Recent Developments in Eco-Friendly Wood-Based Composites II

Recent Developments in Eco-Friendly Wood-Based Composites II

Editor

Pavlo Bekhta

MDPI • Basel • Beijing • Wuhan • Barcelona • Belgrade • Manchester • Tokyo • Cluj • Tianjin

Editor
Pavlo Bekhta
Department of Wood-Based
Composites, Cellulose and
Paper
Ukrainian National Forestry
University
Lviv
Ukraine

Editorial Office
MDPI
St. Alban-Anlage 66
4052 Basel, Switzerland

This is a reprint of articles from the Special Issue published online in the open access journal *Polymers* (ISSN 2073-4360) (available at: www.mdpi.com/journal/polymers/special_issues/Recent_Dev_Eco-Friendly_Wood-Based_Compos_II).

For citation purposes, cite each article independently as indicated on the article page online and as indicated below:

LastName, A.A.; LastName, B.B.; LastName, C.C. Article Title. *Journal Name* **Year**, *Volume Number*, Page Range.

ISBN 978-3-0365-7509-4 (Hbk)
ISBN 978-3-0365-7508-7 (PDF)

Contents

About the Editor

Pavlo Bekhta

Prof. Dr. Pavlo Bekhta works in the Department of Wood-Based Composites, Cellulose and Paper at the Ukrainian National Forestry University, Lviv, Ukraine. He has been the head of the department since 1999. He is involved in research on wood science and technology, wood–polymer composites, lignocellulosic-based composites, and wood modification. He has made important contributions in the field of heat treatment of wood by determining a relationship between color changes and the mechanical strength of thermally modified wood. He also initiated research on the thermo-mechanical densification of a veneer surface before adhesive application by developing technology which has proven to be effective at improving the shear strength of veneer-based products and reducing adhesive consumption. He has contributed significantly to the preparation and modification of wood resin adhesives and has developed techniques for the hot pressing of plywood panels at low temperatures with low emissions. Many of these developments are protected by patents and utility models in Ukraine. He is the author/co-author of 4 textbooks, 8 tutorials, 2 monographs, 4 dictionaries, and over 150 articles, 75 of which are in refereed journals. Under his supervision, 3 dissertations for the degree of Doctor of Technical Sciences and 14 dissertations for the degree of Candidate of Technical Sciences were defended. He was elected as a Fellow of the Ukrainian Academy of Forestry in 2000 and the International Academy of Wood Science in 2022.

Preface to "Recent Developments in Eco-Friendly Wood-Based Composites II"

Today, the most extensively used adhesives in the wood-based materials industry are formaldehyde-based resins. However, the challenge with these resins is that the raw materials come from non-renewable fossil resources and formaldehyde emissions, which restricts their application. Vapors of formaldehyde emitted from many of those wood-based materials and furniture products are one of the main reasons for poor indoor air quality, and hence, cause a negative impact on the indoor environment and human health. Formaldehyde is a human carcinogen that can damage the respiratory, eye, and nervous system. Therefore, the growing environmental problems and strict legal requirements for formaldehyde-free emissions from wood-based materials have posed new challenges for researchers and industry in developing environmentally friendly engineering wood products with close-to-zero formaldehyde emissions. Therefore, developing highly efficient ultra-low formaldehyde emission processes is necessary for the sustainable production of wood-based materials. This reprint contains 12 high-quality original research and reviews papers by 62 authors from 10 countries on 3 continents: Asia (China, India, Indonesia, and Malaysia), Europe (Austria, Czech Republic, Portugal, Romania, and Ukraine), and Africa (South Africa). These papers were published in a Special Issue, "Recent Developments in Eco-Friendly Wood-Based Composites II", of the journal *Polymers*. These papers provide examples of the most recent developments in eco-friendly wood-based composites. The Guest Editor would like to thank all authors who contributed to this Special Issue. The Guest Editor would also like to thank the Special Issue Editor Hunter Jia for his overall professional attitude and kind assistance with the publications.

Pavlo Bekhta

Editor

polymers

MDPI

Editorial

Recent Developments in Eco-Friendly Wood-Based Composites II

Pavlo Bekhta [1,2,3]

[1] Department of Wood-Based Composites, Cellulose and Paper, Ukrainian National Forestry University, 790 57 Lviv, Ukraine; bekhta@nltu.edu.ua
[2] Department of Wood Science and Technology, Faculty of Forestry and Wood Technology, Mendel University in Brno, Zemědělská 3, 613 00 Brno, Czech Republic
[3] Department of Furniture and Wood Products, Technical University in Zvolen, T.G. Masaryka 24, 960 01 Zvolen, Slovakia

Traditional wood-based composites are bonded with synthetic formaldehyde-based adhesives [1,2]. These adhesives bring certain environmental problems because they release formaldehyde emissions, which are a human carcinogen and toxic for the environment [3]. It is difficult to find new uses or new fields for wood-based products because of the lack of proper adhesives which meet the wood industry requirements of being eco-friendly, low-cost, and easy to use. For this reason, growing ecological and environmental consciousness drives efforts for the development of new eco-friendly wood-based composites for various end-use applications. In recent years, significant efforts have been made to reduce formaldehyde emissions from wood-based composites via: (i) the reduction of formaldehyde content in resin formulation [4,5]; (ii) the use of scavengers such as tannins, lignin, starch, wheat and hemp flour, and pulp and paper sludge [6–14] or other compounds (starch derivatives, charcoal, pozzolan, zeolites, and urea) [15–19] that scavenge formaldehyde; (iii) the post-treatment or surface treatment of the wood-based products [18,20]; (iv) the use of natural resins, including soy protein, tannin, lignin, and starch adhesives [21–23]; (v) and the thermal pre-treatment of veneer before bonding [24–26]. Comprehensive information on the reduction of formaldehyde emissions in various ways can be also found in several published reviews [4,20,27–30]. The most acceptable and effective procedure for reducing formaldehyde emissions in wood-based panels is the use of formaldehyde scavengers, which can be classified as synthetic scavengers, bio-based (natural) scavengers, and nano-scavengers [30].

Identifying additives to reduce the total amount of urea-formaldehyde (UF) resin needed without adversely affecting the panel properties is one way to reduce the negative environmental footprint of UF resins caused by the release of formaldehyde. The results provided by Taghiyari et al. [31] showed that small amounts of micron-scale wollastonite could serve as a resin extender. Sugar palm fiber (SPF) was employed as a reinforcement material in a polyvinyl butyral (PVB) polymer matrix to develop SPF-PVB eco-friendly laminated composites through the hot compression method [32]. The laminated composite sample with 80% of PVB and 20% of SPF showed the highest stiffness value. Thermoplastic starch (TPS) and poly (lactic acid) (PLA) are among the most promising biodegradable polymers that have the potential to replace petroleum-based polymers. The study conducted by Nazrin et al. [33] reveals the potential of PLA/TPS blend bionanocomposites for biodegradable packaging applications. The properties of co-extruded wood/polyethylene composites (Co-WPCs) were improved by filling the shell and wood fiber layers with low-cost nano-silica (nSiO_2) and micro-silica (mSiO_2) [34].

One of the possible directions to achieve this goal is the creation of wood composites based on environmentally friendly products, where thermoplastics and their copolymers (low- and high-density polyethylene, polypropylene, co-polyamide, and co-polyester, etc.) are used as adhesives [35–39]. Bark flours obtained from different tree species having a high

Citation: Bekhta, P. Recent Developments in Eco-Friendly Wood-Based Composites II. *Polymers* **2023**, *15*, 1941. https://doi.org/10.3390/polym15081941

Received: 10 April 2023
Revised: 14 April 2023
Accepted: 18 April 2023
Published: 19 April 2023

polyphenol content also exhibited formaldehyde-scavenging properties [40–45]. Equally exciting and revolutionary was the development of the use of citric acid (CA) as a green modifying agent and adhesive for wood [2,46]. There is an excellent review [47] whereby the bonding mechanism and types of wood composites bonded with CA are presented. The authors also discussed the best working conditions for the CA in the fabrication of wood composites. The environmental impacts and future outlook of CA-treated wood and bonded composite are also considered. Another alternative to the use of synthetic formaldehyde-based adhesives is to manufacture binderless wood composites [48], since wood is a natural polymer material which is rich in lignocellulosic compounds such as cellulose, hemicellulose, and lignin.

This Special Issue, entitled "Recent Developments in Eco-Friendly Wood-Based Composites II", comprises 12 high-quality original research and reviews papers by 62 authors from 10 countries on three continents: Asia (China, India, Indonesia, Malaysia), Europe (Austria, Czech Republic, Portugal, Romania, Ukraine), and Africa (South Africa). The papers provide examples of the most recent developments in eco-friendly wood-based composites.

In their paper, Xiao et al. [49] applied a saturated steam heat treatment in a useful way to effectively enhance the dimensional stability and mold-resistance property of bamboo and bamboo-based products. By promoting greenhouse gas sequestration, bamboo and bamboo-based products can improve carbon storage, thereby helping to reduce greenhouses gas emission through replacing traditional products such as concrete, steel, and alloys. The authors observed the decrement of hemicellulose and cellulose after thermal modification, whereby the bamboo samples exhibited improved dimensional stability and anti-fungal properties. The hardness and modulus of elasticity (MOE) of the thermally modified bamboo were 0.75 and 20.6 GPa, respectively.

A very interesting study by Bekhta et al. [50] aimed to evaluate the possibility of using wood particles from deadwood in the production of particleboards. The authors investigated the physical and mechanical properties as well as the formaldehyde content of UF-bonded particleboards with different content of deadwood particles (0%, 25%, 50%, 75%, 100%). It was found that replacing conventional health wood particles with deadwood particles led to the deterioration of the mechanical properties of the boards. In addition, the boards from deadwood particles absorbed more water and swelled more. However, it was shown that adding 3% of MUF resin to UF adhesive increased the bending strength (MOR), MOE, and internal bond strength (IB) by 44.1%, 43.3%, and 294.4%, respectively, while decreasing the water absorption (WA) 24 h and thickness swelling (TS) 24 h by 18.2% and 42.9%, respectively. Moreover, a significant advantage was that boards made from 100% deadwood particles are characterized by 34.5% less formaldehyde content than reference boards made from conventional health wood.

In the study by Ismail et al. [51], a new approach to fabricate the coconut shell nano-biocomposites using waste polypropylene plastic packaging as a matrix was proposed. Coconut shell, an agricultural waste, was bonded with waste plastic to form a biocomposite with a coupling agent. The authors investigated the optimum percentage composition and the effect of coconut shell ball milling time on the physical, mechanical, and thermal properties of the biocomposite. They found that the properties of the biocomposite could be improved by reducing the particle size of the coconut shell (increasing the duration of milling). As the milling time increased from 0 to 40 h, the density increased from 0.9 to 1.02 g/cm^3; TS decreased from 3.4 to 1.8%; porosity decreased from 7.0 to 3.0%; MOR increased from 8.19 to 12.26 MPa; MOE increased from 1.67 to 2.87 GPa; and compressive strength increased from 16.00 to 27.20 MPa. The thermal properties of the biocomposite also improved as the particle size reduced. The authors also found that the performance of the biocomposite improved significantly with a lower percentage matrix and filler nanoparticle rather than when increasing the percentage of the matrix. The finding of this research also indicates that the properties of the biocomposite can be improved by reducing the particle size of the filler to nanometers without having to increase the adhesive composition.

The findings of another study [52] demonstrated that eco-friendly plywood samples using four various wood species (beech, birch, hornbeam and poplar) bonded with LDPE film of different thicknesses (50, 80, 100 and 150 μm) showed satisfactory physical–mechanical properties. Poplar veneer provided the lowest values for MOR, MOE and TS of all the plywood samples, but the bonding strength was at the same level as birch and hornbeam veneer. Beech plywood samples had the best mechanical properties. An increase in LDPE film thickness improved the physical–mechanical properties of plastic-bonded plywood.

In another paper, low-cost wood–plastic composites (WPCs) without any additives were developed from invasive trees without prior processing and low-grade recycled low-density polyethylene [53]. The authors evaluated different biomass/plastic ratios, particle sizes, and press settings to determine the optimum processing parameters to obtain WPCs with adequate properties. The dimensional stability, WA, MOR, MOE, tensile strength, and tensile moduli were improved at longer press times and higher temperatures for all blending ratios. An increased biomass ratio and particle size were positively correlated with WA and TS and inversely related with MOR, tensile strength, and density due to an incomplete encapsulation of the biomass by the plastic matrix.

In another interesting study, the first attempt to investigate low-density insulation boards made of spruce bark fibers in a wet process was conducted [54]. The insulation boards with densities between 160 and 300 kg/m^3 were self-bonded. The authors found that widely available bark residues could be successfully utilized as an innovative raw material for efficient eco-friendly thermal insulation products. The thermal conductivity values of the boards were comparable to established insulation boards based on cork or wood fibers. Based on the measured thermal conductivity and zero formaldehyde content, bark fiber insulation panels might be able to compete with conventional insulations if the density can be further reduced, and applications regarding acoustic insulation are also a possibility.

Oil palm trunks (OPT) are considered significant waste products. Usually, the trunks remain on the plantation site for nutrient recycling or burning, which increases insect and fungi populations. This causes environmental problems for the new palm generation or air pollution due to fire. Therefore, the comprehensive review conducted by Nuryawan et al. [55] summarizes the utilization of OPT into products made of oil palm fibers mainly derived from OPT, and its application for the substitution of wood panel products. Some research works have also analyzed oil palm fibers derived from OPT for the exploitation of their potential as raw material to process into various conventional composite panel products, such as plywood and laminated board, particleboard, or binderless and cement board.

Nanocellulose aerogels are a new category of high-efficiency adsorbents for treating oil spills and water pollution. The review provided by Iskandar et al. [56] presents an introduction to nanocellulose-based aerogel and its fabrication approaches. Different applications of nanocellulose aerogel in environmental, medical and industrial fields are presented. Different strategies for the modification of nanocellulose-based aerogel are also critically discussed in this review, presenting the most recent works in terms of enhancing the aerogel performance in oil absorption in addition to the potential of these materials in near future.

In their comprehensive review paper, Ramesh et al. [57] focused on the processing of WPCs along with additives such as wood flour and various properties of WPCs such as mechanical, structural, and morphological properties. Applications of wood-based composites in various sectors such as automotive, marine, defense, and structural applications are also highlighted in this review.

The processing technology, bonding mechanism and performance of thermoplastic-bonded wood-based panels are comprehensively summarized and reviewed in another interesting paper [58]. Meanwhile, the existing problems for this new kind of panel and their future development trends are also highlighted, which can provide the wood industry with foundations and guidelines for using thermoplastics as environmentally friendly adhesives and effectively solving indoor pollution problems.

In recent years, different types of thermoplastic films such as polyethylene, polypropylene, polyvinyl chloride, co-polyamide and co-polyester have been widely used for wood veneer bonding owing to their excellent water resistance, flexibility, easy processing, and secondary melting characteristics. The findings of another study [59] demonstrated that plastic plywood can be produced using an ethylene–vinyl acetate (EVA) film as a wood adhesive via hot press and cold press processes. The results showed that the EVA film featured good gluing ability, and the EVA plywood could be used in indoor environments.

A very promising direction is the use of lignin and its derivatives as an ecological alternative to petroleum-based adhesives. However, being the most common renewable source of phenolic compounds of natural origin, only 1–2% of the huge annual production volume (50–70 million tons) is actually used for the production of value-added products. Lignosulphonates (LS) account for 90% of the total market of commercial lignin. In their review paper, Gonçalves et al. [60] carried out a comprehensive overview of the methods to improve the reactivity of lignin molecules, and techniques to extract, characterize, and improve the reactivity of LS. The most recent advances in the application of LS in wood adhesives with and without their combination with formaldehyde, are also discussed.

The papers from this Special Issue represent only some of the recent developments in eco-friendly wood-based composites. The utilization of recycled plastics, lignin and their derivatives, wood (bark) and agricultural wastes to manufacture wood composites as well as traditional WPCs is highly viable concerning eco-friendliness, and contributes to the improvement of the circular economy. It also saves the usage of virgin materials, thus enhancing sustainability in the production of composite materials. However, most of the proposed methods to manufacture high-performance, eco-friendly wood composites with a lower environmental impact have been studied only in laboratory conditions, can only find use in some nonstructural applications, and have not been introduced in large-scale industrial production as yet. Hence, further research is still needed in order to develop methods for improving reactivity and the selection of suitable crosslinkers for lignin-based adhesives, and modification methods to improve the interfacial adhesion between hydrophilic wood and hydrophobic thermoplastics in order to expand their use in some exterior and structural applications.

Acknowledgments: This work was supported by the EU NextGenerationEU through the Recovery and Resilience Plan for Slovakia under the project No. 09I03-03-V01-00124.

Conflicts of Interest: The authors declare no conflict of interest.

References

1. Mantanis, G.I.; Athanassiadou, E.T.; Barbu, M.C.; Wijnendaele, K. Adhesive systems used in the European particleboard, MDF and OSB industries. *Wood Mater. Sci. Eng.* **2018**, *13*, 104–116. [CrossRef]
2. Pizzi, A.; Papadopoulos, A.N.; Policardi, F. Wood Composites and Their Polymer Binders. *Polymers* **2020**, *12*, 1115. [CrossRef] [PubMed]
3. Formaldehyde, 2–Butoxyethanol and 1–tert–Butoxypropan–2–ol. In *IARC Monographs on the Evaluation of Carcinogenic Risk to Humans;* World Health Organization—International Agency for Research on Cancer: Lyon, France, 2006; Volume 88.
4. Myers, G.E. How mole ratio of UF resin affects formaldehyde emission and other properties: A literature critique. *For. Prod. J.* **1984**, *34*, 35–41.
5. Dunky, M. Urea–formaldehyde (UF) adhesive resins for wood. *Int. J. Adhes. Adhes.* **1998**, *18*, 95–107. [CrossRef]
6. Eom, Y.-G.; Kim, J.-S.; Kim, S.; Kim, J.-A.; Kim, H.-J. Reduction of formaldehyde emission from particleboards by bio–scavengers. *J. Korean Wood Sci. Technol.* **2006**, *34*, 29–41.
7. Kim, S.; Kim, H.; Kim, H.; Lee, H. Effect of bio-scavengers on the curing behavior and bonding properties of melamine-formaldehyde resins. *Macromol. Mater. Eng.* **2006**, *291*, 1027–1034. [CrossRef]
8. Kim, S. Environment-friendly adhesives for surface bonding of wood-based flooring using natural tannin to reduce formaldehyde and TVOC emission. *Bioresour. Technol.* **2009**, *100*, 744–748. [CrossRef]
9. Moubarik, A.; Allal, A.; Pizzi, A.; Charreir, B.; Carreir, F. Characterization of a formaldehyde–free cornstarch–tannin wood adhesive for interior plywood. *Eur. J. Wood Prod.* **2010**, *68*, 427–433. [CrossRef]

10. Boran, S.; Usta, M.; Ondaral, S.; Gümüşkaya, E. The efficiency of tannin as a formaldehyde scavenger chemical in medium density fiberboard. *Compos. Part B Eng.* **2012**, *43*, 2487–2491. [CrossRef]
11. Gangi, M.; Tabarsa, T.; Sepahvand, S.; Asghari, J. Reduction of formaldehyde emission from plywood. *J. Adhes. Sci. Technol.* **2013**, *27*, 1407–1417. [CrossRef]
12. Bekhta, P.; Sedliačik, J.; Kačík, F.; Noshchenko, G.; Kleinová, A. Lignocellulosic waste fibers and their application as a component of urea-formaldehyde adhesive composition in the manufacture of plywood. *Eur. J. Wood Wood Prod.* **2019**, *77*, 495–508. [CrossRef]
13. Kawalerczyk, J.; Siuda, J.; Mirski, R.; Dziurka, D. Hemp flour as a formaldehyde scavenger for melamine-urea-formaldehyde adhesive in plywood production. *BioResources* **2020**, *15*, 4052–4064. [CrossRef]
14. Bekhta, P.; Noshchenko, G.; Réh, R.; Krišťák, L.; Sedliačik, J.; Antov, P.; Mirski, R.; Savov, V. Properties of Eco-Friendly Particleboards Bonded with Lignosulfonate-Urea-Formaldehyde Adhesives and pMDI as a Crosslinker. *Materials* **2021**, *14*, 4875. [CrossRef] [PubMed]
15. Park, B.D.; Kang, E.C.; Park, J.Y. Thermal curing behavior of modified urea-formaldehyde resin adhesives with two formaldehyde scavengers and their influence on adhesion performance. *J. Appl. Polym. Sci.* **2008**, *110*, 1573–1580. [CrossRef]
16. Kim, S. The reduction of indoor air pollutant from wood-based composite by adding pozzolan for building materials. *Constr. Build. Mater.* **2009**, *23*, 2319–2323. [CrossRef]
17. Kmec, S.; Sedliacik, J.; Smidriakova, M.; Jablonski, M. Zeolite as a filler of UF resin for lower formaldehyde emission from plywood. *Ann. Wars. Univ. Life Sci.* **2010**, *70*, 161–165.
18. Costa, N.A.D.; Pereira, J.; Ferra, J.; Cruz, P.; Martins, J.; Magalhaes, F.D.; Mendes, A.; Carvalho, L.H. Scavengers for Achieving Zero Formaldehyde Emission of Wood-Based Panels. *Wood Sci. Technol.* **2013**, *47*, 1261–1272. [CrossRef]
19. Kowaluk, G.; Zajac, M.; Czubak, E.; Auriga, R. Physical and mechanical properties of particleboards manufactured using charcoal as additives. *iForest* **2016**, *10*, 70–74. [CrossRef]
20. Myers, G.E. Effects of post–manufacture board treatments on formaldehyde emission: A literature review (1960–1984). *For. Prod. J.* **1986**, *36*, 41–51.
21. Lorenz, L.F.; Conner, A.H.; Christiansen, A.W. The effect of soy protein additions on the reactivity and formaldehyde emissions of urea-formaldehyde adhesive resin. *For. Prod. J.* **1999**, *49*, 73–78.
22. Maulana, M.I.; Lubis, M.A.R.; Febrianto, F.; Hua, L.S.; Iswanto, A.H.; Antov, P.; Kristak, L.; Mardawati, E.; Sari, R.K.; Zaini, L.H.; et al. Environmentally Friendly Starch-Based Adhesives for Bonding High-Performance Wood Composites: A Review. *Forests* **2022**, *13*, 1614. [CrossRef]
23. Neitzel, N.; Hosseinpourpia, R.; Adamopoulos, S. A dialdehyde starch-based adhesive for medium-density fiberboards. *BioResources* **2023**, *18*, 2155–2171. [CrossRef]
24. Aydin, I.; Colakoglu, G. Formaldehyde Emission, Surface Roughness, and Some Properties of Plywood as Function of Veneer Drying Temperature. *Dry. Technol.* **2005**, *23*, 1107–1117. [CrossRef]
25. Murata, K.; Watanabe, Y.; Nakano, T. Effect of Thermal Treatment of Veneer on Formaldehyde Emission of Poplar Plywood. *Materials* **2013**, *6*, 410–420. [CrossRef] [PubMed]
26. Bekhta, P.; Sedliačik, J.; Bekhta, N. Effect of Veneer-Drying Temperature on Selected Properties and Formaldehyde Emission of Birch Plywood. *Polymers* **2020**, *12*, 593. [CrossRef] [PubMed]
27. İstek, A.; Özlüsoylu, I.; Onat, S.M.; Özlüsoylu, Ş. Formaldehyde Emission Problems and Solution Recommendations on Wood-Based Boards: A review. *J. Bartin Fac. For.* **2018**, *20*, 382–387.
28. Kariuki, S.W.; Wachira, J.; Kawira, M.; Murithi, G. Formaldehyde Use and Alternative Biobased Binders for Particleboard Formulation: A Review. *Hindawi J. Chem.* **2019**, *2019*, 5256897. [CrossRef]
29. Antov, P.; Savov, V.; Neykov, N. Reduction of Formaldehyde Emission from Engineered Wood Panels by Formaldehyde Scavengers—A Review. In Proceedings of the 13th International Scientific Conference Wood EMA 2020 and 31st International Scientific Conference ICWST 2020 Sustainability of Forest-Based Industries in the Global Economy, Vinkovci, Croatia, 28–30 September 2020; pp. 7–11, ISBN 978-953-57822-8-5.
30. Kristak, L.; Antov, P.; Bekhta, P.; Lubis, M.A.R.; Iswanto, A.H.; Réh, R.; Sedliačik, J.; Savov, V.; Taghiayri, H.R.; Papadopoulos, A.N.; et al. Recent Progress in Ultra-Low Formaldehyde Emitting Adhesive Systems and Formaldehyde Scavengers in Wood-Based Panels: A Review. *Wood Mater. Sci. Eng.* **2022**, *18*, 763–782. [CrossRef]
31. Taghiyari, H.R.; Esmailpour, A.; Majidi, R.; Morrell, J.J.; Mallaki, M.; Militz, H.; Papadopoulos, A.N. Potential Use of Wollastonite as a Filler in UF Resin Based Medium-Density Fiberboard (MDF). *Polymers* **2020**, *12*, 1435. [CrossRef]
32. Syaqira, S.N.; Leman, Z.; Sapuan, S.M.; Dele-Afolabi, T.T.; Azmah Hanim, M.A. Tensile Strength and Moisture Absorption of Sugar Palm-Polyvinyl Butyral Laminated Composites. *Polymers* **2020**, *12*, 1923. [CrossRef]
33. Nazrin, A.; Sapuan, S.M.; Zuhri, M.Y.M. Mechanical, Physical and Thermal Properties of Sugar Palm Nanocellulose Reinforced Thermoplastic Starch (TPS)/Poly (Lactic Acid) (PLA) Blend Bionanocomposites. *Polymers* **2020**, *12*, 2216. [CrossRef] [PubMed]
34. Sun, L.; Zhou, H.; Zong, G.; Ou, R.; Fan, Q.; Xu, J.; Hao, X.; Guo, Q. Effects of SiO_2 Filler in the Shell and Wood Fiber in the Core on the Thermal Expansion of Core–Shell Wood/Polyethylene Composites. *Polymers* **2020**, *12*, 2570. [CrossRef]
35. Bekhta, P.; Sedliačik, J. Environmentally-Friendly High-Density Polyethylene-Bonded Plywood Panels. *Polymers* **2019**, *11*, 1166. [CrossRef] [PubMed]
36. Mirski, R.; Bekhta, P.; Dziurka, D. Relationships between Thermoplastic Type and Properties of Polymer-Triticale Boards. *Polymers* **2019**, *11*, 1750. [CrossRef]

37. Bekhta, P.; Müller, M.; Hunko, I. Properties of Thermoplastic-Bonded Plywood: Effects of the Wood Species and Types of the Thermoplastic Films. *Polymers* **2020**, *12*, 2582. [CrossRef]
38. Mirski, R.; Banaszak, A.; Bekhta, P. Selected Properties of Formaldehyde-Free Polymer-Straw Boards Made from Different Types of Thermoplastics and Different Kinds of Straw. *Materials* **2021**, *14*, 1216. [CrossRef]
39. Bekhta, P.; Pizzi, A.; Kusniak, I.; Bekhta, N.; Chernetskyi, O.; Nuryawan, A. A Comparative Study of Several Properties of Plywood Bonded with Virgin and Recycled LDPE Films. *Materials* **2022**, *15*, 4942. [CrossRef]
40. Aydin, I.; Demirkir, C.; Colak, S.; Colakoğlu, S. Utilization of bark flours as additive in plywood manufacturing. *Eur. J. Wood Wood Prod.* **2017**, *75*, 63–69. [CrossRef]
41. Ružiak, I.; Igaz, R.; Krišťák, L.; Réh, R.; Mitterpach, J.; Očkajová, A.; Kučerka, M. Influence of Urea-Formaldehyde Adhesive Modification with Beech Bark on Chosen Properties of Plywood. *BioResources* **2017**, *12*, 3250–3264. [CrossRef]
42. Réh, R.; Igaz, R.; Krišťák, Ľ.; Ružiak, I.; Gajtanska, M.; Božíková, M.; Kučerka, M. Functionality of Beech Bark in Adhesive Mixtures Used in Plywood and Its Effect on the Stability Associated with Material Systems. *Materials* **2019**, *12*, 1298. [CrossRef] [PubMed]
43. Tudor, E.M.; Barbu, M.C.; Petutschnigg, A.; Réh, R.; Krišťák, L. Analysis of Larch-Bark Capacity for Formaldehyde Removal in Wood Adhesives. *Int. J. Environ. Res. Public Health* **2020**, *17*, 764. [CrossRef]
44. Réh, R.; Krišťák, Ľ.; Sedliačik, J.; Bekhta, P.; Božiková, M.; Kunecová, D.; Vozárová, V.; Tudor, E.M.; Antov, P.; Savov, V. Utilization of Birch Bark as an Eco-Friendly Filler in Urea-Formaldehyde Adhesives for Plywood Manufacturing. *Polymers* **2021**, *13*, 511. [CrossRef]
45. Bekhta, P.; Sedliačik, J.; Noshchenko, G.; Kačík, F.; Bekhta, N. Characteristics of beech bark and its effect on properties of UF adhesive and on bonding strength and formaldehyde emission of plywood panels. *Eur. J. Wood Wood Prod.* **2021**, *79*, 423–433. [CrossRef]
46. Del Menezzi, C.; Amirou, S.; Pizzi, A.; Xi, X.; Delmotte, L. Reactions with Wood Carbohydrates and Lignin of Citric Acid as a Bond Promoter of Wood Veneer Panels. *Polymers* **2018**, *10*, 833. [CrossRef] [PubMed]
47. Lee, S.H.; Md Tahir, P.; Lum, W.C.; Tan, L.P.; Bawon, P.; Park, B.-D.; Osman Al Edrus, S.S.; Abdullah, U.H. A Review on Citric Acid as Green Modifying Agent and Binder for Wood. *Polymers* **2020**, *12*, 1692. [CrossRef]
48. Zhang, D.; Zhang, A.; Xue, L. A review of preparation of binderless fiberboards and its self-bonding mechanism. *Wood Sci. Technol.* **2015**, *49*, 661–679. [CrossRef]
49. Xiao, X.; Liang, X.; Peng, H.; Wang, K.; Liu, X.; Li, Y. Multi-Scale Evaluation of the Effect of Thermal Modification on Chemical Components, Dimensional Stability, and Anti-Mildew Properties of Moso Bamboo. *Polymers* **2022**, *14*, 4677. [CrossRef]
50. Bekhta, P.; Kozak, R.; Gryc, V.; Sebera, V.; Tippner, J. Effects of Wood Particles from Deadwood on the Properties and Formaldehyde Emission of Particleboards. *Polymers* **2022**, *14*, 3535. [CrossRef] [PubMed]
51. Ismail, I.; Aini, Q.; Jalil, Z.; Olaiya, N.G.; Mursal, M.; Abdullah, C.K.; H.P.S., A.K. Properties Enhancement Nano Coconut Shell Filled in Packaging Plastic Waste Bionanocomposite. *Polymers* **2022**, *14*, 772. [CrossRef] [PubMed]
52. Bekhta, P.; Chernetskyi, O.; Kusniak, I.; Bekhta, N.; Bryn, O. Selected Properties of Plywood Bonded with Low-Density Polyethylene Film from Different Wood Species. *Polymers* **2022**, *14*, 51. [CrossRef]
53. Mohammed, A.S.; Meincken, M. Properties of Low-Cost WPCs Made from Alien Invasive Trees and rLDPE for Interior Use in Social Housing. *Polymers* **2021**, *13*, 2436. [CrossRef]
54. Gößwald, J.; Barbu, M.-C.; Petutschnigg, A.; Tudor, E.M. Binderless Thermal Insulation Panels Made of Spruce Bark Fibres. *Polymers* **2021**, *13*, 1799. [CrossRef]
55. Nuryawan, A.; Sutiawan, J.; Rahmawaty; Masruchin, N.; Bekhta, P. Panel Products Made of Oil Palm Trunk: A Review of Potency, Environmental Aspect, and Comparison with Wood-Based Composites. *Polymers* **2022**, *14*, 1758. [CrossRef]
56. Iskandar, M.A.; Yahya, E.B.; Abdul Khalil, H.P.S.; Rahman, A.A.; Ismail, M.A. Recent Progress in Modification Strategies of Nanocellulose-Based Aerogels for Oil Absorption Application. *Polymers* **2022**, *14*, 849. [CrossRef] [PubMed]
57. Ramesh, M.; Rajeshkumar, L.; Sasikala, G.; Balaji, D.; Saravanakumar, A.; Bhuvaneswari, V.; Bhoopathi, R. A Critical Review on Wood-Based Polymer Composites: Processing, Properties, and Prospects. *Polymers* **2022**, *14*, 589. [CrossRef] [PubMed]
58. Mo, X.; Zhang, X.; Fang, L.; Zhang, Y. Research Progress of Wood-Based Panels Made of Thermoplastics as Wood Adhesives. *Polymers* **2022**, *14*, 98. [CrossRef] [PubMed]
59. Zhang, Y.; He, Y.; Yu, J.; Lu, Y.; Zhang, X.; Fang, L. Fabrication and Characterization of EVA Resins as Adhesives in Plywood. *Polymers* **2023**, *15*, 1834. [CrossRef]
60. Gonçalves, S.; Ferra, J.; Paiva, N.; Martins, J.; Carvalho, L.H.; Magalhães, F.D. Lignosulphonates as an Alternative to Non-Renewable Binders in Wood-Based Materials. *Polymers* **2021**, *13*, 4196. [CrossRef]

polymers

MDPI

Article

Binderless Thermal Insulation Panels Made of Spruce Bark Fibres

Jakob Gößwald [1], Marius-Cătălin Barbu [1,2], Alexander Petutschnigg [1,3] and Eugenia Mariana Tudor [1,2,*]

[1] Forest Products Technology and Timber Construction Department, Salzburg University of Applied Sciences, Markt 136a, 5431 Kuchl, Austria; jgoesswald.htw-m2020@fh-salzburg.ac.at (J.G.); cmbarbu@unitbv.ro (M.-C.B.); alexander.petutschnigg@fh-salzburg.ac.at (A.P.)

[2] Faculty of Furniture Design and Wood Engineering, Transilvania University of Brasov, B-dul. Eroilor nr. 29, 500036 Brasov, Romania

[3] Institute of Wood Technology and Renewable Materials, University of Natural Resources and Life Sciences (BOKU), Konrad Lorenz-Straße 24, 3340 Tulln, Austria

* Correspondence: eugenia.tudor@fh-salzburg.ac.at

Abstract: Tree bark is a by-product of the timber industry available in large amounts, considering that approximately 10% of the volume of a tree stem is bark. Bark is used primarily for low-value applications such as heat generation or as mulch. To the best of our knowledge, this study is the first one that scrutinises thermal insulation panels made from spruce bark fibres with different densities and fibre lengths manufactured in a wet process. The insulation boards with densities between 160 and 300 kg/m^3 were self-bonded. Internal bond, thermal conductivity, and dimensional stability (thickness swelling and water absorption), together with formaldehyde content, were analysed. The thermal properties of the boards were directly correlated with the density and reached about 0.044 W/m*K, while the internal bond was rather influenced by the fibre length and was relatively low (on average 0.07 N/mm^2). The water absorption was high (from 55% to 380%), while the thickness swelling remained moderate (up to 23%). The results of this study have shown that widely available bark residues can be successfully utilised as an innovative raw material for efficient eco-friendly thermal insulation products.

Keywords: tree bark fibre; thermal insulation panels; thermal conductivity; self-bonded boards; zero formaldehyde content

Citation: Gößwald, J.; Barbu, M.-C.; Petutschnigg, A.; Tudor, E.M. Binderless Thermal Insulation Panels Made of Spruce Bark Fibres. *Polymers* **2021**, *13*, 1799. https://doi.org/10.3390/polym13111799

Academic Editor: Adriana Kovalcik

Received: 8 May 2021
Accepted: 27 May 2021
Published: 29 May 2021

1. Introduction

Bark is the outer layer of trees, divided into two anatomically different layers, the outer bark, whose primary purpose is the protection of the underlaid tissues, and the inner bark that transports the assimilation products from leaves to the root with active tissues close to cambium [1]. Various types of extractives (especially carbohydrates) are included in the tree bark [2,3].

Today, in Europe, the majority of the available bark is used for bioenergy production or is used for even less value-added purposes like composting and incinerating [4]. Tree bark can be superiorly utilised as raw material, for example, as a filler in urea formaldehyde adhesives [5–8] to replace wheat flour, reducing in this way the formaldehyde emissions. Extractives and chemical compounds of the bark offer applications as medicine, plastics, or aggregates [9]. Some tree species also allow utilisations as tissue [10–12].

For particleboards (PB), the use of recycled wood or other lignocellulose materials is well established [13]. Due to its availability, bark, which encompasses about 10% of the stem volume [14], has interesting potential in bark-based composites. Muszynski and McNatt [15] showed that with up to 30% spruce and pine bark, it is possible to produce particleboards with acceptable properties. Blanchet et al., 2000 [15] indicated that even larger wood particle proportions could be replaced, especially when using the inner bark of birch trees [16]. With treatments in hot water, the mechanical properties of particleboards containing tree bark can be modified [17].

Because bark is a fibrous tissue, the production and utilisation of black spruce bark fibres in the core layer of medium density fibreboards (MDF) were evaluated by Xing et al. [18,19]. According to this study, it can be stated that black spruce bark can be an auxiliary material for the core layers of MDF.

Bark fibres can be produced by thermo-mechanical refining [19,20], while bark particles can be obtained with various methods, for example, shredding or hammer-milling [15,21].

One purpose of bark is the protection of the tree from external influences like moisture loss and temperature changes; therefore, a natural optimisation of bark toward these insulation properties has already been the subject of study of many research teams [22]. Bark composites with clay as a binder open a new means of the manufacture of panels with enhanced resistance to fire [23]

The thermal conductivity (TC) of bark-based panels was evaluated in various studies, for example, in insulation boards bonded with natural tannins [21,24,25] or with larch, pine, spruce, fir, and oak tree bark resinated with urea-formaldehyde, melamine urea-, and tannin-based adhesives. The evaluation of the effects of particle orientation in insulation panels for larch bark showed lambda (λ) values (TC) between 0.056 to 0.1 W/m*K [25].

Regarding the differences between inner and outer bark with regard to thermal properties, some studies indicate that the inner bark is able to insulate better compared to the outer bark, especially in trees that contain large fibres [12]. Another important issue for the implementation of tree bark in added-value applications is its comminution type [26].

With reinforced surfaces, the mechanical properties of bark insulation boards can be enhanced [27]. Lower TC (0.045 W/m*K) was achieved after alkaline extraction of poplar bark [14]. Apart from thermal insulation, bark-based composites can be used as sound-absorbing panels. At densities below 500 kg/m^3, the bark composites had better sound absorption than most other wood-based products [28,29].

Another advantage of bark might be its suitability for self-bonded boards. The studies of [15] and [30] that dealt with the manufacture of self-agglomerated PB based on bark observed the effect of particle plasticisation and extractive polymerisation on bark particles' self-bonding.

Burrows (1960) [31] studied the properties of self-bonded Douglas-fir PB and suggested that the plasticisation mechanism may occur due to the lignocellulosic character of bark and due to the presence of water as a plasticiser. This premise is complemented by the conclusion of [32] regarding the lignocellulosic materials with lightweight molecules (lignin polymers, non-crystalline cellulose, and hemicellulose) that permit softening at a convenient temperature for producing a plasticised matrix that can connect particles in self-bonding panels.

In the present study, the suitability of spruce bark fibres for use in low-density insulation panels was analysed. It was assumed that the reduction in density would decrease the TC of the boards. Due to extractives and the fibrous nature of the bark material, stable boards can be produced in a wet process, without supplementary resins. Subsequently, the influence of the fibre length and density was examined, together with formaldehyde content.

2. Materials and Methods

The bark was sourced from fresh spruce trees (*Picea abies*) from a sawmill in Altötting (Germany) with a diameter over 20 cm. The logs were debarked using high-pressure water jets provided by a Kärcher HD 1090 (Winnenden, Germany). Due to the high water pressure (200 bar), the breakdown of bark resulted in larger pieces and fibrous material, as observed by Krivo et al. (1983) [33]. Since spruce bark in contact with air dries and oxidates quickly, visible due to a change in colour from white to brown within some hours, air contact was avoided where possible.

The wet fibre material was fractionated using sieves of 7, 4, and 1.6 mm. The wet sieving provides the advantage that the fibres can be easily processed, because the fibres clump together during drying. Three types of fibre bundle lengths were chosen for this experimental design, 1.6, 4, and 7 mm, which correspond to the mesh size of the sieves. Due to anatomical differences between phloem and phellem, the brittle parts of the bark degraded into small particles, while the fibrous parts of the bark formed fibre bundles and single fibres. An example of the composition of 1.6 mm fibres is presented in Figure 1.

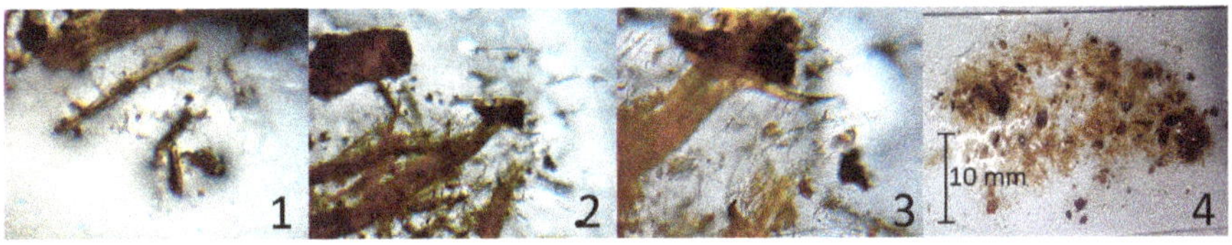

Figure 1. Bark fibres included in the length class 1.6 mm; magnification: 4×, 4×, 10×, reference picture; dependencies: **1** and **2** are details of **4**; **3** is a detail of **2**.

The wet process was applied for the low-density bark fibre boards without using additives or adhesive. At the laboratory scale, the fibres were mixed with 8 L water and were subsequently dehydrated on a fabric supported by a sieve and further squeezed using an overlaying sieve with 9 kPa. The wet board with the size of 30 cm × 30 cm was dried in a Binder (Tuttlingen, Germany) oven at 103 °C for 24 h. Table 1 shows the density and fibre length class of the bark fibre insulation boards. The panel type is coded as follows: the letter A is for bark fibre length 1.6 mm, letter B for 4 mm fibre length, and C for 7 mm fibre length. To the codification belongs also the target density (200 and 250 kg/m^3).

Table 1. Experimental design of bark fibre boards with three fibre bundle lengths (1.6, 4, and 7 mm) and two density levels (200 and 250 kg/m^3).

Insulation Panel (Target Density)	Density (kg/m^3)	Fibre Length (mm)	Boards Number
A200	277	1.6	3
B250	245	4	3
B200	185	4	4
C200	204	7	3

Due to the shrinking during drying, the boards need to be calibrated (milled to obtain a homogeneous and constant thickness) and cut to size to measure the thermal conductivity, carried out using the single-plate λ-Meter EP 500e of the Lambda Messtechnik GmbH (Dresden, Germany) according to EN 12677:2001 [34]. After testing, 50 mm × 50 mm samples were used to determine the internal bond according to EN 1607:2013 [35] with a Zwick Roel Z250 universal testing machine (Ulm, Germany). The dimensional stability (thickness swelling/water absorption after 24 h) was determined according to EN 317:2005 [36], as well as the free formaldehyde content with the perforator method EN ISO 12460-5:2015 [37]. All boards were cut in compliance with EN 326: 1994 [38] and conditioned at 20 °C and 65% relative air humidity for one week, until constant mass was reached, before the testing. The results were analysed using Python software. A regression analysis with all variables at 5% significance was performed in combination with an ANOVA and a test of heteroskedasticity.

3. Results and Discussion

Before the first samples were tested, the easy processability of bark fibre boards was observed, due to the efficient grindability and the cuttability with the cutter knife, especially at fibre bundles length of 1.6 mm.

3.1. Formaldehyde Content

The formaldehyde content of the boards with fibre bundle length of 4 mm and 180 kg/m^3 density was determined, according to EN ISO 12460:5:2015 at the company Kaindl (Wals, Austria), to be 0 mg/100 g. Since bark has the ability to bind formaldehyde [6], values significantly under 1 mg were expected; however, similar studies based on larch bark panels showed slightly higher formaldehyde contents [6,39]. With no formaldehyde content, these boards are included in the super E0 classification (<1.5 mg/100 g). The zero value for the formaldehyde content may be attributed to the high amount of lignin in the chemical composition of tree bark [40]. The lignin content of spruce bark ranges from 26% [41] to 37% [42]. Due to their phenolic nature, bark tannins can react with formaldehyde as a substitute for phenol in the formation of wood adhesives, which can be confirmed by the low formaldehyde content [43].

3.2. Physical Properties

The results of the physical and mechanical properties of the insulation panels made of spruce bark fibres are shown in Table 2.

Table 2. Physical and mechanical properties of the spruce bark fibre insulation boards (values with the same letter (a, b, c, d) are not significantly different ANOVA, post-hoc Tukey HSD, p = 0.05; standard deviation in parentheses).

Sample	TS %	WA %	TC 10 °C mW/(m*K)	TC 25 °C mW/(m*K)	TC 40 °C mW/(m*K)	IB N/mm^2	Density kg/m^3
A200	10.0 ^{b} (5.4)	207 ^{b} (108)	59.9 ^{c} (3.4)	62.2 ^{c} (3.3)	64.5 ^{c} (3.1)	0.129 ^{c} (0.035)	277 ^{d} (19)
B250	18.6 ^{d} (3.8)	301 ^{c} (32)	58.9 ^{c} (3.9)	60.8 ^{c} (3.7)	62.4 ^{c} (3.5)	0.069 ^{b} (0.011)	245 ^{c} (27)
B200	14.4 ^{c} (3.0)	305 ^{c} (57)	47.5 ^{a} (2.8)	49.3 ^{a} (2.7)	51.1 ^{a} (2.6)	0.034 ^{b} (0.006)	185 ^{a} (21)
C200	8.0 ^{a} (3.8)	172 ^{a} (45)	50.6 ^{b} (0.9)	53.4 ^{b} (1.4)	56.1 ^{b} (1.9)	0.009 ^{a} (0.013)	204 ^{b} (4)

3.3. Thickness Swelling and Water Absorption after 24 h

For the analysis of the thickness swelling (TS) and water absorption (WA) after 24 h measured according to EN 317:1993, two cases need to be considered. In some samples for all fibre bundles lengths (1.6, 4, and 7 mm), dry spots could be located, which show a different behaviour compared to the wet samples (Figure 2). Dry spots seem to occur at the lowest densities of each fibre bundle length. Such an effect could not be observed by similar studies, which rather show a linear corelation [21]. To the best of our knowledge, this study is the first attempt to investigate the low-density bark fibre boards, so the incidence of dry spots is specific to this material. Additionally, the values for 7 mm fibre bundles' length need to be considered less precise due to the very low values of internal bond (Subchapter 3.4) that influenced the measuring error of TS and WA.

Figure 2. Water penetration over the cross section of the tested sample: fibre length 1.6 mm, wet and dry (**left**); fibre length 4 mm, wet and dry (**middle**); and fibre length 7 mm, wet and dry (**right**).

For TS and WA, after 24 h, a multiple polynomial regression (MPR) analysis was performed with the significant variables: intercept, density, fibre length, and fibre length squared. The regression of TS for the wet samples was highly significant and positively correlated, while R^2 showed a value of 0.61 (Figure 3). With a polynomial regression of the fibre length, it was possible to model the thickness swelling for all fibre bundle classes. A typical TS for wood-based panels shows a positive correlation with the fibre thickness [13]. Since with longer fibres an increased fibre thickness can be expected, when using the described defibration method, the 7 mm fibre bundle length (C) did not follow this prediction. Due to the reduced IB and density of this panel, combined with the low slenderness ratio, large holes and less felted regions can be expected, compared with 1.6 (A) and 4 mm (B) fibre lengths. This characteristic can affect the TS, since holes are a favourable field for the swelling fibres. For the dry samples, the regression was still significant; however, R^2 was 0.47, most likely due to grouped density range of the dry samples. Additionally, the dry samples of fibre length 1.6 (A) and 4 mm (B) seem to have the lowest density of their classes. The difference in TS between wet and dry samples was not as strong as in the WA. It is assumed that the core fibres soaked up just enough water to stay under the equivalent of the fibre saturation point in wood. The highest TS 24 h value of 23% could be found for 4 mm fibre length (B), whereas the lowest values with 3.3% and 1.6% could be found for 1.6 mm (A) and 7 mm (C) fibres. Medium-sized fibres seem to have a disadvantage in terms of TS. The smaller density variation for 1.6 mm (A) and 7 mm (C) fibre length makes it hard to predict whether the TS and WA trends foreseen by the model are stable at other density levels, too.

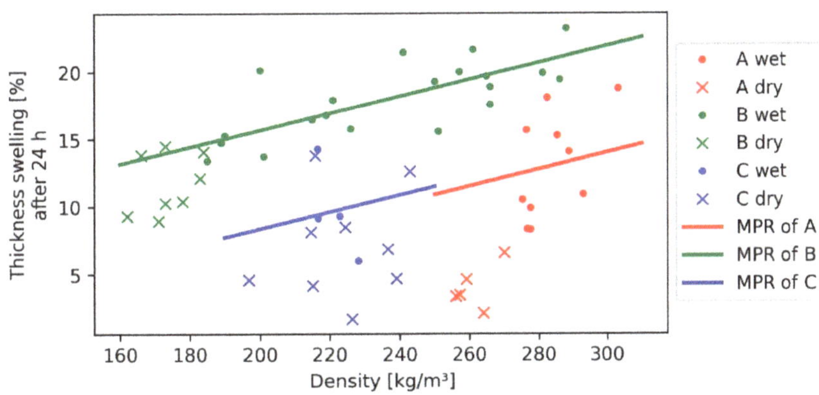

Figure 3. Multiple polynomial regression (MPR) of thickness swelling after 24 h for each bark fibre length class with the density; only the regression for wet samples is depicted.

The water absorption (WA) after 24 h for the wet sample values varied from 217% to 380%, according to the production method and the absence of adhesive or hydrophobic additives. In contrast, in dry samples with 1.6 mm fibre length, values under 55% were achieved. Figure 4 shows relatively similar WA of wet samples of 1.6 mm (A) and 4 mm (B). As explained in previous subchapter TS, the WA of the C samples (7 mm) might be underestimated. If the WA of 7 mm fibre bundles is excepted from the measurement series, it could be stated that the fibre length seems to have little effect on the water absorption. Between the dry and wet samples of each fibre length, a sudden drop can be identified, which increases with decreased fibre length, indicating a polynomial corelation to the density, caused by the dry samples. In that case, it can be expected that with smaller fibre length and lower densities, the water absorption can be decreased, which in most cases is a favourable material property. The decreased water absorption of dry samples is considered to be caused by the increased pore size of lower density samples, which come along with a reduced capillary force, so the water could not be soaked into the sample. Assuming that, compared to wood, bark contains more hydrophobic substances, such as suberin [14], an improvement of water repellence can be observed in such composites. The regression of the WA shows a negative correlation with the density, which is also more significant and shows less prediction error due to an R^2 of 0.73. Since similar studies [25] also show a negative correlation, the measured data correspond to the expectations. For the dry samples, a highly significant model with an R^2 of 0.88 was obtained. As Figure 4 depicts, the dry samples appear grouped and do not show overlapping density areas like the wet samples. The accuracy of the model regarding the negative correlation of the dry samples model needs to be questioned. On the one hand, fibre bundles of 7 mm (C) and (A) indicate the negative correlation. On the other hand, the fibre bundles of 4 mm (B) seemed to follow more of a polynomial behaviour towards the density and therefore a positive correlation with the density in the corresponding interval.

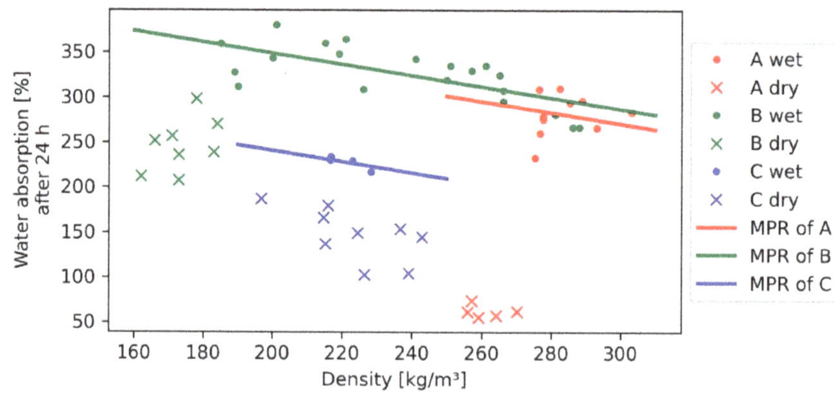

Figure 4. Polynomial regression (MPR) of water absorption after 24 h for each bark fibre length class (FLC) with the density; only the regression for wet samples is depicted.

3.4. Internal Bond

The internal bond ranged between 0.2 and 0.0 N/mm^2; however, only boards with 1.6 mm fibre bundle length (A) were able to achieve values over 0.1 N/mm^2 (Figure 5). One reason is the higher density of those boards (277 kg/m^3 average), caused by the proportionately higher shrinking of the board during the drying process.

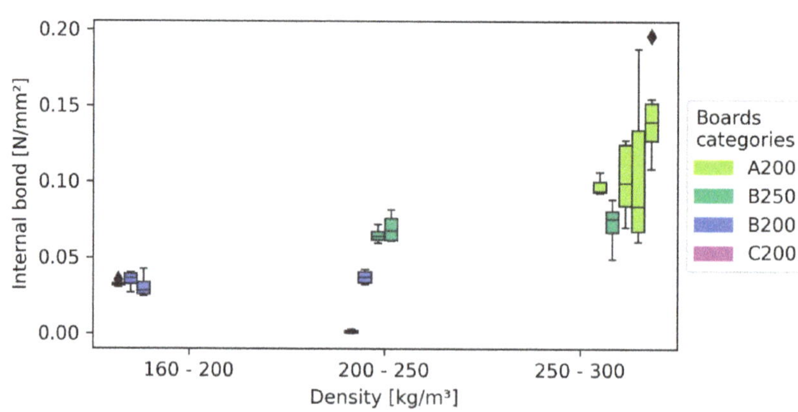

Figure 5. Internal bond of the insulation panels with 1.6, 4, and 7 mm bark fibre length as a function of density.

As depicted in Figure 3, the internal bond of C200 boards (7 mm fibre bundle length) shown in comparison to the other boards' values close to 0.0 N/mm^2. This behaviour can be explained due to thicker fibres that decreased the homogeneity and, as a consequence, the self-bonding capacity [13]. Within the other board categories (B200, B250, and A200), the IB showed almost no correlation with the density. Because such a correlation is typically for wood-based panels [13], this indicates together with the asymmetry of some boxplots (due to the scattering of internal bond, especially for the fibre length 1.6 mm (A)), that this effect is caused by variation within the panels. Because this variation increases with the density, heteroscedastic effects were detected by the "White-test" for heteroscedasticity within the regression model. Because a constant amount of process water was used during the panel manufacturing, the solid content of the fibre-water suspension of boards with higher density was subsequently higher than those of panels with lower density, therefore inducing more variation to the higher density boards. Simple linear regressions (SLR) involving only one fibre class do not show heteroscedasticity, and it is therefore a result of the larger dataset of the multiple linear regression (MLR). However, the variation of fibre length 1.6 mm seems to be higher than that of fibre length of 4 mm (Figure 6), slightly indicating that smaller fibres might be more prone to irregular fibre distribution during the forming process than longer fibres at the same density.

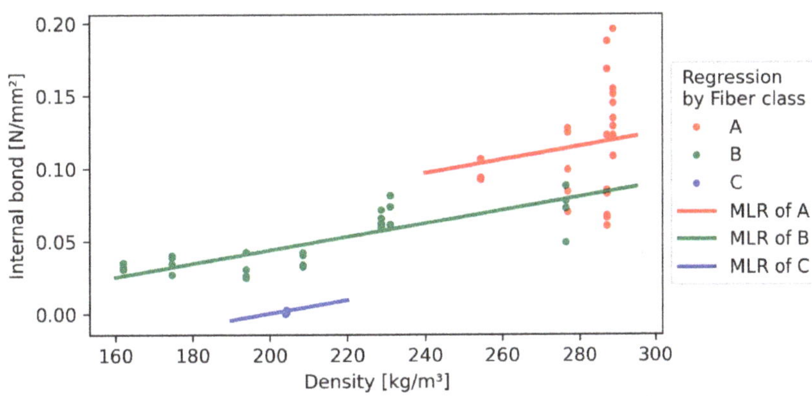

Figure 6. Multiple linear regression (MLR) of internal bond towards the independent variables density and bark fibre length (lines) and measured values (points).

The MLR shows an R^2 of 0.7, while both variables (fibre length and density) are highly significant. Due to the uneven distribution of the samples of fibre length of 1.6 mm (A) and 7 mm (C), the results of the regression are only valid in areas, where data points are available. Additionally, its explanatory power is decreased by its heteroscedasticity. Irregular fibre distribution can lead to a decrease in the average performance of the board and an increased variation within the board [44]. Therefore, heteroskedasticity in the model indicates an underestimation of the slope of the model.

3.5. Thermal Conductivity (TC)

The lowest TC was measured with 0.044 W/(m*K) and a density of 162 kg/m^3 at the fibre length of 4 mm (B), whereas the highest value of 0.063 W/(m*K) was measured in the same fibre length at 276 kg/m^3. In general, boards with a lower variation in density also showed a lower TC and a larger asymmetry in the boxplots (Figure 7), as also the coefficient of variation indicates, since both vary in similar intervals of 2–7% for TC and 2–11% for the density. Bark fibre boards performed in terms of TC around 8% better at lower temperature (10 °C) than at higher temperature (40 °C), as can be seen in Figure 7.

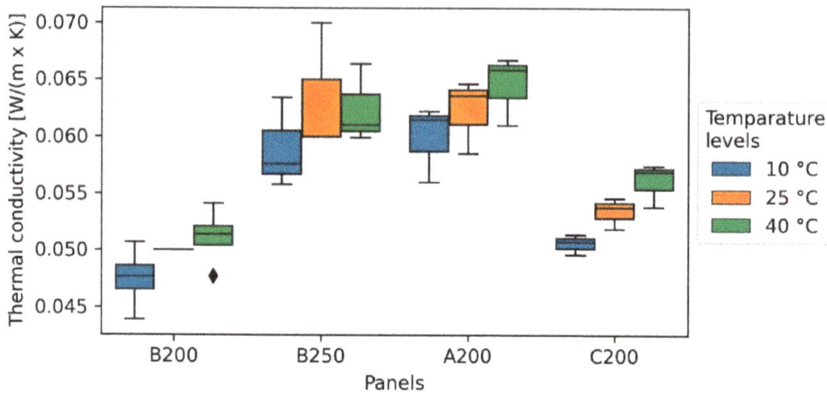

Figure 7. Thermal conductivity of bark fibreboards under different temperature levels (10, 25, and 40 °C).

Similar to the IB, the TC also correlates highly significantly with the density, but in contrast to it, the fibre length was not significant, resulting in a simple linear model. However, some studies indicate a slight influence of the fibre length [45]. To exclude any influence of the fibre length, a larger data set is required. However, as Figure 8 presents, the regression model fits the data with R^2 of 0.94 in comparison to the other models quite well, most likely because density variation within the boards does not influence the outcome strongly. For the same reason, the model is not affected by heteroscedasticity like the IB model, even if it is possible for irregularities in the panel forming to also influence the TC to some extent.

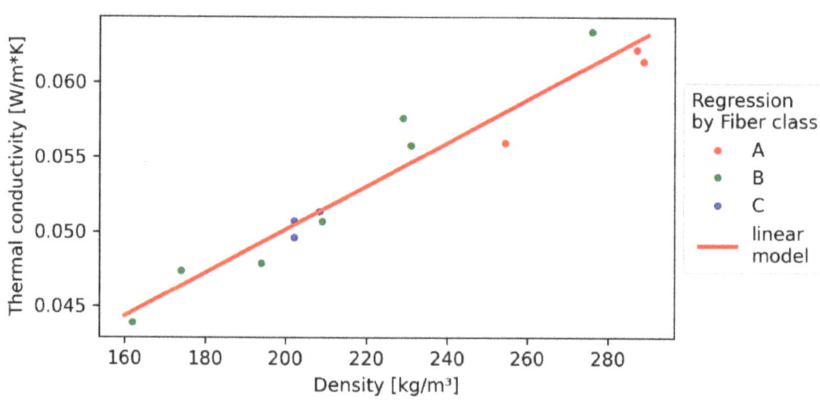

Figure 8. Simple linear regression (SLR) of thermal conductivity towards density (line) and measured values (points).

Compared to similar studies, the slope of the model (0.013 W/m*K/100 kg/m^3) is 54% higher (Figure 8) [21]. The TC of the spruce bark fibre insulation panels is at least 15% higher compared to mineral wool and polystyrene (approximately 0.03 W/m*K) but seemed to have a benefit over particle-based insulations, as reported by Kain et al., 2020 [21].

4. Conclusions

To the best of our knowledge, this research is the first attempt to investigate low-density insulation boards made of bark fibres.

The results of this study showed that the thermal insulation properties of bark fibre insulation boards can reach thermal conductivity from 0.044 W/m*K (at a density of 164 kg/m^3) to 0.063 W/m*K (276 kg/m^3), being significantly influenced by the density. These TC values are comparable to established insulation boards based on cork or wood fibres [46]. The effect of the fibre length was not significant for the TC, as observed in previous studies [45]. However, with spruce bark fibres, it is possible to achieve lower thermal conductivity than with particle-based bark panels [21].

The internal bond was furthermore influenced significantly by the length of the bark fibres bundles. However, the used model has a large variation and is therefore less reliable. The variances are most likely caused by an insufficiently low solid content during the board production, but the wet process has still proved its ability to produce bark fibre-based insulation boards without adding resins, therefore indicating a sufficient self-agglomeration and sticking of bark fibres.

However, without additional hydrophobic additives, the water absorption after 24 h can rise up to 380%, while the thickness swelling after 24 h remains under 25%. At lower density, bark fibre boards did not show a complete wetting anymore, which goes along with a drop in water absorption down to 55% and a reduced thickness swelling.

Based on the measured thermal conductivity and zero formaldehyde content, bark fibre insulation panels might be able to compete with conventional insulations if the density can be further reduced, but also applications regarding its acoustic insulation are thinkable [28]. To answer these questions, further research is necessary regarding fields such as the impacts of bark species and bark quality as well as other production methods or properties crucial to certain applications such as its protection capability towards structure-borne noise or fire.

Author Contributions: Conceptualisation, J.G. and E.M.T.; methodology, J.G.; validation, E.M.T., M.-C.B., and A.P.; formal analysis, A.P.; investigation, J.G.; resources, J.G.; data curation, J.G.; writing—original draft preparation, J.G and E.M.T.; writing—review and editing, J.G., E.M.T., and M.-C.B.; visualisation, J.G.; supervision, M.-C.B.; project administration, M.-C.B. and A.P.; funding acquisition, A.P. All authors have read and agreed to the published version of the manuscript.

Funding: This research received no external funding.

Institutional Review Board Statement: Not applicable.

Informed Consent Statement: Not applicable.

Data Availability Statement: It is not the case, no datasets.

Acknowledgments: The authors want to express their thankfulness to Thomas Wimmer from "Forest Products Technology and Timber Construction Department" at Campus Kuchl of the Salzburg University of Applied Sciences for his support during the sample testing.

Conflicts of Interest: The authors declare no conflict of interest.

References

1. Pásztory, Z.; Mohácsiné, I.R.; Gorbacheva, G.; Börcsök, Z. The utilization of tree bark. *Bioresources* **2016**, *11*, 7859–7888. [CrossRef]
2. Chow, P.; Nakayama, F.S.; Blahnik, B.; Youngquist, J.A.; Coffelt, T.A. Chemical constituents and physical properties of guayule wood and bark. *Ind. Crop. Prod.* **2008**, *28*, 303–308. [CrossRef]
3. Jablonsky, M.; Nosalova, J.; Sladkova, A.; Haz, A.; Kreps, F.; Valka, J.; Miertus, S.; Frecer, V.; Ondrejovic, M.; Sima, J.; et al. Valorisation of softwood bark through extraction of utilizable chemicals. A review. *Biotechnol. Adv.* **2017**, *35*, 726–750. [CrossRef]
4. Borysiuk, P.; Boruszewski, P.; Auriga, R.; Danecki, L.; Auriga, A.; Rybak, K.; Nowacka, M. Influence of a bark-filler on the properties of PLA biocomposites. *J. Mater. Sci.* **2021**, *56*, 9196–9208. [CrossRef]
5. Aydin, I.; Demirkir, C.; Colak, S.; Colakoglu, G. Utilization of bark flours as additive in plywood manufacturing. *Eur. J. Wood Prod.* **2017**, *75*, 63–69. [CrossRef]
6. Barbu, M.C.; Lohninger, Y.; Hofmann, S.; Kain, G.; Petutschnigg, A.; Tudor, E.M. Larch bark as a formaldehyde scavenger in thermal insulation panels. *Polymers* **2020**, *12*, 2632. [CrossRef]
7. Sutrisno Alamsyah, E.M.; Syamsudin, T.S.; Purwasasmita, B.S.; Suzuki, S.; Kobori, H. The potential using of organic nanoparticles synthesized from Gmelina (Gmelina arborea Roxb.) wood bark as nanofiller of wood adhesive: Physical, chemical and thermal properties. *J. Indian Acad. Wood Sci.* **2020**, *17*, 165–175. [CrossRef]

8. Réh, R.; Krišťák, L'.; Sedliačik, J.; Bekhta, P.; Božiková, M.; Kunecová, D.; Vozárová, V.; Tudor, E.M.; Antov, P.; Savov, V. Utilization of birch bark as an eco-friendly filler in urea-formaldehyde adhesives for plywood manufacturing. *Polymers* **2021**, *13*, 511. [CrossRef]
9. Feng, S.; Cheng, S.; Yuan, Z.; Leitch, M.; Xu, C. Valorization of bark for chemicals and materials: A review. *Renew. Sustain. Energy Rev.* **2013**, *26*, 560–578. [CrossRef]
10. Bortenschlager, S.; Oeggl, K. *The Iceman and His Natural Environment: Palaeobotanical Results*; Springer: Vienna, Austria, 2000.
11. Pásztory, Z.; Ronyecz Mohácsiné, I.; Börcsök, Z. Investigation of thermal insulation panels made of black locust tree bark. *Constr. Build. Mater.* **2017**, *147*, 733–735. [CrossRef]
12. Pásztory, Z.; Ronyecz, I. The thermal insulation capacity of tree bark. *Acta Silv. Lignaria Hung.* **2013**, *9*, 111–117. [CrossRef]
13. Paulitsch, M.; Barbu, M.C. *Holzwerkstoffe der Moderne, 1. Aufl.*; DRW-Verlag: Leinfelden-Echterdingen, Germany, 2015.
14. Busquets-Ferrer, M.; Czabany, I.; Vay, O.; Gindl-Altmutter, W.; Hansmann, C. Alkali-extracted tree bark for efficient bio-based thermal insulation. *Constr. Build. Mater.* **2021**, *271*, 121577. [CrossRef]
15. Blanchet, P.; Cloutier, A.; Riedl, B. Particleboard made from hammer milled black spruce bark residues. *Wood Sci. Technol.* **2000**, *34*, 11–19. [CrossRef]
16. Pedieu, R.; Riedl, B.; Pichette, A. Properties of mixed particleboards based on white birch (*Betula papyrifera*) inner bark particles and reinforced with wood fibres. *Eur. J. Wood Prod.* **2009**, *67*, 95–101. [CrossRef]
17. Yemele, M.; Koubaa, A.; Diouf, P.N.; Blanchet, P.; Cloutier, A.; Stevanovic, T. Effects of hot-water treatment of black spruce bark and trembling aspenbark rawmaterialon the physical and mechanical properties of bark particleboards. *Wood Fiber Sci.* **2008**, *40*, 339–351.
18. Xing, C.; Zhang, S.Y.; Deng, J.; Wang, S. Investigation of the effects of bark fiber as core material and its resin content on three-layer MDF performance by response surface methodology. *Wood Sci. Technol.* **2007**, *41*, 585–595. [CrossRef]
19. Xing, C.; Deng, J.; Zhang, S.Y. Effect of thermo-mechanical refining on properties of MDF made from black spruce bark. *Wood Sci. Technol.* **2007**, *41*, 329–338. [CrossRef]
20. Gao, Z.; Wang, X.M.; Wan, H.; Brunette, G. Binderless panels made with black spruce bark. *Bioresources* **2011**, *6*, 3960–3972.
21. Kain, G.; Tudor, E.M.; Barbu, M.-C. Bark thermal insulation panels: An explorative study on the effects of bark species. *Polymers* **2020**, *12*, 2140. [CrossRef]
22. Rosell, J.A.; Gleason, S.; Méndez-Alonzo, R.; Chang, Y.; Westoby, M. Bark functional ecology: Evidence for tradeoffs, functional coordination, and environment producing bark diversity. *New Phytol.* **2014**, *201*, 486–497. [CrossRef] [PubMed]
23. Tudor, E.M.; Scheriau, C.; Barbu, M.C.; Réh, R.; Krišťák, L'.; Schnabel, T. Enhanced resistance to fire of the bark-based panels bonded with clay. *Appl. Sci.* **2020**, *10*, 5594. [CrossRef]
24. Kain, G.; Güttler, V.; Barbu, M.-C.; Petutschnigg, A.; Richter, K.; Tondi, G. Density related properties of bark insulation boards bonded with tannin hexamine resin. *Eur. J. Wood Prod.* **2014**, *72*, 417–424. [CrossRef]
25. Kain, G.; Lienbacher, B.; Barbu, M.-C.; Richter, K.; Petutschnigg, A. Larch (Larix decidua) bark insulation board: Interactions of particle orientation, physical–mechanical and thermal properties. *Eur. J. Wood Prod.* **2018**, *76*, 489–498. [CrossRef]
26. Tudor, E.M.; Zwickl, C.; Eichinger, C.; Petutschnigg, A.; Barbu, M.C. Performance of softwood bark comminution technologies for determination of targeted particle size in further upcycling applications. *J. Clean. Prod.* **2020**, *269*, 122412. [CrossRef]
27. Tsalagkas, D.; Börcsök, Z.; Pásztory, Z. Thermal, physical and mechanical properties of surface overlaid bark-based insulation panels. *Eur. J. Wood Prod.* **2019**, *77*, 721–730. [CrossRef]
28. Tudor, E.M.; Dettendorfer, A.; Kain, G.; Barbu, M.C.; Réh, R.; Krišťák, L. Sound-Absorption coefficient of bark-based insulation panels. *Polymers* **2020**, *12*, 1012. [CrossRef] [PubMed]
29. Li, M.; van Renterghem, T.; Kang, J.; Verheyen, K.; Botteldooren, D. Sound absorption by tree bark. *Appl. Acoust.* **2020**, *165*, 107328. [CrossRef]
30. Yemele, M.C.; Blanchet, P.; Cloutier, A.; Koubaa, A. Effect of bark content and particle geometry on the physical and mechanical properties of particleboard made from black spruce and trembling aspen bark. *For. Prod. J.* **2008**, *58*, 48–56.
31. Burrows, C.H. Particleboard from Douglas-fir bark without additives. In *Forest Products Research Laboratories, Report, No.15 pp.40 Ref.15.*; United States Department of Agriculture: Madison, WI, USA, 1960.
32. Almusawi, A.; Lachat, R.; Atcholi, K.E.; Gomes, S. Proposal of manufacturing and characterization test of binderless hemp shive composite. *Int. Biodeterior. Biodegrad.* **2016**, *115*, 302–307. [CrossRef]
33. Krilov, A. Debarking of fibrous-barked hardwoods by ultra-high pressure water jets. *Wood Sci.Technol.* **1983**, *17*, 145–158. [CrossRef]
34. *EN 12667:2001-Thermal Performance of Building Materials and Products—Determination of Thermal Resistance by Means of Guarded Hot Plate and Heat Flow Meter Methods—Products of High and Medium Thermal Resistance*; CEN, European Committee for Standardization: Brüssel, Belgium, 2001.
35. *EN 1607:2013-Plattenebene Thermal Insulating Products for Building Applications—Determination of Tensile Strength Perpendicular to Faces*; CEN, European Committee for Standardization: Brüssel, Belgium, 2013.
36. *EN 317:2005-Particleboards and Fibreboards—Determination of Swelling in Thickness after Immersion in Water*; CEN, European Committee for Standardization: Brüssel, Belgium, 2005.
37. *ISO 12460-5:2015-Wood-Based Panels—Determination of Formaldehyde Release—Part 5: Extraction Method (Called the Perforator Method)*; CEN, European Committee for Standardization: Brüssel, Belgium, 2015.

38. *EN 326-1:2005-Wood-Based Panels—Sampling, Cutting and Inspection—Part 1: Sampling and Cutting of Test Pieces and Expression of Test Results*; CEN, European Committee for Standardization: Brüssel, Belgium, 2005.
39. Tudor, E.M.; Barbu, M.C.; Petutschnigg, A.; Réh, R.; Krišťák, Ľ. Analysis of larch-bark capacity for formaldehyde removal in wood adhesives. *Int. J. Environ. Res. Public Health* **2020**, *17*, 764. [CrossRef]
40. Medved, S.; Gajšek, U.; Tudor, E.M.; Barbu, M.C.; Antonović, A. Efficiency of bark for reduction of formaldehyde emission from particleboards. *Wood Res.* **2019**, *64*, 307–316.
41. Burhenne, L.; Messmer, J.; Aicher, T.; Laborie, M.-P. The effect of the biomass components lignin, cellulose and hemicellulose on TGA and fixed bed pyrolysis. *J. Anal. Appl. Pyrolysis* **2013**, *101*, 177–184. [CrossRef]
42. Kemppainen, K.; Siika-aho, M.; Pattathil, S.; Giovando, S.; Kruus, K. Spruce bark as an industrial source of condensed tannins and non-cellulosic sugars. *Ind. Crop. Prod.* **2014**, *52*, 158–168. [CrossRef]
43. Li, J.; Li, C.; Wang, W.; Zhang, W. Reactivity of larch and valonia tannins in synthesis of tannin-formaldehyde resins. *Bioresources* **2016**, *11*, 2256–2268. [CrossRef]
44. Blechschmidt, J. (Ed.) *Taschenbuch der Papiertechnik*; Carl Hanser Verlag GmbH & Co. KG: München, Germany, 2010.
45. Medved, S.; Lesar, B.; Tudor, E.M.; Humar, M. Thermal insulation panels from cellulosic fibres. *For. Prod. J.* **2015**, *65*, 554–558.
46. Sprengard, C.; Treml, S.; Holm, H.A. Technologien und Techniken zur Verbesserung der Energieeffizienz von Gebäuden durch Wärmedämmstoffe: Metastudie Wärmedämmstoffe—Produkte—Anwendungen—Innovationen. 2013. Available online: http://www.fiw-muenchen.de/media/pdf/metastudie_waermedaemmstoffe.pdf (accessed on 9 January 2020).

polymers

MDPI

Review

Lignosulphonates as an Alternative to Non-Renewable Binders in Wood-Based Materials

Sofia Gonçalves [1], João Ferra [2], Nádia Paiva [2], Jorge Martins [1,3], Luísa H. Carvalho [1,3] and Fernão D. Magalhães [1,*]

1 LEPABE–Faculdade de Engenharia da Universidade do Porto, Rua Dr. Roberto Frias, 4200-465 Porto, Portugal; up201808942@edu.fe.up.pt (S.G.); jmmartins@estgv.ipv.pt (J.M.); lhcarvalho@estgv.ipv.pt (L.H.C.)
2 Sonae Arauco Portugal S.A., Lugar do Espido—Via Norte, 4470-177 Porto, Portugal; joao.ferra@sonaearauco.com (J.F.); nadia.paiva@sonaearauco.com (N.P.)
3 DEMad–Departamento de Engenharia de Madeiras, Instituto Politécnico de Viseu, Campus Politécnico de Repeses, 3504-510 Viseu, Portugal
* Correspondence: fdmagalh@fe.up.pt

Abstract: Lignin is a widely abundant renewable source of phenolic compounds. Despite the growing interest on using it as a substitute for its petroleum-based counterparts, only 1 to 2% of the global lignin production is used for obtaining value-added products. Lignosulphonates (LS), derived from the sulphite pulping process, account for 90% of the total market of commercial lignin. The most successful industrial attempts to use lignin for wood adhesives are based on using this polymer as a partial substitute in phenol-formaldehyde or urea-formaldehyde resins. Alternatively, formaldehyde-free adhesives with lignin and lignosulphonates have also been developed with promising results. However, the low number of reactive sites available in lignin's aromatic ring and high polydispersity have hindered its application in resin synthesis. Currently, finding suitable crosslinkers for LS and decreasing the long pressing time associated with lignin adhesives remains a challenge. Thus, several methods have been proposed to improve the reactivity of lignin molecules. In this paper, techniques to extract, characterize, as well as improve the reactivity of LS are addressed. The most recent advances in the application of LS in wood adhesives, with and without combination with formaldehyde, are also reviewed.

Keywords: lignosulphonates; lignin; formaldehyde; wood adhesives

Citation: Gonçalves, S.; Ferra, J.; Paiva, N.; Martins, J.; Carvalho, L.H.; Magalhães, F.D. Lignosulphonates as an Alternative to Non-Renewable Binders in Wood-Based Materials. *Polymers* **2021**, *13*, 4196. https://doi.org/10.3390/polym13234196

Academic Editor: Pavlo Bekhta

Received: 15 November 2021
Accepted: 22 November 2021
Published: 30 November 2021

1. Introduction

Lignin is a complex, amorphous, natural polymer, and one of the most abundant in nature, only behind cellulose [1]. Although lignin is the main by-product of the paper pulping processes, it is usually burned as fuel [2].

Out of the 50 to 70 million tons of the lignin that is produced annually, only 1 to 2% is actually used for the production of value-added products. Therefore, it can be concluded that lignin is an underutilized material [3].

In plants, cellulose is present in the form of bundled fibrils that are bound together by hemicellulose and lignin, providing rigidity to the cell wall [4]. Lignin performs the following functions in plants: providing mechanical support, slowing down decomposition, forming a barrier towards water evaporation, and helping to transport water to vital areas of the plant [2]. In vitro studies have shown that lignin and lignin extracts display antimicrobial and antifungal activity, act as antioxidants, absorb UV radiation, and exhibit flame-retardant properties [5].

Lignin is the main renewable source of phenolic compounds of natural origin. This has been increasingly researched as a more eco-friendly alternative to the petroleum-based counterparts [6]. The molecule is comprised of several types of methoxylated phenylpropanoid units (C9). The three primary precursors of lignin are: *p*-coumaryl, coniferyl, and sinapyl alcohols, as shown in Figure 1. These monolignols are also known as *p*-hidroxyphenyl (H), guaiacyl (G), and syringyl (S) units, respectively [6–8].

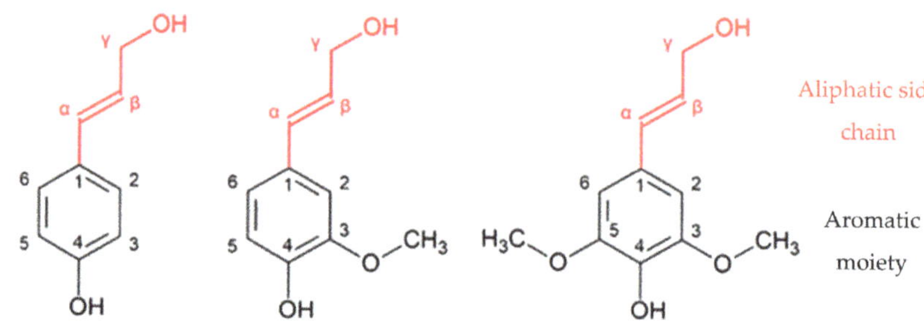

Figure 1. Monomeric lignin precursors: *p*-coumaryl alcohol, coniferyl alcohol, and sinapyl alcohol (adapted from [6]).

These monolignols differ in the number of methoxy groups that are attached to the aromatic moiety: sinapyl alcohol has two methoxy groups, coniferyl alcohol has one methoxy group, and *p*-coumaryl alcohol has none [6]. Different types of plants display different degrees of participation of the major monolignols. As shown in Table 1, the main monolignols in softwood and hardwoods are coniferyl and sinapyl alcohol, respectively [9].

Table 1. Percentages of the monolignols in lignin in different plants [9].

Linkage Type	*p*-Coumaryl Alcohol (%)	Coniferyl Alcohol (%)	Sinapyl Alcohol (%)
Coniferous; softwoods	<5 [a]	>95	0 [b]
Eudicotyledonous; hardwoods	0–8	25–50	45–75
Monocotyledonous; grasses	5–35	35–80	20–55

[a] Higher amount in compression wood. [b] Some exceptions exist.

Researchers admit that no plants contain lignin that is only derived from the three primary precursors [6]. Noncanonical subunits that have been identified include caffeoyl alcohol, which was discovered in the seeds of some Vanilla and Cactaceae species [10]. Other subunits are ferulic acid, ferulates (which form linkages between hemicellulose and lignin), coniferaldehyde, sinapaldehyde, and acylated monoglinols [6]. Intermediate free radicals are generated from these lignin precursors through the dehydrogenation of phenolic OH groups by the plant's peroxidase and laccase enzymes. The polymerization process then occurs as follows: firstly, two radicals are coupled forming a dimer, and the process progresses with the coupling of monomeric radicals with dimer, trimers, and oligomers resulting in a complex branched polymer (lignification) [11,12].

The great complexity of the lignin structure is due to the variability of the linkages found in it, such as ether, esters, and carbon-carbon, the most abundant being the bond β-O-4, β-β, and β-5, as seen in Figure 2 [6,13].

Lignin has a great variability of functional groups in its complex structure. The main ones are shown in Figure 3. Their abundance in the structure also depends on the source of lignin [13–15]. The frequency and type of the most common linkages in softwood and hardwood lignins are described in Table 2.

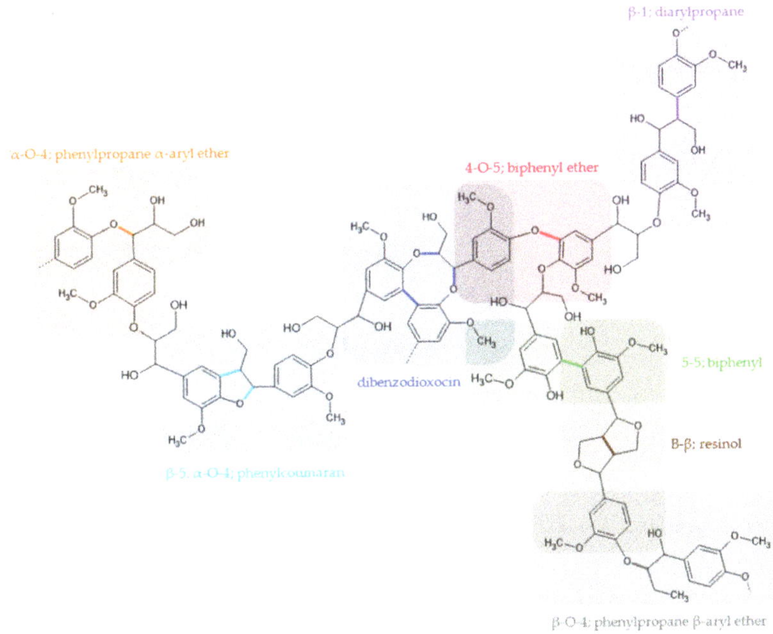

Figure 2. Model of lignin's structure with the important linkages and units (adapted from [7]).

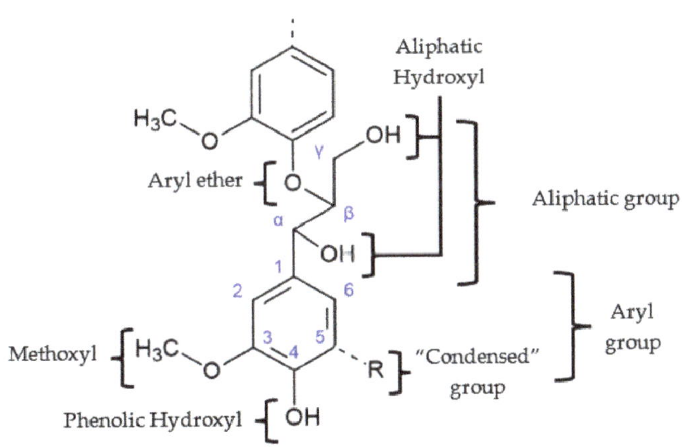

Figure 3. Main functional groups in lignin's structure (adapted from [14]).

Table 2. Types and frequencies of linkages in softwood and hardwood lignins [11].

Linkage Type	Softwood (Spruce) (%)	Hardwood (Birch) (%)
β-O-4-Aryl ether	46	60
α-O-4-Aryl ether	6–8	6–8
4-O-5-Diaryl ether	3.5–4	6.5
β-5-Phenylcoumaran	9–12	6
5-5-Biphenyl	9.5–11	4.5
β-1-(1,2-Diarylpropane)	7	7
β-β-(Resinol)	2	3
Others	13	5

2. Technical Lignins

Technical lignins are mostly obtained as co-products of the manufacture of cellulose pulp for paper, through wood pulping (delignification) processes [16,17]. The main objective of wood pulping is to liberate the cellulose fibers from the lignin binder and other non-fibrous compounds. The most common commercial processes can be grouped into four types: chemical, semichemical, chemimechanical, and mechanical [18].

In chemical pulping, delignification occurs until most of the lignin in the middle lamella of the woody cell is removed. This results in an easy separation of the fibers. Most of these processes are currently based predominantly on the sulfate (kraft) process and, less frequently, on the sulphite process [18].

On the other hand, in semichemical processes only a partial dissolution of lignin is achieved, as the wood chips are cooked during shorter periods of time or under milder conditions. Chemimechanical pulps are produced by pretreating the wood chips usually at elevated temperatures in alkaline solutions of sodium sulfite before defibration [18]. Lastly, mechanical pulping uses no chemicals, only mechanical abrasion combined with water or steam. Thermomechanical pulping has become the most important method of this kind. In this procedure, the refiners are first pressurized with steam at high temperatures in order to promote fiber liberation and operate at ambient temperature in a second stage [19].

During pulping lignin's structure is inevitably modified. Therefore, the type of pulping process determines the type of lignin that is industrially obtained. Some examples of chemical pulping that will be addressed in this study are sulphite (lignosulphonates), kraft, soda, and organosolv lignins [6,16].

Although there is a great diversity of technical lignins, this article will focus mainly on lignosulphonates.

2.1. Lignosulphonates

Most of the pulp in the world was produced through the sulphite process until the 1950s. After that, the kraft process has been the dominating method. However, the sulphite process is still important in some countries and for certain pulp qualities [18].

Currently, the total annual global production of lignosulphonates is approximately 1.8 million tons. These technical lignins account for 90% of the total market of commercial lignin since the kraft process yields relatively small quantities of usable lignin [3,16].

This consists in the digestion of wood at 130–180 °C with an aqueous solution of a sulphite or bisulphite salt of sodium, ammonium, magnesium, or calcium [16,18]. There are several modifications of the sulphite method, which are designated according to the pH of the cooking liquor, as shown in Table 3 [18].

Table 3. Different modifications of the sulphite method [13,18].

	Acid (bi)Sulphite	Bisulphite	Neutral Sulphite	Alkaline Sulphite
pH range	1–2	3–5	6–9	9–13
Base alternatives	Ca^{2+}, Mg^{2+}, Na^+, NH_4^+	Mg^{2+}, Na^+, NH_4^+	Na^+, NH_4^+	Na^+
Active reagents	HSO_3^-, H^+	HSO_3^-, H^+	HSO_3^-, SO_3^{2-}	SO_3^{2-}, HO^-
Max. temp. (°C)	125–145	150–170	160–180	160–180
Time at max. temp. (h)	3–7	1–3	0.25–3	3–5
Softwood pulp yield (%)	45–55	50–65	75–90 [a]	45–60

[a] Hardwood.

During the digestion process the linkages between lignin and carbohydrates are cleaved, as well as the carbon–oxygen bonds that connect lignin units. However, the sulphonation of the lignin aliphatic chain is the most important reaction as it lends the resulting LS their main characteristics [16].

This reaction consists in the attack on lignin's structure by the negatively charged sulphite or bisulphite ions. The targets of this nucleophilic attack depend on the pH of the sulphite delignification. At high pH, quinone methide structures of phenolic units are the targets. However, at low pH, carbenium (benzylium) ions of phenolic or non-phenolic units are sulphonated at the position of the side chain, as shown in Figure 4 [16].

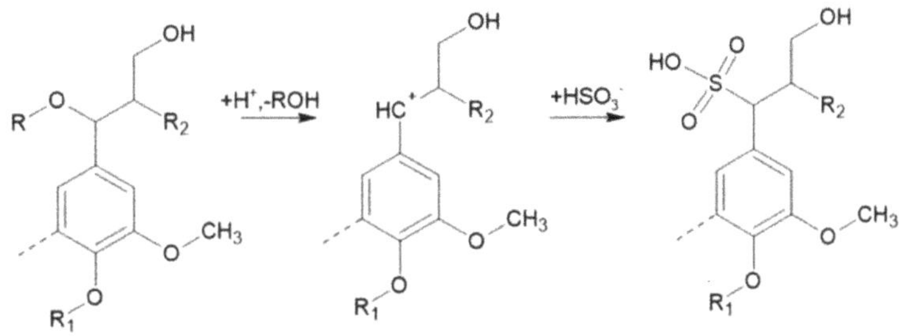

R = H, alkyl, aryl
R_1 = H, neighbouring lignin unit
R_2 = neighbouring lignin unit

Figure 4. Main reactions for lignosulphonate formation during acid sulphite pulping (adapted from [16]).

During pulping, about 4–8% sulphur is incorporated into lignin's structure, making lignin water-soluble and preventing its recondensation, which would result in the redeposition of lignin on the cellulose fibers [16,20].

The resulting fiber pulp is separated from the spent pulping liquor through filtration and washing. This liquor contains 50 to 80 wt% of lignosulphonates, as well as hemicelluloses and residual pulping chemicals. The structures of these lignosulphonates may vary significantly due to the wide range of conditions under which sulphite pulping can be operated [3,16].

Purified forms of lignosulphonates are more valuable and have broader applications than crude spent sulphite liquor. The purification process can be performed according to a variety of techniques [5,16].

Ultrafiltration has been applied on an industrial scale to recover lignosulphonates from spent sulphite liquors. The higher molecular weight of lignosulphonates in comparison to other components in the spent liquors allows this method to be applied effectively. The result is a sugar-rich permeate and a retentate with up to 95% of lignosulphonates [3,5,16].

Alternatively, the spent sulphite liquor can be purified through alcoholic fermentation of the sugars. Lignosulphonates exhibiting a purity over 90% can then be recovered resorting to ultrafiltration [3,5,16].

The cation (or base) originally present in the pulping liquor can influence the physicochemical properties of lignosulphonates. For example, sodium sulphite produces longer lignin chains that are more suitable as dispersants. However, the use of calcium sulphite results in a more compact lignin [5]. Changing this cation may therefore be required for certain applications and can be achieved through the use of ion-exchange resins [16].

Lignosulphonates are soluble in water, have high molecular weights with a broad distribution, and a high ash content. These compounds also contain a variety of functional groups including phenolic hydroxyl and carboxylic groups and sulphur containing groups [20].

2.2. *Kraft Lignin*

The kraft process is the leading pulping process worldwide. Kraft pulp mills have a highly engineered incorporated system for recovery of pulping chemicals and energy. This system is crucial for the economic and environmental performance of these mills and relies on the combustion of the black liquor. Thus, the quantity of kraft lignin recovered for chemical use is low in comparison to lignosulphonates [12,16].

Kraft pulping uses a solution composed of sodium hydroxide and sodium sulphide, named white liquor, to cleave lignin's ether bonds [5,18]. This process has a total duration of approximately 2 h and the temperatures range from 150 to 170 °C [21].

Kraft lignin is hydrophobic, highly modified, and displays a lower molecular mass than native lignin [22]. It is also not soluble in water and mostly solvent insoluble except for in highly alkaline mediums (pH > 11) [23].

2.3. Soda Lignin

Soda pulping was introduced in the 1850s as the first chemical pulping method [5,16,18]. Kraft pulping originated from this process and has almost completely replaced it, due to having better delignification selectivity which results in pulp with higher quality [18]. However, the soda process is currently becoming the main chemical pulping method of non-wood fibers such as bagasse, wheat straw, hemp, flax, kenaf, and sisal [5,16,20].

The soda process differs from the kraft process mainly in the cooking liquor that is sulphur-free [16,20]. Both kraft lignin and lignosulphonates can be classified as sulphur containing lignins, which are associated with environmental concerns [23]. Soda lignin, on the other hand, is sulphur-free, meaning that its chemical composition is closer to that of native lignin [20,24]. Soda lignins also differ from lignosulphonates in their low molecular weight, low levels of sugar and ash contaminants, and in being water insoluble. Thus, they are more similar to kraft lignins [16].

Soda lignins from non-wood plants differ from wood lignins since they contain more *p*-hydroxyl units, as well as high silicate and nitrogen [16,20].

2.4. Organosolv Lignin

With the environmental concerns associated with sulphur, many sulphur-free extraction processes have been developed, including the organosolv process [23]. Thus, lignin is extracted from plant-tissues with an aqueous solution of organic solvents at high temperature and pressure [12,22]. These solvents include methanol, ethanol, acetic acid, butanol, phenol, peroxiorganic acids, ethyl acetate, and formic acid [5,12,20,23].

The main advantages of this type of lignin is the absence of sulphur, as well as a less modified and more hydrophobic structure than kraft lignin, combined with lower ash content, higher purity, and usually lower molecular weight [5]. Organosolv lignins also contain many reactive side chains available for further chemical reactions [20]. However, this delignification process is not used widely, since it is expensive, requires a high amount of organic solvents, produces pulp fiber with low quality, and causes extensive corrosion of the plant equipment [5,12,23].

2.5. Comparison of Technical Lignins

As discussed, the lignin's final structure and composition is highly influenced by its origin and extraction method [23]. Therefore, technical lignins display a wide range of properties [20].

Applications of these compounds require them to possess tailored properties. These include molecular weight, purity, homogeneity, and the presence of certain functional groups. The main chemical properties of technical lignins are summarized in Table 4 [20].

Table 4. Main chemical properties of technical lignins [20].

Parameter	Lignosulphonates	Kraft Lignin	Soda Lignin	Organosolv Lignin
Ash content (%)	4.0–8.0	0.5–3.0	0.7–2.3	1.7
Sulphur content (%)	3.5–8.0	1.0–3.0	0	0
Molecular weight, Mw	1000–50,000 (up to 150,000)	1500–5000 (up to 25,000)	1000–3000 (up to 15,000)	500–5000
Polydispersity	4.2–7.0	2.5–3.5	2.5–3.5	1.5

It is important to note that the information collected in Table 4 includes lignin from different sources. As seen in the table, lignosulphonates have the highest ash, sulphur content, molecular weight, and polydispersity [20].

Studies have shown that the molecular weight of LS also depends on wood species. Thus, hardwood LS displayed a lower weight than softwood LS [25].

All the lignins described in Table 4 contain a low amount of β-O-4 linkages (below 10%) and high amounts of C-C bonds, unlike native lignin. However, LS are water-soluble for most of the pH-range, which is not the case for kraft lignin [23].

3. Physico-Chemical Characterization of Lignosulphonates

As previously mentioned, the chemical structure and composition of lignin varies with the source and isolation method. Therefore, it is important to carry out a detailed characterization before it can be included in the production of value-added products [26]. However, this is not an easy task due to lignin's three-dimensional architecture combined with a diversity of chemical linkages and functional groups [27].

Currently there are no uniform or standardized methods for the characterization of lignin. Nevertheless, efforts are being made to develop a series of ISO methods for the characterization of the following lignin features: general composition; functional groups; size and morphology; thermal properties; structural features; and safe handling and processability [28].

Thus, there is little information on fundamental analysis and characterization of technical lignins in literature, especially when it comes to lignosulphonates. This may be due to the declining use of sulphite pulping, the greater complexity of lignosulphonates, the lack of widely accepted standard protocols for the purification of these type of lignins, as well as the fact that the existing methods need to be adapted before they can be applied to lignosulphonates [16].

3.1. Chemical Composition

3.1.1. Lignosulphonate Content

Measuring the amount of lignin in biomass is one of the most important steps in developing lignocellulosic biomass for bio-based chemicals. However, while attempting to achieve this task researchers have been presented with several challenges [29].

One of the most used wet chemical techniques to measure lignin content is the Klason or acid-insoluble lignin method. In this analysis the insoluble residues, after hydrolysis with H_2SO_4 72% (*w*/*w*), are filtered, dried, and weighed. This acid-insoluble lignin is referred to as "Klason lignin". A small portion of lignin is dissolved during hydrolysis. The remainder acid-soluble lignin is determined through ultraviolet (UV) spectroscopy, from the absorbance at 205 nm of the filtrated solution. The total lignin is determined as the sum of these two fractions. The Klason method is described thoroughly in ISO 21,436 [29,30]. Although this process is well established, the sample preparation and analysis are time consuming, costly and may destroy the sample [29]. This method may also be inapplicable when the biomass undergoes multiple transformations and therefore, unable to correctly determine lignin content. This is the case for all processes in which covalent bonds between lignin and secondary components, such as carbohydrates and their condensation products, are formed [31].

For commercial LS samples and spent sulphite liquors, LS content is frequently determined through the direct measurement of absorbance at selected wavelengths using UV spectroscopy. However, determining the extinction coefficient is not a simple task [29]. It should also be noted the wavelength choice may influence the obtained results. Measurements at 205 and 280 nm can be influenced by other substances, such as SO_2 and carbohydrate degradation products, like furfural which also absorbs strongly at 280 nm [32,33]. Solvents for UV measurements of LS include water, and an ethanol/water mixture (2:8 *v*/*v*). Lin summarizes several absorptivity values at 280 nm found in literature for several types of lignin, including lignosulphonates [33]. Some studies also resort to the absorptivity value at 205 nm for acid-soluble lignin described in TAPPI UM 250 and ISO 21,436 [30,34]. Alonso and co-workers, on the other hand, determined the purity of commercial LS samples at 232.5 nm [35]. A procedure for the determination of lignin through the UV method is described by Lin [33].

Lignosulphonate content may also be determined through the Pearl-Benson method which is based on a chemical reaction involving nitrosation. Thus, the phenolic units in lignin react with acidified sodium nitrite forming a nitrosophenol. Then, upon addition of alkali, an intensely colored quinone mono-oxime structure is formed whose absorbance is measured at 430 nm and related to lignin concentration. However, a previous calibration with standard lignin is needed. This procedure is described in detail by Dence [32].

Near-infrared spectroscopy (NIR) has been used to successfully measure lignin content in samples. Firstly, a set reference samples must be prepared and the lignin content of these samples must be determined through a reference method. These data can then be used to calibrate the NIR signal. Thus, the lignin content in unknown samples can be rapidly quantified resorting to their NIR spectra [9].

3.1.2. Total Ash Content

For polymer applications, lignosulphonates with low ash content are desired [36]. Therefore, it is important to determine the total ash content in samples. This parameter can be determined gravimetrically after incineration at 525 °C according to the procedure described in ISO 1762 (2019). In this method the samples are weighed in a heat-resistant crucible and ignited in a muffle furnace at 525 ± 25 °C. Then, the ash content is determined on a dry basis, using the mass of residue after ignition and the dry matter content of the sample [37,38].

3.2. *Molecular Weight*

Gel Permeation Chromatography (GPC)

Molecular weight (MW) is an important property of a polymer, that can provide information, such as degree of polymerization, even before investigating the chemical structure [26,39]. Gel permeation chromatography (GPC) or size exclusion chromatography (SEC) is the method of choice for the determination of the molecular weight distribution (MWD) of technical lignins [39–41]. However, it has been noted in several studies that secondary separation effects in GPC columns can interfere significantly with the obtained results [42].

These effects are caused by interactions between lignin molecules (association), lignin and solvent (solvatation), and lignin column packing material (adsorption), which should all be minimized in order to obtain absolute MWDs. Additionally, studies have found extremely high inter-laboratory deviations of calculated molar masses from similar lignin samples [42].

This method requires a previously made calibration from a series of standards of different MW, that display the relationship between molecular weight and elution time [39]. After calibration, the result can be converted into molar mass distribution. Studies have performed this calibration using polymers such as polystyrene sulfonates, pullulans, or proteins with mixed results. Thus, if the standards present do not present a similar molecular configuration to lignosulphonates, very large calculation errors may occur. Promising results have been obtained using lignosulphonate fractions with molecular weights determined previously by analytical ultracentrifuge for calibration [43].

Tetrahydrofuran (THF) is a widely used eluent, but most lignins are not entirely soluble in this compound. Therefore, a previous acetylation and/or methylation step is required [39,42]. Studies have attempted to acetylate lignosulphonates. However, these were only soluble up to 4.5% in the acetylation mixture. Even after this process, only the low MW compounds were soluble in THF. Therefore, it was concluded that the THF system is not suitable for lignosulphonates [44].

Various GPC systems have been applied for lignosulphonates and spent sulphite liquors analysis. Some are reviewed in Table 5 [39,42]. Some studies have achieved accurate results with in-line multi-angle laser light scattering (MALLS). This system solved previous problems caused by unrepresentative calibration standards, variations in refractive index, and fluorescence [45]. The complex mobile phase used by Fredheim and co-workers was chosen in order to prevent formation of aggregates and absorption to the column material [43,46].

Table 5. GPC systems for lignosulphonates and spent sulphite liquors (adapted from [39]).

Sample	Column Type	Eluent	Standards	Detectors	Reference
		Organic Solvent			
LS and LS–QAM complex	3 SDVB (styrene–divinylbenzene) columns	THF + QAM (quaternary amine methyltrioctylammonium chloride)	PS and biphenyl	UV	[47]
SSLs fractions	PSS GRAM 30, 2 columns and a guard column	LiBr (0.05 M) in DMSO/ water (90:10)	Pullulan	UV RI Viscosimetric	[48]
SSLs fractions	2 Polyacrylate methacrylate Columns	DMSO:H_2O (9:1) and 0.05 M LiBr	Pullulan (high Mw) glucose/cellobiose (low Mw)	RI Viscosimetric	[42]
	2 PFG-PRO (silica)	DMAc and 0.11 M LiCl	Polyethylene glycol Polyethylene oxide		
Purified LS	3 Agilent PolarGel M colums 1 guard column	DMSO/LiBr (0.5% w/v)	PSS	UV RI	[41]
		Aqueous System			
SSL and purified LS	2 PL aquagel And pre-column	$NaNO_3$ 0.1 M	PS	RI	[49]
LS and fractioned LS	Jordi Glucose-DVB And pre-column	Water/DMSO/ Na_2HPO_4-$4H_2O$/SDS	-	DAWN-F MALLS RI	[46]
LS	Ultrahydrogel or Ultrastyragel	Na_2NO_3 or NaCl solutions	Pulluan	RI	[50]

3.3. Chemical Structure Characterization

Methods used for the characterization of lignin's structure include spectroscopic methods such as Fourier transform infrared (FTIR), UV/vis and Nuclear Magnetic Resonance (NMR), and Raman spectroscopy [51]. Wet chemistry methods may also be used [9].

3.3.1. Ultraviolet (UV) Spectroscopy

Ultraviolet (UV) spectroscopy refers to absorption spectroscopy in the UV region (200–400 nm). This method is one the most useful for the quantitative and qualitative analyses of lignin in solution. Because of its aromatic nature, lignin strongly absorbs UV light and exhibits characteristic maxima in the ultraviolet light region. Samples are most often liquids, thus lignin is typically dissolved in a solvent. Potential solvents for lignosulphonates include water and an ethanol/water mixture (2:8 v/v) [33,51].

The type of lignin, its chemical modifications, and the solvent used determine the location and intensity of the maxima. The spectrum of softwood lignin presents a maximum of absorbance at 280 nm, a shoulder at 230 nm, and a sharp peak at 200–210 nm. These bands are designated B, E_2, and E_1, respectively. On the other hand, the spectra of hardwood lignins exhibits B bands in the 268–277 nm range. Additionally, the B band absorptivity values for softwood lignins (18–21 L g^{-1} cm^{-1}) are significantly higher than those for hardwood lignins (12–14 L g^{-1} cm^{-1}). Lignosulphonates also present a lower light absorptivity when compared to kraft lignin, as sulphite pulping has a bleaching effect on lignin [33]. An example of an UV spectra of hardwood magnesium-based thick spent sulphite liquor (HLS) and softwood sodium lignosulfonates (SLS) in water is shown in Figure 5.

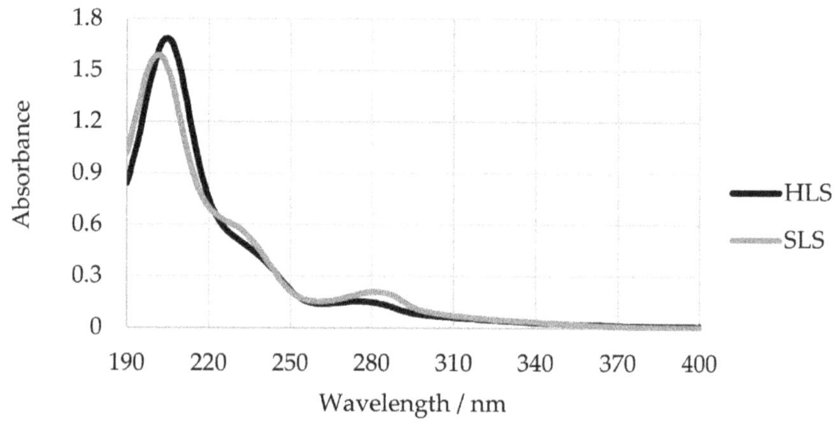

Figure 5. UV spectra of hardwood magnesium-based thick spent sulphite liquor (HLS) and softwood sodium lignosulfonates (SLS) in water.

3.3.2. FTIR Spectroscopy

Infrared (IR) spectroscopy has been used to characterize lignin since the early 1950s. This is due to the simplicity of the technique along with the fact that the samples do not need to be dissolved in any solvent and only small quantities are needed. Fourier transform infrared (FTIR) spectrometers are the most used equipment [52,53].

The mid-IR region is normally used to characterize lignin (4000–400 cm^{-1}) [51]. Lignin IR spectra are easy to obtain. If possible, the preferred method for spectra acquisition should the KBr pellet transmission technique. Reflection techniques such as attenuated total reflectance (ATR) or diffuse reflectance (DR), as well as photo acoustic (PA) methods, should be used for analyzing liquids or high consistency pastes, and for surface analysis in the solid state [52].

IR spectroscopy is mostly used for the analysis of lignin in solid state. For liquid samples, UV and NMR spectroscopy are more frequently used. However, the need to measure IR spectra in solution exists when there is no time to isolate lignin from a solution. An example is the monitoring of spent pulping liquors for process control, which must be performed rapidly [52].

Due to the differences between the spectra of softwood and hardwood lignins, studies have used the IR spectrum as an indicator of the ratio of syringyl to guaiacyl units in lignins [53].

An example of a FTIR spectra of hardwood magnesium-based thick spent sulphite liquor and softwood sodium lignosulfonates is shown in Figure 6. The assignments for the main bands of milled-wood lignins have been explained in detail by Faix. Hemmilä and co-workers have also characterized two samples of ammonium and sodium lignosulphonates through FTIR spectroscopy and summarized their main absorption bands [36].

Advantages of FTIR spectroscopy include high signal to-noise ratio and linearity, high accuracy in frequency, the mechanical simplicity the spectrometer, and the easy data treatment due to well developed software for spectral data manipulation [52].

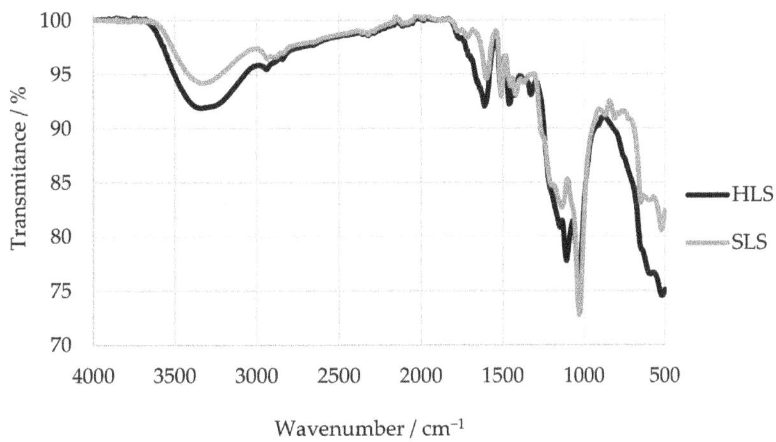

Figure 6. FTIR spectra of HLS and SLS.

3.3.3. Raman Spectroscopy

Much like in IR spectroscopy, in Raman spectroscopy an incident IR beam makes contact with the sample. However, in Raman spectroscopy, the photons involved are neither absorbed nor emitted but instead shifted in frequency. This shift is equal to the energy of the vibrational transition [54]. Raman spectra are determined through the difference between the frequency of the incident and refracted light. Each variation is related to one of the normal vibration modes of the molecules [55].

Raman spectroscopy and IR spectroscopy both provide information on the vibration and rotations of molecules and, in some cases, electronic transitions. Although the mechanism of these methods is different, the obtained information is complementary. The active vibrations in Raman may not be active in IR and vice versa. It should also be noted that in Raman spectra each band has a characteristic polarization which is very useful for the characterization of the molecular structure of a compound [55].

The first spectra of lignin samples revealed that the Raman signal was greatly obscured by laser-induced fluorescence. This problem was solved for most samples when a new Raman instrument based on NIR excitation was created. Thus, NIR Raman spectroscopy or NIR FT-Raman is rapidly becoming a promising technique in the analysis of lignin. However, the study of commercial lignins remains a challenge, as these samples produce a significant amount of fluorescence even when excited at 1064 nm. Therefore, further research is needed in other to develop more adequate approaches [56].

Advantages of the Raman spectroscopy include the ability to choose frequencies that allow selective lignin excitation and the easy characterization of heterogeneous samples. The latter case is a difficult task in IR spectroscopy due to the Rayleigh scatter of infrared photons [54].

Agarwal and Atalla summarized the main bands in the FT-Raman spectra of softwood and hardwood milled-wood lignins [56]. Ertani and co-workers also summarized the observed main absorption bands of the samples of two specialty lignosulphonates [57].

3.3.4. Nuclear Magnetic Resonance (NMR) Spectroscopy

Nuclear Magnetic Resonance (NMR) spectroscopy is widely used for the characterization, classification, and detailed structural analysis of lignin because it provides information that cannot be obtained by chemical analysis. Nuclei of interest in NMR spectroscopy include: ^{1}H, ^{13}C, ^{31}P. The most abundant carbon isotope, ^{12}C, is not NMR-active, and the ^{13}C isotope is only present in 1.11% relative abundance [58].

In an early stage, ^{1}H NMR was frequently applied for lignin analysis due to fact that the proton nucleus is of 100% natural abundance and of high sensitivity. This technique is used almost entirely on acetylated lignins as this provides better signal resolution. ^{1}H NMR is able to quantify a number of notable lignin structural features. Nevertheless, it has drawbacks such as a limited range of chemical shifts, extensive signal overlapping, and proton-coupling effects [53].

^{13}C NMR has contributed greatly to the current knowledge of lignin's structure. However, this technique has an even lower sensitivity than ^{1}H NMR, due to the even closer energy levels than and the low natural abundance of ^{13}C. To obtain good spectra in acceptable times sample sizes in the milligram range are required. Despite apparent sensitivity issues, NMR spectra contain more information than those of other spectral techniques. NMR alone can often fully identify compounds as well as determine their structure and bonding patterns, even in complex molecules [58].

Other approaches with increased sensitivity have been developed [58]. In two-dimensional (2D) NMR, particularly heteronuclear single quantum coherence (HSQC) NMR, a ^{13}C-^{1}H correlation spectrum is obtained where each signal corresponds to a unique C–H bond in the analyzed sample. Currently, this technique is commonly used for the elucidation of lignin structure and bonding [4,53]. This approach leads to a much easier analysis since the overlapping of peaks is completely avoided or largely minimized [4].

Unlike other types of "non-polar" or "non-sulfonated" technical lignins, LS have not been studied extensively by NMR spectroscopy. A major cause for this is that LS are insoluble in almost all organic solvents. LS contain some paramagnetic ions which cause signal broadening due to an increased relaxation rate. Another issue is that LS are usually used as metal salts. When these are dissolved in water the currents in the conducting solution and sample heating can difficult NMR analysis [43].

The methods present in literature for making lignin soluble in organic solvents are not applicable to LS. However, when LS is ion exchanged to the acid form it becomes soluble in methanol. This treatment also removes most of the metal ions from the solution, thus eliminating previously mentioned difficulties. Using methanol-d4 instead of D_2O improves spectra resolution [43].

Lebo and co-workers successfully studied eight different commercial LS samples using this method. Thus, important structural information was obtained [43].

On the other hand, Marques and co-workers studied purified LS from thin and thick spent liquors by 1D/2D NMR, using deuterated water with Sodium 3-(trimethylsilyl)propionate as an internal standard. In this study, the major obtained carbon signals in the ^{13}C NMR spectra of LS were summarized [59].

3.3.5. Wet Chemistry Methods

Wet chemistry methods have also been extensively used to characterize the structures of different lignins. Contrary to the spectroscopy methods previously described, these methods degrade the sample and the information is obtained through the resulting products [9]. Chemical degradation reactions of lignins were the only available techniques for structural characterization before the appearance of NMR methods [60].

Frequently used wet chemical techniques used for quantifying lignin monomers are acidolysis, cupric oxide, NBO, permanganate oxidation, and thioacidolysis [29]. Each method can provide a piece of structural information but in most cases, the information is semi-quantitative at best [9]. Therefore, these methods will not be addressed further in this study.

3.4. Functional Group Analysis

3.4.1. Determination of Phenolic Hydroxyl Groups

One the most important factors which influences the physical and chemical properties of lignin are the phenolic hydroxyl groups. The chemical reactivity of lignin is deeply affected by its phenolic hydroxyl content, namely in the reaction with formaldehyde for the production of adhesives. Therefore, determining the content of these groups provides insights on the structure and reactivity of lignin samples [61].

Methods that are frequently used to measure phenolic hydroxyl content in lignin include potentiometric and conductometric titration, ionization UV spectroscopy, and NMR spectroscopy [61]. However, when it comes to lignosulphonates, this quantification has been challenging for researchers [62].

The UV method is quick, simple, and applicable to lignosulphonates [62]. In alkaline solutions, phenolic hydroxyl groups are ionized and the absorption shifts towards longer wavelengths and higher intensities. Therefore, it is useful to examine the ionization spectrum ($\Delta\varepsilon_i$), that is determined though the subtraction of the neutral solution spectrum by the spectrum derived from an alkaline solution. Thus, the quantification of the phenolic hydroxyl groups is based on this principle [51]. Lin provided a procedure for this method in detail [33]. However, lignin model compounds or a lignin of known phenolic hydroxyl content need to be used for calibration. This may introduce errors in the obtained values since the structure of these model compounds can be simple in comparison to technical lignins [61,62].

Alonso and co-workers determined the content of phenolic hydroxyl groups in lignosulphonates by the Goldschmid method (UV absorption). The spectra of an alkaline lignosulphonate solution (boric acid pH = 12) were measured against a neutral lignosulphonate solution (potassium dihydrogen phosphate pH = 6). The phenolic hydroxyl content was calculated from the absorptivity at the maximum of 250 and 400 nm [63,64].

The titration method can be used to quantify this group along with carboxyl or sulfonate groups. This procedure is based on the acidity of phenolic hydroxyl groups and relies on an internal standard. However, this method does not provide information on the distribution of the different types of phenolic OH groups. The obtained results may also be influenced by other substances such as methoxyl groups, high sugar contents, and sulphites [61,62].

When this analysis was attempted via ^{1}H or ^{13}C NMR spectroscopy, a previous acetylation step was required. However, as previously mentioned, this reaction could not be completed with lignosulphonates, resulting in inadequate results [62].

^{31}P NMR has also been used to quantify hydroxyl and phenolic groups in lignin, but a functionalization step where these groups are phosphitylated is still required [4]. Stücker and co-workers analyzed several softwood lignosulphonates using this spectroscopic method. Firstly, the LS samples were converted into their corresponding lignosulphonic acid by ion exchange treatment and later lyophilized. These samples were then dissolved in a mixture of anhydrous N,N-dimethylformamide (DMF), deuterated DMF and pyridine. Endo-N-hydroxy-5-norbornene-2,3-dicarboxylic acid imide was used as an internal standard. The relaxation agent and phosphitylation reagent were chromium (III) acetylacetonate and 2-chloro-4,4,5,5-tetramethyl-1,3,2-dioxaphospholane, respectively. The authors concluded that the proposed method was accurate for the determination of hydroxyl groups in LS samples [62].

3.4.2. Determination of Methoxyl Groups

The methoxyl content of lignin is frequently determined according to Zeisel procedures. ^{13}C NMR spectroscopy has also been used as well as on ^{1}H NMR, although in the latter case for a rough estimation based on spectral examinations [65].

The original Zeisel procedure cannot be used to determine methoxyl groups in sulfur containing compounds. Therefore, it is not suited for lignosulphonates. In order to apply this method these compounds, modifications were proposed by Vieböch and co-workers. Through this method, it is possible to express the methoxyl content of lignin on a particular basis, such as equivalents per C_9 unit but, lignin must be previously purified. If the Klason lignin content of the sample inferior to 95%, further purification should be carried out. Chen and co-workers have explained this method in detail [66].

Marques and co-workers determined the content of methoxyl groups in purified and unpurified sulphite liquor using the Zeisel-Vieböch-Schwappach method and ^{13}C NMR. For the former method, 0.5 g of phenol, 1 g of KI, and 2 mL of H_3PO_4 (85%) were added sequentially to 20 mg of the samples. The obtained mixture was then heated to 145 °C for 40 min and then dragged to a solution of Br_2 by a current of N_2. Next, the Br_2 was dragged with a solution of sodium acetate (20%) and destroyed with formic acid (4%). Then, 20 mL of H_2SO_4 (10%) and 5 mL of KI 10% were added. Lastly the mixture was titrated with $Na_2S_2O_3$ 0.1 M until it turned pink. Starch was used as an indicator. The quantification through the ^{13}C NMR method was achieved through the integral of the centered signal at 56.8 ppm. Although different results were obtained with these two methods, this difference was not considered to be significant [13].

More recently, another method has been proposed by Sumerskii and co-workers with higher precision, accuracy, and sample throughput than the Zeisel-Vieböch-Schwappach method. This new approach is based on headspace-isotope dilution Gas chromatography–mass spectrometry and can be applied to any type of lignin [31].

3.5. Thermal Properties

Thermogravimetry (TGA) and differential scanning calorimetry (DSC) are the most frequently used techniques in lignin chemistry for determining its thermal properties. Thermal analysis can be applied to study the physical properties of lignin as well as its derivatives. In such studies, it provides information on molecular arrangements, phase transitions, and the interaction between lignin and low molecular weight substances such as water. For more specific applications, these techniques may be used determine the durability of lignin and its respective glass transition and degradation temperature. A great advantage is that only a small amount of the sample is used and any kind of material may be analyzed, including powders and liquids [67].

DSC is often used to determine the glass transition temperature and heat capacity. Variations in these values can be induced by low molecular weight contaminants, molecular weight, thermal history, cross-linking, and pressure. Therefore, thus, a glass transition temperature value for any specific lignin type cannot be precisely reported [51,67].

TGA is primarily conducted dynamically and provides information about degradation temperatures and other phase transition temperatures [51].

A disadvantage of thermal analysis is that although it may provide accurate data on transition and degradation temperature, it only gives general information on the molecular arrangements of polymers can be obtained through these techniques [67].

4. Approaches to Increase Reactivity

Lignin's guaiacyl (G) units and phenol present a similar structure, as shown in Figure 7. However, in the aromatic ring of lignin's G units positions 1 and 3 are blocked. On the other hand, in the aromatic ring of lignin's syringyl (S) all of the ortho and para positions are blocked. Therefore, the methoxyl groups are the cause of lignin's low reactivity. This factor, alongside its high dispersity in molecular weight, has hindered the application of industrial lignins in resin synthesis [23,68].

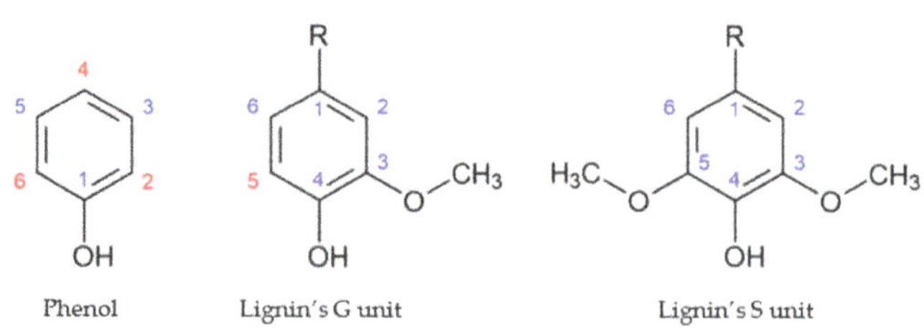

Figure 7. Reactive sites of Phenol and lignin's phenolic units (adapted from [23]).

However, it should be noted that, in recent studies, it was suggested that hardwood lignins, although containing a lower number of reactive sites in their aromatic rings, could still be used as a phenol substitute in phenol-formaldehyde (PF) resins. It was suggested that other factors may also influence the final performance of the obtained resin such as: molecular branching and conformity, steric availability, molar mass, and its distribution [69]. The amount of phenolic and aliphatic hydroxyl groups is also highly important for all adhesive applications that require reaction of lignin with different aldehydes, phenols, tannins, or isocyanates [70].

Several methods to improve the reactivity of lignin molecules have been reviewed in literature and will be discussed in this study.

4.1. Phenolation

Phenolation or phenolysis is one of the most promising treatments for lignin modification. This method allows the reduction of lignin's molecular weight as well as an increase of its phenolic hydroxyl groups, therefore increasing its reactivity. Phenolation reactions result in the attachment of phenol onto lignin. These reactions can occur in alkaline or acidic mediums [71].

In the first step of the reaction, the protonation of the benzyl hydroxyl group occurs. The next step is the dehydration at the α-carbon, forming a carbonium ion. An electrophilic attack to the phenol molecule by the carbonium ion produces a phenol condensation product, adduct. Then occurs the incorporation of ortho or para-phenyl to the α-hydroxyl groups of the propane side chains of lignin. Lastly, adduct fragmentation takes place. This results in a decrease in the molecular weight of the reaction products, thus facilitating their incorporation in resins. Depending on the reaction conditions, side reactions may occur. Thus, a self-condensation product could be produced by the reaction of the carbonium ion in a lignosulphonate molecule. The treated lignin can react in an acidic or alkaline medium with formaldehyde to produce resol or novolac resins [72].

A recent study has suggested that lignin substructures such as β-O-4′, β-5′/α-O-4′, β-β′, and α-carbonyl react with phenol, thus increasing the amount of phenolic hydroxyl groups present in the structure. This work also considered the ortho and para positions for lignin phenolation and the presence of more substructures, shown in Figure 8, because of the elimination of formaldehyde from the γ-carbon [71].

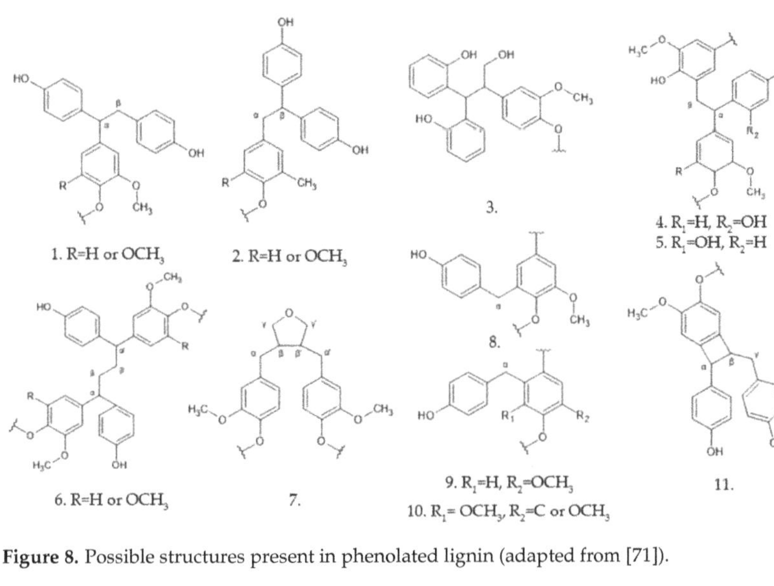

Figure 8. Possible structures present in phenolated lignin (adapted from [71]).

Phenolation is frequently used for the modification of lignosulphonates since it increases the content of phenolic hydroxyl groups, while reducing molecular weight, thus simplifying their structure. The removal of the sulfonic acid group after phenolation has also been reported [71].

Alonso and co-workers studied the phenolation method to modify the structure of softwood ammonium lignosulphonates. The lignosulphonates reacted with phenol using oxalic acid as a catalyst. The methods used to characterize the final products were GPC, FTIR, and ^{1}H NMR. It was concluded that high temperatures, a long reaction time, and low lignosulphonate concentration favored high phenol conversion rates. Therefore, the chosen ideal reaction conditions were 120 °C, 160 min, and 30% lignosulphonate content [73].

Hu and co-workers used phenolation to modify lignosulphonates for the synthesis of a phenolic resol with phenol and formaldehyde. For the phenolation process, 20 g of lignosulphonate were added to 100 g of molten phenol. The pH of the mixture was adjusted to be between 9 and 10 through the addition of sodium hydroxide aqueous solution. The resulting mixture was stirred and heated slowly to 100–120 °C in an oil bath, then refluxed for 1 h. The reaction was quenched by cooling to 70 °C [74].

A major advantage of phenolation is that the final product is soluble in phenol. Therefore, direct incorporation in PF resin production may be possible without prior purification [3].

4.2. Hydroxymethylation

Lignin methylolation, or hydroxymethylation, introduces hydroxymethyl (–CH_2OH) onto lignin molecules. The obtained lignin can be directly incorporated in the synthesis of PF resol resins for wood adhesives as a phenol substitute. Lignin methylolation usually occurs in an alkaline medium with formaldehyde [72].

Three reactions occur. In the main reaction, the Lederer–Manasse reaction, lignin's reactivity is increased through the incorporation of hydroxymethyl groups into the aromatic rings, as shown in Figure 9-reaction a. The Tollens reaction may also take place, but it is not desirable. This involves the substitution of lignin's side chains by aliphatic methylol groups, Figure 9-reaction b [75]. With the increase in temperature, this reaction may be followed by a condensation in which hydroxymethyl groups react at free positions of other lignin units to form methylene bonds (Figure 9 reaction c), thus decreasing hydroxyl group content. Another undesirable side reaction is the Cannizzarro reaction, in which formaldehyde reacts with itself [35,72,75].

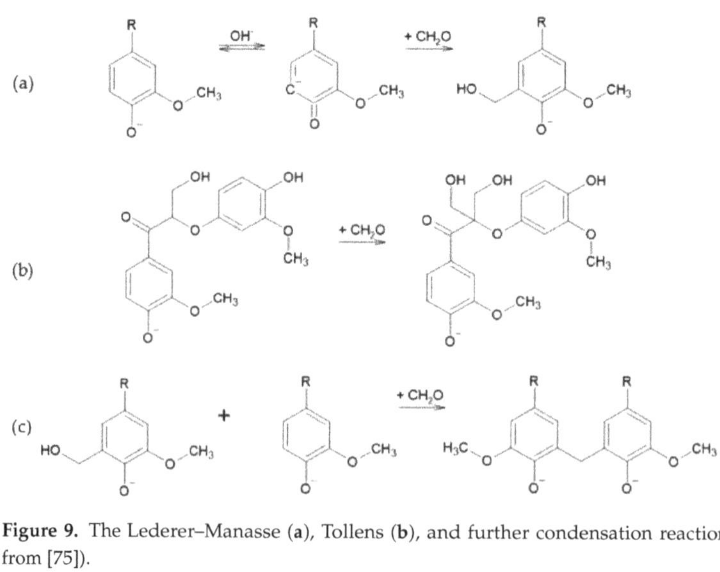

Figure 9. The Lederer–Manasse (**a**), Tollens (**b**), and further condensation reactions (**c**) (adapted from [75]).

Lignin's reactivity in hydroxymethylation or cross-linking depends on the reaction conditions but, also on lignin's source and pulping process [75].

Alonso and co-workers studied the methylolation of softwood and hardwood LS. FTIR and ^{1}H NMR were used to evaluate these samples before and after methylolation. Softwood ammonium LS were used to optimize the operation conditions to promote the Lederer–Manasse reaction, as they displayed the most promising characteristics. The reaction was followed by the changes in the concentration of free formaldehyde. The chosen optimum operating conditions for methylolation were 1.0 of formaldehyde-to-lignin molar ratio, 45 °C, and a sodium hydroxide-to-lignin molar ratio of 0.80. Hardwood LS exhibited lower reactivity due to their predominant structural unit, syringyl, having a higher degree of substitution on the aromatic ring [35].

It should be noted that the reaction of formaldehyde at the meta position of phenolic hydroxyl groups has been reported to occur when all the ortho and para positions are blocked, as in syringyl units. However, it is significantly slower and incapable of saturating all of the available meta positions [76].

Pang et al. used a different procedure for the methylolation of calcium LS. Initially, the solution of calcium LS (30%) was heated to 80 °C. The pH was adjusted to 11 with NaOH, under stirring. Then, a 37% formaldehyde solution was added slowly to the mixture until a mass ration of formaldehyde to lignosulphonate of 0.35. After a reaction time of 2 h, the process was ended [77].

Aro and Fatehi suggested the purification of LS after methylolation through drying or membrane filtration [3].

However, as formaldehyde is a known carcinogenic, studies have proposed the total substitution of formaldehyde (LD50 rat > 800 mg/kg) in lignin-based wood adhesives with glyoxal, a less toxic aldehyde (LD50 rat > 2960 mg/kg), which is non-volatile with regard to the aqueous phase in the lignin hydroxymethylation step [78–80].

Mansouri and co-workers proposed a procedure for the glyoxalation of LS for later incorporation in wood adhesives. Firstly, calcium LS powder (96% solid) was slowly added to water in the proportion of 29.5 to 38.4 parts by mass. Sodium hydroxide solution (30%) was added when necessary, in order to maintain the pH between 12 and 12.5. This step, alongside vigorous overhead stirring, allowed for adequate dissolution of LS. In total, 18.1 parts by mass of sodium hydroxide solution (30%) were added, resulting in a final pH of 12.5. The solution was then transferred to a 250 mL flat bottom flask equipped with a condenser, thermometer and magnetic stirrer bar and heated to 58 °C. Then, 17.15 parts of glyoxal (40% in water) were added. Lastly, the mixture was continuously stirred with a magnetic stirrer for 8 h [78].

4.3. Oxidation

Oxidative lignin conversion is a promising way to increase lignin's value by converting it into chemicals. In this process, lignin's structure is disassembled into its phenolic building blocks, which can be converted into targeted end products. This method can be used to produce highly functionalized, valuable structures such as vanillin and is, therefore, of great interest [81].

The mechanism of lignin depolymerization depends on the oxidants and catalysts used. The most used oxidants include oxygen, hydrogen peroxide, ozone, and peroxyacids of chlorine (including chlorine dioxide). The involved mechanisms are electrophilic, radical, and nucleophilic [82].

Oxygen is the most promising oxidant, both in terms operability and costs. When oxygen is used as an oxidant, radical and electrophilic mechanisms are responsible lignin's degradation. The main end products are vanillaldehyde/acid, syringaldehyde/acid, and *p*-hydrobenzaldehyde/acid [82]. The structure of these aldehydes is shown in Figure 10.

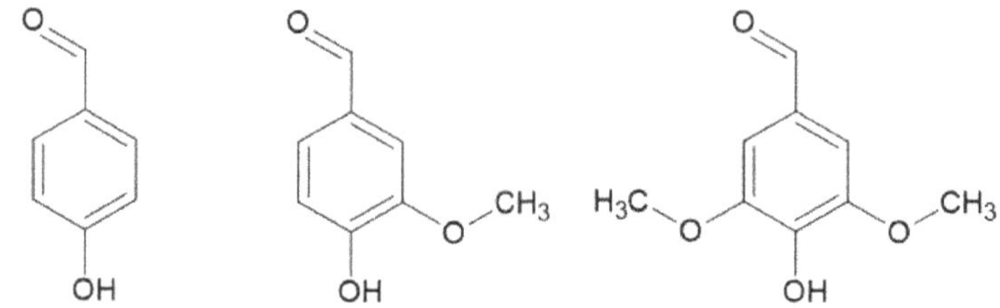

Figure 10. Aromatic aldehydes derived from lignin's oxidation (adapted from [83]).

Lignin oxidative conversion follows three steps: side chain cleavage, ring opening, and condensation [82].

In case hydrogen peroxide is used, the mechanisms involved are of radical and nucleophilic nature, resulting in aldehydes, acids, and quinines as the main products [82].

The catalysts used in oxidative lignin conversion can generally be divided into three types: metal-inorganic catalysts, metal-organic catalysts, and organic catalysts. Metal oxide presents high activity, low cost, and is easy to manipulate; therefore, it is seen as one of the most promising catalysts for oxidative biomass conversion [82].

Santos et al. studied the oxidation kinetics of hardwood magnesium LS with oxygen under alkaline conditions. The reaction system O_2/NaOH resulted mainly in the production of vanillin and syringic aldehyde. Studies with the addition of a catalyst (copper salt, 20% *w*/*w*) were also conducted resulting in an increase in yield of 25 to 50%. The highest obtained yields of syringic aldehyde (16.1%) and vanillin (4.5%) were obtained using the following reaction conditions: 150 °C for 20 min in 0.9 M alkaline solution and oxygen pressure of 10 bar. It was noted that the sugars present in the sulphite liquor significantly reduced the yield of the aromatic aldehydes. Thus, the removal of these compounds prior to LS oxidation was highly advised [83].

Yuan and co-workers oxidized ammonium LS with H_2O_2 for subsequent incorporation in a binder for a wood-based green composite. Firstly, 50 g of LS was mixed with 100 g of water. The pH of the resulting solution was adjusted to 10 using a sodium hydroxide solution (30% *w*/*w*). Next, a 30% H_2O_2 dosage based on the dry weight to LS was added. The mixture was heated to 60 °C and stirred for 30 min. Lasty, the solution was kept at 80 °C in an oven for 6 h and then cooled to ambient temperature [84].

Other procedures were suggested in different studies, namely for the incorporation of LS in a foaming resol resin as a partial phenol substitute [77,85].

Oxidation with hydrogen peroxide has been widely used in the pulping industry and is environmentally friendly. However, the decomposition of H_2O_2 forms molecular oxygen as well as several different radical species. These compounds react with lignin in a variety of ways, thus creating a complicated reaction mechanism [3].

4.4. Hydrolysis

In the hydrolysis process, water is used to cleave the ether bonds that connect lignin's phenyl-propane units, usually in the presence of an acid or an alkaline catalyst. The final product has a lower molecular weight. In lignin's case, hydrolysis can occur in an alkaline or acidic medium under sub or supercritical conditions [15,71].

In acidic hydrolysis, lignin is reacted with a strong acid and catalyst, resulting in the demethylation of the phenyl-propane unit and in an ortho substitution by hydroxyl groups. Hydrolyzed lignin is subjected to high temperatures in order to cleave the ether bonds of the phenyl-propane units in the lignin structure. This results in more reactive monomeric units [71].

In an alkaline medium, –OH groups act as a primary catalytic agent and are commonly used in pulping reactions for severing lignin. These groups attack different lignin bonds and produce ether bonds that can be cleaved easily through other treatments [71].

Whether it was subjected to an acidic or alkaline treatment, the treated lignin is susceptible to degradation, thus producing new phenolic hydroxyl-containing units with low molecular weight. In an alkaline treatment, the products of this degradation can be of value in condensation reactions to produce phenolic products. Hydrolysis in an acidic medium usually occurs in non-aqueous media; therefore, the obtained product must be previously extracted from an organic solvent before its use [71].

Mansouri and co-workers increased the reactivity of softwood LS using alkaline hydrolysis with sodium hydroxide. The modified samples were tested according to the content of phenolic hydroxyl groups, aromatic protons, as well as weight-average MW, number-average MW, and LS content. UV spectroscopy, ^{1}H NMR, and aqueous GPC were used for this characterization. The powder LS was used with a ratio of $1/10$ (w/w) to the sodium hydroxide solution, 2% (w/w). The optimum operating conditions were 170 °C and a reaction time of 90 min. The LS modified under these conditions presented a high concentration of phenolic hydroxyl groups, aromatic protons, as well as high LS content and low MW. The reactivity towards formaldehyde was also tested through a methylolation reaction. In this test, the modified LS displayed an increase in reactivity in more than 50% when compared to the original lignin. Thus, the authors concluded that increasing the severity of the operating conditions produced largely improved formaldehyde reactivity [86].

However, studies have indicated that lignin yields can decrease at extreme reaction conditions due to re-polymerization, especially under acidic conditions. This reaction can be controlled and the conversion rates can be increased resorting to alkaline catalysts. Catalysts can also increase selectivity by selectively cleaving bonds [15].

4.5. Ionic Liquid (ILs) Treatment

Although glyoxilation, phenolation, and hydroxymethylation significantly influence the reactivity of lignin in resin synthesis, they rely on volatile organic reagents. Thus, there is still a need for more eco-friendly methods [3,68].

Ionic liquids (ILs) present chemical and thermal stability, low vapor pressure, and high ionic conductivity [68]. Their low melting point does not exceed 100 °C and, due to their negligible vapor pressure, these liquids obey the principles of green chemistry [87]. These properties have resulted in an increasing interest in these compounds in different sectors as "green solvents" [68].

In lignin chemistry, ILs can be modifiers or depolymerizing compounds [87]. One of the most recent methods to improve lignin's reactivity is the pretreatment with ILs such as 1-Butyl-3-methylimidazolium chlorine ([Bmim]Cl) and 1-ethyl-3-methylimidazolium acetate([Emim][OAc]) [68].

These ILs were applied to treat industrial lignin by Qu and co-workers. In this study, ([Bmim]Cl) or ([Emim][OAc]) were mixed with industrial lignin at a ratio of 20:1, respectively. Then the mixture has heated at 120 °C for 30 min. In order to recover the lignin by vacuum filtration, deionized water was added at a ratio of 5:1. The recovered solid was washed repeatedly with deionized water until the filtrate was colorless. Lastly, the obtained lignin was dried overnight in a vacuum oven at 40 °C. Each treatment was carried out twice. This procedure reduced the average MW and polydispersity of the treated lignin. When [Emim][OAc] was used the phenolic hydroxyl and carboxyl groups were preserved but the methoxyl groups were reduced by 45% [88].

In a later study by Younesi-Kordkheili et al., soda bagasse lignin was modified through this method for incorporation in a UF resin. Modified lignin, when compared to unmodified lignin, had a better performance as an additive in these resins [68]. In a following study, Younesi-Kordkheili and co-workers proposed the mechanism shown in Figure 11 for the reaction between [Emim][OAc] and lignin [89].

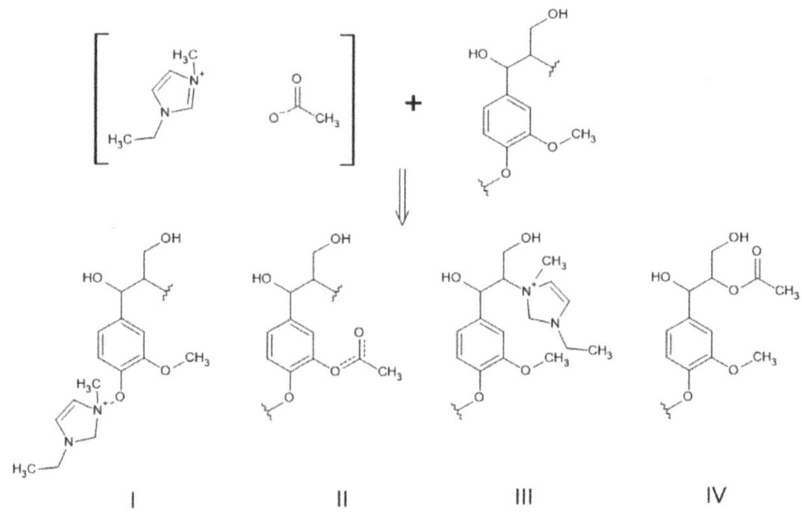

Figure 11. Possible reactions between lignin and [Emim][OAc] (adapted from [89]).

In another study, ionic liquid lignins were prepared from sodium LS by a cation exchange to modify its thermophysical properties. When sodium was replaced by the tris-[2-(2-methoxyethoxy)ethyl]amine cation a flowable ionic liquid lignin was produced with a Tg of −13 °C. Through this approach, the properties of both LS and ILs were combined to create a dispersant and binder for cellulose and gluten bio composites. Panels with fewer defects and improved toughness were produced with modified LS in comparison to those with unmodified LS or free ionic liquid. The authors concluded that the retention of OH groups and the low Tg value of the modified LS was essential to produce composites resistant to high stress. Hence, this treatment introduced new functionalities while maintaining the properties of LS [90].

5. Lignosulphonates in Wood Adhesives

As previously mentioned, despite being produced in high quantities, lignin is usually burned for energy production [2,3,91].

Nevertheless, lignin displays several characteristics which make it a promising polymer for the development of new products. These characteristics include good rheological and viscoelastic properties, good film-forming ability, small particle size, and compatibility with a wide range of industrial chemicals. Depending on its origin, lignin can be of hydrophilic or hydrophobic nature allowing for the production of a wide range of blends. The presence of aromatic rings in its structure also provides thermal stability, good mechanical properties, and the possibility of a broad range of chemical transformations [5].

Lignosulphonates are currently used for a variety of value-added products, as shown in Figure 12; however, here we will focus only on wood adhesives [5].

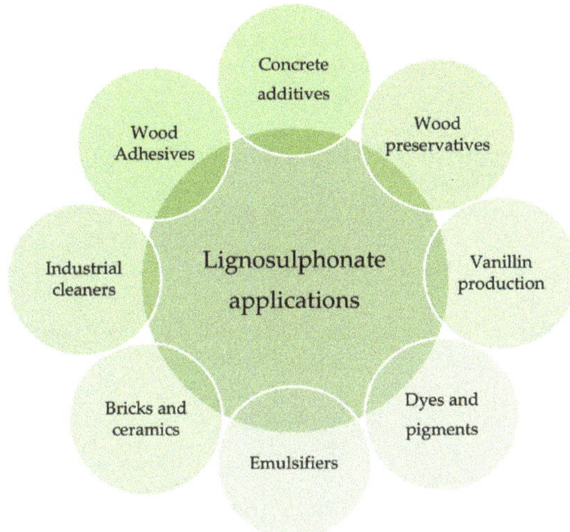

Figure 12. LS applications.

Since the start of wood pulping, the large quantities of generated lignin waste have been proposed for the preparation of adhesives. Many studies have been conducted on the use of lignins to produce wood adhesives, including lignosulphonates [23,78].

The most successful industrial attempts to use lignin for wood adhesives consisted in using this polymer as a partial substitute in phenol-formaldehyde (PF) resins or urea-formaldehyde resins [23,78].

Nevertheless, these resins include formaldehyde which is a known carcinogenic. Nowadays, concerns about formaldehyde emissions and the sustainability of the used raw materials and final products are reducing the popularity of formaldehyde-based synthetic resins for wood-based panels [23,68,92]. Thus, formaldehyde-free systems which combine LS with polyethylenimine, furfuryl alcohol, polymeric isocyanate, chitosan, tannins, glyoxal, and wheat flour have been proposed [70,78,84,93–95].

5.1. *Formaldehyde Adhesives*

5.1.1. Lignin-Urea-Formaldehyde (LUF) Resins

In the second half of the 20th century, a considerable number of studies were developed which report the incorporation of 5 to 20% of LS in UF resins, after or during its synthesis. The produced UF resins were reported by some works to display increased cold adhesion to wood particles and higher storage times. On the other hand, the resultant particleboards had improved mechanical properties, as well as lower formaldehyde emissions [92].

One of these studies added hardwood calcium-based spent sulphite liquor from the acid sulphite process (55–60% total solids, pH 3.0 to 3.5) during the UF synthesis. This type of LS was selected as calcium was suspected to catalyze the reaction. To produce the resin, acidic methylolation was accomplished with 3.0 to 5.0 parts of formaldehyde at 90 to 95 °C for 12 to 24 h. The acidic medium was preferred as it allowed the reaction to take place preferably at the meta positions, which are more readily available. Next, the temperature was decreased to 70–80 °C and 20 parts of urea were added and left to react for two hours. Then, 22.5 parts of formaldehyde were added for another reaction period of four to six hours at 70–85 °C. Lastly, 5.8 parts of ammonia were added providing a final pH of 8.0–8.5. The authors claimed that the obtained resin bonded wood particles when used at concentrations of 1 to 5% by weight while displaying a superior water resistance to the unmodified UF resin. It was concluded that all the urea reacted with the methylol lignosulfonate [96].

A recent study attempted to incorporate unmodified or hydroxymethylated thick spent sulphite liquor (TSSL) into a standard UF resin, with less promising results. TSSL was added in different amounts (10 and 20%) during the LUF resin synthesis alongside the second urea or after the synthesis. It was found that wood particleboards with 90% the amount of UF resin alone displayed similar IB strengths, as well as improved thickness swelling (TS) than those produced with any of the modified resins. The authors concluded that unmodified or hydroxymethylated TSSL acted as either inert or detrimental compounds in the performance of the UF resin, depending on the TSSL modification and incorporation stage. However, incorporation of unmodified or hydroxymethylated TSSL during the synthesis led to lower condensation times of the resins, lowering production costs [92].

These authors noted a higher acid buffer capacity of the LUF resins, which was more pronounced when LS was added during synthesis. Bekhta and coworkers also reported this effect. Thus, the sulfite and bisulfite ions in the sulphite liquor create a buffer system, therefore increasing the amount of acid needed to change the pH and slowing down resin cure [92,97]. This factor alongside the presence of ashes and sugars may significantly increase gel time [92].

Antov et al. combined an UF resin with commercial ammonium LS for the production of high-density fiberboard (HDF) panels from hardwood fibers. HDF panels were produced with low content UF, 3%, and ammonium LS content of 6, 8, and 10%, based on the dry fibers. The physical and mechanical properties of the fiberboards, such as water absorption (WA), thickness swelling, modulus of elasticity (MOE), bending strength (MOR), internal bond strength (IB), as well as formaldehyde content, were examined. The produced HDF panels complied with the requirements for use in loadbearing applications in humid conditions. The obtained formaldehyde content was also extremely low, and equivalent to that of natural wood. Notably, panels produced with more than 8% of ammonium LS displayed better physical and mechanical properties than the control panels with only UF resin (at 6%). To conclude, the authors suggested that future studies should add suitable cross-linking agents and study the bonding interaction of all the components in these panels, in order to decrease the hot-pressing factor [98].

It should be noted that the low formaldehyde release reported with the study while using ammonium LS may be caused by the reaction between the ammonium ion and formaldehyde under the conditions applied to the panels [76].

Gao and co-workers tried a different approach: incorporating sodium LS in UF resins as an anionic surfactant or stabilizer in order to improve colloidal stability and extend storage stability. When LS was added at the methylolation stage, the storage stability improved from 30 to 200 days. The nature of the interaction between LS and UF resins was studied using techniques such as FTIR, ^{13}C cross polarized magic angle spinning NMR. It confirmed that LS could increase the electrostatic repulsion of UF resins thus avoiding aggregation and aging. However, no chemical reaction between UF resins and LS was observed. As previously noted in other studies, LS addition slightly reduced the curing rate; but the thermal stability of the UF resin was improved. The shear strength and formaldehyde release of the obtained resins were also comparable to standard UF resins [99].

5.1.2. Lignin-Phenol-Formaldehyde (LPF) Resins

Ghorbani and co-workers studied four commercial spruce LS for different sulphite pulping processes as partial (40% *w*/*w*) phenol in lignin-phenol-formaldehyde (LPF) resole resins. The incorporation of the studied LS resulted in a faster viscosity gain during resole cooking compared to the lignin-free reference resin. Although displaying the lowest average MW and dispersity, sodium LS, when incorporated into the LPF resin for gluing beech veneer strips, resulted in the best curing and tensile shear strength development under hot pressing. On the other hand, calcium and magnesium LS were less suited as phenol replacements, since the obtained LPF adhesives had a poor performance. In this study, phenolation of sodium and ammonium LS, which had the most promising characteristics, did not significantly improve the performance of the obtained LPF resins [100].

Studies have also shown that impurities in crude lignin, specifically elemental sugars, lowered the reactivity of lignin towards formaldehyde for the incorporation in LPF resins, thus leading to extended pressing times. The strength and water resistance of the obtained resins were also hindered [101,102].

Alonso and co-workers used methylolated softwood ammonium LS as a phenol substitute in the synthesis of a LPF resin. The optimum operating conditions for this synthesis were studied. The obtained LPF resins were characterized according to free phenol and formaldehyde contents, gel time, alkaline number, viscosity, pH, solid content, and chemical structure changes (^{13}C NMR). The optimum operating conditions were sodium hydroxide to phenol-modified LS molar ratio of 0.6; formaldehyde to phenol-modified LS molar ratio of 2.5; and 35% *w*/*w* of replaced phenol by LS. The obtained LPF resin under these conditions complied with the specifications necessary for its utilization in plywood. Its characteristics were also similar to those of the commercial PF resol resin used as a reference [63].

In a later study, the structural (FTIR), thermal (TGA), and rheological properties of a LPF resin 30 wt% of replaced phenol by methylolated softwood ammonium LS were studied and compared to those of a commercial PF resin. While the PF resin displayed a higher reactivity, the TG and DTG thermograms revealed that the addition of LS improved the thermal decomposition temperature of the LPF resin. The DSC profile of these resins was similar, indicating that LS functioned as an extender. Lastly, it was concluded that the flow behavior of both resins was different, as PF exhibited a behavior close to a Newtonian flow, while the LPF resin displayed a pseudoplastic behavior. This was explained by the increased branching obtained for the LPF resin. The authors concluded that this change in behavior may be beneficial for certain applications and suggests that a better mechanical performance can be obtained for LPF resins [103].

Antov et al. assessed the viability of calcium LS as an eco-friendly additive in a PF resin, at different ratios, for medium-density fiberboards (MDF). It was concluded that addition of LS should not exceed 10% (on the dry fibers) to avoid deterioration in the mechanical properties. The authors recommended a content of 3.5% PF resin to produce LS-PF bonded MDF panels that obey EN standard requirements [104].

5.2. Formaldehyde-Free Adhesives

Spent sulphite liquor alone has been used as a wood adhesive; however, it was associated with high pressing temperatures and long pressing times. Combinations with strong mineral acids and hydrogen peroxide have been proposed; however, neither have found industrial success [76].

Unmodified magnesium LS alone has been studied as a binder for industrial waste fibers by Antov and co-workers. Composites were produced using a hot press temperature of 210 °C, a pressing time of 16 min, and a 15% gluing content of magnesium LS (based on the dry fibers). The obtained MOR and MOE values were higher than the minimum required for PBs for interior fitments for use in dry conditions and equivalent to the value for MDF panels. However, these composites showed deteriorated moisture properties [101].

Ferreira et al. also confirmed the adhesive properties of TSSL alone, allowing the production of PBs with IB values above the requirements of standard EN 312 for general purpose boards for use in dry conditions (type P1). The improvement of PBs performance was achieved by adding wheat flour (WF) to TSSL, particularly when the mixture was pre-heated at 94 °C [95].

In a later study, manufacturing conditions for the particleboards using TSSL and WF were investigated. It was possible to produce PBs with densities ranging from 682 kg m^{-3} to 783 kg m^{-3}, temperatures from 180 to 210 °C, and pressing times between 8 to 10 min. All the particleboards produced in these conditions were in accordance with the IB requirements of standard EN 312 for PBs type P2. These standards were also obeyed when the

PBs were manufactured 20 days after binder preparation. However, the authors concluded that further studies need to be conducted to improve water resistance and decrease the pressing time, allowing the increase in the range of application of these PBs [91].

In other studies, the use of unpurified LS has also been reported to result in a resin with lower reactivity, resulting in increased pressing times. Therefore, LS, as the main component of adhesives, should be accompanied by a suitable chemical crosslinker in order to compensate for this lower reactivity. Another approach is the modification of the parameters for the production of new composites [101]. The use of a crosslinker could also increase water resistance [91].

Oxidized ammonium LS combined with polyethylenimine (PEI) have been studied with satisfying results as a binder for wood-based green composites. The effects of hot-pressing temperature, hot-pressing time, binder content, and MAL/PEI weight ratio on the physico-mechanical properties of the composites were investigated. The optimum operating conditions were 170 °C, 7 min, 20% *w*/*w*, 7:1, respectively. Under these conditions, the composites met the mechanical property requirements for furniture grade medium density fiberboard (MDF-FN REG). Dynamic mechanical analysis (DMA), scanning electron microscopy (SEM), FTIR, and X-ray diffraction analysis (XRD) were used to determine the influence of LS oxidation. FTIR analysis of unmodified and oxidized LS confirmed the higher reactivity of the latter. The authors also obtained evidence for the formation of amide linkages in the modified LS-PEI mixture under optimum conditions. SEM analysis confirmed the superior bonding strength of the composites obtained with modified LS. The higher crystallinity of these composites and good viscoelastic properties were also verified by XRD and DMA, respectively [84].

Composites which met the Chinese requirements for MDF have also been produced using chitosan and ammonium LS as a binder. The 6% *w*/*w* of chitosan-lignin adhesive content was used with a lignin/chitosan weight ratio of 1:2. The SEM analysis confirmed the adhesive's high performance while the FTIR analysis indicated that hydrogen bonding between chitosan and LS occurred. Lastly, TGA proved that these linkages would decompose below 230 °C [93].

Hemmilä and co-workers also highlighted the need for need for a suitable crosslinker for LS adhesives for PBs, when LS is used without modification. Both bio-based furfuryl alcohol (FOH) and pMDI were studied as crosslinkers for ammonium LS. The addition of mimosa tannin to LS before crosslinking was also studied. A significant decrease in curing temperature and curing heat with crosslinker addition was confirmed. The best results were obtained for the LS-pMDI sample. FOH resulted in PBs with inferior mechanical properties when compared to those with pMDI, both with and without tannin. Formaldehyde emissions of pMDI crosslinked particleboards were at the level of natural wood, unlike FOH crosslinked boards which emitted beyond this level. Regarding the addition of mimosa tannin (10% to LS amount), no significant effects on the mechanical properties of PBs were observed. However, a decrease in the formaldehyde emissions of FOH crosslinked LS samples was noted. The authors concluded that, until more suitable crosslinkers are found, pMDI is a suitable option [70].

Polymeric isocyanate (pMDI) has been used as a crosslinker for glyoxylated LS. For methylolated lignin, the reaction mechanism with pMDI in aqueous solution has been elucidated by Pizzi. Thus, pMDI reacts preferably with the methylol groups in methylolated lignin, forming urethane bonds. Additional crosslinking occurs through the reaction of the isocyanate groups with the -NH- of the urethane bridges [76,78].

As previously mentioned, Mansouri and co-workers have replaced formaldehyde (LD50 rat > 800 mg/kg) in lignin-based wood adhesives with glyoxal, a less toxic aldehyde (LD50 rat > 2960 mg/kg), which is non-volatile with regard to the aqueous phase [78–80]. For this purpose, an adhesive composed of glyoxalated calcium LS and pMDI was proposed. The optimum operating conditions for PBs production were resin load on dry wood of 8%, pressing time of 28.1 s/mm board thickness at temperatures between 195–200 °C. The ratio of glyoxalated LS to pMDI was of 60 to 40. These PBs yielded good IB strength results that comfortably passed those required in international standard specifications for exterior-grade panels. Their performance was also similar to PBs manufactured with formaldehyde-based commercial adhesives [78].

LS and tannin adhesives have also been suggested. Rhazi et al. combined glyoxylated ammonium LS with natural tannin extracted using different processes to produce an adhesive for plywood panels. The optimized operating conditions for plywood production were ratio of tannin solution to glyoxylated LS of 75:25, pressing time of 4 min, temperature of 150 °C, and pressure of 12 bar. The produced panels showed similar mechanical properties to those made with the commercial UF and PF resins. However, formaldehyde emission levels were also significantly lower than when commercial PF resins were used. These panels also followed the requirements of EN 314-2:1993, and therefore presented good bonding quality [94].

6. Conclusions and Future Challenges

Although lignin is the most abundant renewable source of phenolic compounds of natural origin, it is usually treated as an industrial sub-product and burned as fuel. Lignosulphonates, obtained from the sulphite pulping process, alone account for 90% of the total market of commercial lignin with a total annual global production of approximately 1.8 million tons.

As the chemical structure and composition of lignin varies with the source and isolation method, it is important to carry out a detailed characterization before it can be included in the production of value-added products. Currently, there are no uniform or standardized methods for the characterization of lignin. Nevertheless, efforts are being made to develop a series of ISO methods for the characterization of the following lignin features: general composition; functional groups; size and morphology; thermal properties; structural features; safe handling; and processability. Thus, there is little information on fundamental analysis and characterization of technical lignins in the literature, especially when it comes to lignosulphonates. Additionally, the methods that are currently applied need to be adapted before they can be used on lignosulphonates.

One of the possible applications of lignins and lignosulphonates in the production of wood adhesives. The most successful industrial attempts to use lignin for this purpose are based on using this polymer as a partial substitute in phenol-formaldehyde resins or urea-formaldehyde resins. Although lignin and phenol present a similar structure, lignin presents a lower number of reactive sites for polymerization reactions.

In order to address this issue, several methods to improve the reactivity of lignin molecules have been studied with the most promising being phenolation and hydroxymethylation. While these methods influence the reactivity of lignin in resin synthesis, they rely on organic reagents. Thus, there is still a need for more eco-friendly methods like oxidation and ionic liquid treatment.

So far LPF resins with 35% w/w of replaced phenol by methylolated softwood ammonium LS which complied with the standards for plywood application and displayed similar characteristics to commercial PF resins have been developed. In other studies, high-density fiberboard panels were produced with low content UF, 3%, and ammonium LS content of 8%, based on the dry fibers which complied with the requirements for use in loadbearing applications in humid conditions.

However, these resins include formaldehyde which is a known carcinogenic. Thus, concerns about formaldehyde emissions and the sustainability of the used raw materials and final products are reducing the popularity of formaldehyde-based synthetic resins for wood-based panels. Formaldehyde emissions from panels, especially in indoor applications, are raising great concerns.

Therefore, bio-based adhesives have been increasingly studied. Systems which combine LS with polyethylenimine, furfuryl alcohol, polymeric isocyanate, chitosan, tannins, glyoxal, and wheat flour have been proposed.

Nevertheless, lignin adhesives are associated with long pressing times due to their low reactivity, which increases production costs in panel manufacturing. Studies have also reported the low water resistance of the obtained wood panels. Thus, to address these problems and increase the industrial success of lignin and LS adhesives, the interaction between these compounds needs to be further elucidated and more suitable crosslinkers need to be studied.

To conclude, although several studies have been conducted on lignosulphonates, important advances are still needed to warrant success for lignin-based adhesives, namely in their characterization techniques, methods for improving reactivity, and selection of suitable crosslinkers.

Author Contributions: Conceptualization, S.G., J.F., N.P., F.D.M., L.H.C. and J.M.; funding acquisition, J.F., N.P., F.D.M., L.H.C. and J.M.; investigation, S.G., J.F., N.P., F.D.M., L.H.C. and J.M.; methodology, S.G.; project administration, J.F.; supervision, N.P., F.D.M. and L.H.C.; validation, J.F., N.P., F.D.M., L.H.C. and J.M.; writing—original draft, S.G.; writing—review and editing, N.P., F.D.M. and L.H.C. All authors have read and agreed to the published version of the manuscript.

Funding: This work was financially supported by: Base Funding–UIDB/00511/2020 of the Laboratory for Process Engineering, Environment, Biotechnology and Energy–LEPABE–funded by national funds through the FCT/MCTES (PIDDAC).

Conflicts of Interest: The authors declare no conflict of interest.

References

1. Berlin, A. Resin Compositions Comprising Lignin Derivatives. U.S. Patent 9,267,027, 2 December 2010.
2. Hemmilä, V.; Trischler, J.; Sandberg, D. Lignin: An Adhesive Raw Material of the Future or Waste of Research Energy? In Proceedings of the 9th Meeting of the Northern European Network for Wood Science and Engineering, Hannover, Germany, 11–12 September 2013; Brischke, C., Meyer, L., Eds.; Leibniz Universität: Hannover, Germany, 2013; pp. 98–103.
3. Aro, T.; Fatehi, P. Production and Application of Lignosulfonates and Sulfonated Lignin. *ChemSusChem* **2017**, *10*, 1861–1877. [CrossRef]
4. Bertella, S.; Luterbacher, J.S. Lignin Functionalization for the Production of Novel Materials. *Trends Chem.* **2020**, *2*, 440–453. [CrossRef]
5. Doherty, W.O.S.; Mousavioun, P.; Fellows, C.M. Value-Adding to Cellulosic Ethanol: Lignin Polymers. *Ind. Crop. Prod.* **2011**, *33*, 259–276. [CrossRef]
6. Li, C.; Zhao, X.; Wang, A.; Huber, G.W.; Zhang, T. Catalytic Transformation of Lignin for the Production of Chemicals and Fuels. *Chem. Rev.* **2015**, *115*, 11559–11624. [CrossRef]
7. Tribot, A.; Amer, G.; Abdou Alio, M.; de Baynast, H.; Delattre, C.; Pons, A.; Mathias, J.-D.; Callois, J.-M.; Vial, C.; Michaud, P.; et al. Wood-Lignin: Supply, Extraction Processes and Use as Bio-Based Material. *Eur. Polym. J.* **2019**, *112*, 228–240. [CrossRef]
8. Zakzeski, J.; Bruijnincx, P.C.A.; Jongerius, A.L.; Weckhuysen, B.M. The Catalytic Valorization of Lignin for the Production of Renewable Chemicals. *Chem. Rev.* **2010**, *110*, 3552–3599. [CrossRef] [PubMed]
9. Gellerstedt, G.; Henriksson, G. Lignins: Major Sources, Structure and Properties. In *Monomers, Polymers and Composites from Renewable Resources*; Elsevier: Amsterdam, The Netherlands, 2008.
10. Beckham, G.T.; Johnson, C.W.; Karp, E.M.; Salvachúa, D.; Vardon, D.R. Opportunities and Challenges in Biological Lignin Valorization. *Curr. Opin. Biotechnol.* **2016**, *42*, 40–53. [CrossRef]
11. Pandey, M.P.; Kim, C.S. Lignin Depolymerization and Conversion: A Review of Thermochemical Methods. *Chem. Eng. Technol.* **2011**, *34*, 29–41. [CrossRef]
12. Khan, T.A.; Lee, J.-H.; Kim, H.-J. Lignin-Based Adhesives and Coatings. In *Lignocellulose for Future Bioeconomy*; Elsevier: Amsterdam, The Netherlands, 2019.
13. Marques, A.P. Caracterização e Transformação de Lenhosulfonatos de Eucalyptus Globulus. Ph.D. Thesis, Universidade de Aveiro, Aveiro, Portugal, 2011.
14. Dimmel, D. Overview. In *Lignin and Lignans: Advances in Chemistry*; Heitner, C., Dimmel, D., Schmidt, J., Eds.; CRC Press: Boca Raton, FL, USA, 2010.
15. Siddiqui, H. Production of Lignin-Based Phenolic Resins Using De-Polymerized Kraft Lignin and Process Optimization. Master's Thesis, University of Western Ontario, London, ON, Canada, 2013.
16. Lora, J. Industrial Commercial Lignins: Sources, Properties and Applications. In *Monomers, Polymers and Composites from Renewable Resources*; Belgacem, M.N., Gandini, A., Eds.; Elsevier: Amsterdam, The Netherlands, 2008.
17. Becker, J.; Wittmann, C. A Field of Dreams: Lignin Valorization into Chemicals, Materials, Fuels, and Health-Care Products. *Biotechnol. Adv.* **2019**, *37*, 107360. [CrossRef]
18. Sjöström, E. *Wood Chemistry, Fundamentals and Applications*, 2nd ed.; Academic Press: San Diego, CA, USA, 1993.

19. Biermann, J.C. Pulping Fundamentals. In *Handbook of Pulping and Papermaking*; Academic Press: San Diego, CA, USA, 1996; pp. 55–100.
20. Vishtal, A.; Kraslawski, A. Challenges in Industrial Applications of Technical Lignins. *BioResources* **2011**, *6*, 3547–3568. [CrossRef]
21. Chakar, F.S.; Ragauskas, A.J. Review of Current and Future Softwood Kraft Lignin Process Chemistry. *Ind. Crop. Prod.* **2004**, *20*, 131–141. [CrossRef]
22. Calvo-Flores, F.G.; Dobado, J.A. Lignin as Renewable Raw Material. *ChemSusChem* **2010**, *3*, 1227–1235. [CrossRef]
23. Hemmilä, V.; Adamopoulos, S.; Karlsson, O.; Kumar, A. Development of Sustainable Bio-Adhesives for Engineered Wood Panels—A Review. *RSC Adv.* **2017**, *7*, 38604–38630. [CrossRef]
24. Windeisen, E.; Wegener, G. Lignin as Building Unit for Polymers. In *Polymer Science: A Comprehensive Reference*; Elsevier: Amsterdam, The Netherlands, 2012.
25. Braaten, S.M.; Christensen, B.E.; Fredheim, G.E. Comparison of Molecular Weight and Molecular Weight Distributions of Softwood and Hardwood Lignosulfonates. *J. Wood Chem. Technol.* **2003**, *23*, 197–215. [CrossRef]
26. Watkins, D.; Nuruddin, M.; Hosur, M.; Tcherbi-Narteh, A.; Jeelani, S. Extraction and Characterization of Lignin from Different Biomass Resources. *J. Mater. Res. Technol.* **2015**, *4*, 26–32. [CrossRef]
27. Chung, H.; Washburn, N.R. Extraction and Types of Lignin. In *Lignin in Polymer Composites*; Elsevier: Amsterdam, The Netherlands, 2016.
28. Ajao, O.; Jeaidi, J.; Benali, M.; Restrepo, A.; el Mehdi, N.; Boumghar, Y. Quantification and Variability Analysis of Lignin Optical Properties for Colour-Dependent Industrial Applications. *Molecules* **2018**, *23*, 377. [CrossRef]
29. Lupoi, J.S.; Singh, S.; Parthasarathi, R.; Simmons, B.A.; Henry, R.J. Recent Innovations in Analytical Methods for the Qualitative and Quantitative Assessment of Lignin. *Renew. Sustain. Energy Rev.* **2015**, *49*, 871–906. [CrossRef]
30. ISO. *Pulps—Determination of Lignin Content—Acid Hydrolysis Method*; ISO 21436; ISO: Geneva, Switzerland, 2020.
31. Sumerskii, I.; Zweckmair, T.; Hettegger, H.; Zinovyev, G.; Bacher, M.; Rosenau, T.; Potthast, A. A Fast Track for the Accurate Determination of Methoxyl and Ethoxyl Groups in Lignin. *RSC Adv.* **2017**, *7*, 22974–22982. [CrossRef]
32. Dence, C.W. The Determination of Lignin. In *Methods in Lignin Chemistry*; Lin, S.Y., Dence, C.W., Eds.; Springer: Berlin/Heidelberg, Germany, 1992.
33. Lin, S.Y. Ultraviolet Spectrophotometry. In *Methods in Lignin Chemistry*; Lin, S.Y., Dence, C.W., Eds.; Springer: Berlin/Heidelberg, Germany, 1992.
34. Fatehi, P.; Gao, W.; Sun, Y.; Dashtban, M. Acidification of Prehydrolysis Liquor and Spent Liquor of Neutral Sulfite Semichemical Pulping Process. *Bioresour. Technol.* **2016**, *218*, 525–528. [CrossRef]
35. Alonso, M.V.; Rodríguez, J.J.; Oliet, M.; Rodríguez, F.; García, J.; Gilarranz, M.A. Characterization and Structural Modification of Ammonic Lignosulfonate by Methylolation. *J. Appl. Polym. Sci.* **2001**, *82*, 2661–2668. [CrossRef]
36. Hemmilä, V.; Hosseinpourpia, R.; Adamopoulos, S.; Eceiza, A. Characterization of Wood-Based Industrial Biorefinery Lignosulfonates and Supercritical Water Hydrolysis Lignin. *Waste Biomass Valorization* **2020**, *11*, 5835–5845. [CrossRef]
37. TAPPI. *Ash in Wood, Pulp, Paper and Paperboard: Combustion at 525 °C*; T 211 Om-02; TAPPI: Peachtree Corners, GA, USA, 2002.
38. ISO. *Paper, Board, Pulps and Cellulose Nanomaterials—Determination of Residue (Ash Content) on Ignition at 525 °C*; ISO 1762; ISO: Geneva, Switzerland, 2019.
39. Ji, L. Characterization of Lignin Molar Mass and Molecular Conformation by Multi-Angle Light Scattering. Master's Thesis, University of British Columbia, Vancouver, BC, Canada, 2019.
40. Gellerstedt, G. Gel Permeation Chromatography. In *Methods in Lignin Chemistry*; Lin, S.Y., Dence, C.W., Eds.; Springer: Berlin/Heidelberg, Germany, 1992.
41. Sulaeva, I.; Zinovyev, G.; Plankeele, J.-M.; Sumerskii, I.; Rosenau, T.; Potthast, A. Fast Track to Molar-Mass Distributions of Technical Lignins. *ChemSusChem* **2017**, *10*, 629–635. [CrossRef]
42. Ringena, O.; Lebioda, S.; Lehnen, R.; Saake, B. Size-Exclusion Chromatography of Technical Lignins in Dimethyl Sulfoxide/Water and Dimethylacetamide. *J. Chromatogr. A* **2006**, *1102*, 154–163. [CrossRef]
43. Lebo, S.E.; Brten, S.M.; Fredheim, G.E.; Lutnaes, B.F.; Lauten, R.A.; Myrvold, B.O.; McNally, T.J. Recent Advances in the Characterization of Lignosulfonates. In *Characterization of Lignocellulosic Materials*; Blackwell Publishing Ltd.: Oxford, UK, 2008; pp. 189–205.
44. Brudin, S.; Schoenmakers, P. Analytical Methodology for Sulfonated Lignins. *J. Sep. Sci.* **2010**, *33*, 439–452. [CrossRef]
45. Ruwoldt, J. A Critical Review of the Physicochemical Properties of Lignosulfonates: Chemical Structure and Behavior in Aqueous Solution, at Surfaces and Interfaces. *Surfaces* **2020**, *3*, 42. [CrossRef]
46. Fredheim, G.E.; Braaten, S.M.; Christensen, B.E. Molecular Weight Determination of Lignosulfonates by Size-Exclusion Chromatography and Multi-Angle Laser Light Scattering. *J. Chromatogr. A* **2002**, *942*, 191–199. [CrossRef]
47. Majcherczyk, A.; Hüttermann, A. Size-Exclusion Chromatography of Lignin as Ion-Pair Complex. *J. Chromatogr. A* **1997**, *764*, 183–191. [CrossRef]
48. Ringena, O.; Saake, B.; Lehnen, R. Isolation and Fractionation of Lignosulfonates by Amine Extraction and Ultrafiltration: A Comparative Study. *Holzforschung* **2005**, *59*, 405–412. [CrossRef]
49. Marques, A.P.; Evtuguin, D.V.; Magina, S.; Amado, F.M.L.; Prates, A. Chemical Composition of Spent Liquors from Acidic Magnesium–Based Sulphite Pulping of *Eucalyptus Globulus*. *J. Wood Chem. Technol.* **2009**, *29*, 322–336. [CrossRef]

50. Chen, F.; Li, J. Aqueous Gel Permeation Chromatographic Methods for Technical Lignins. *J. Wood Chem. Technol.* **2000**, *20*, 265–276. [CrossRef]
51. Stark, N.M.; Yelle, D.J.; Agarwal, U.P. Techniques for Characterizing Lignin. In *Lignin in Polymer Composites*; Elsevier: Amsterdam, The Netherlands, 2016.
52. Faix, O. Fourier Transform Infrared Spectroscopy. In *Methods in Lignin Chemistry*; Lin, S.Y., Dence, C.W., Eds.; Springer: Berlin/Heidelberg, Germany, 1992.
53. Calvo-Flores, F.G.; Dobado, J.A.; Isac-García, J.; Martín-Martínez, F.J. Functional and Spectroscopic Characterization of Lignins. In *Lignin and Lignans as Renewable Raw Materials*; John Wiley & Sons, Ltd.: Chichester, UK, 2015.
54. Atalla, R.H.; Agarwal, U.P.; Bond, J.S. Raman Spectroscopy. In *Methods in Lignin Chemistry*; Lin, S.Y., Dence, C.W., Eds.; Springer: Berlin/Heidelberg, Germany, 1992.
55. Carvalho, L. Estudo Da Operação de Prensagem Do Aglomerado de Fibras de Média Densidade (MDF). Ph.D. Thesis, Faculdade de Engenharia da Universidade do Porto, Porto, Portugal, 1999.
56. Agarwal, U.; Atalla, R. Vibrational Spectroscopy. In *Lignin and Lignans: Advances in Chemistry*; Heitner, C., Dimmel, D., Schmidt, J., Eds.; CRC Press: Boca Raton, FL, USA, 2010.
57. Ertani, A.; Nardi, S.; Francioso, O.; Pizzeghello, D.; Tinti, A.; Schiavon, M. Metabolite-Targeted Analysis and Physiological Traits of Zea Mays, L. in Response to Application of a Leonardite-Humate and Lignosulfonate-Based Products for Their Evaluation as Potential Biostimulants. *Agronomy* **2019**, *9*, 445. [CrossRef]
58. Ralph, J.; Landucci, L. NMR of Lignins. In *Lignin and Lignans*; CRC Press: Boca Raton, FL, USA, 2010.
59. Marques, A.P.; Evtuguin, D.V.; Magina, S.; Amado, F.M.L.; Prates, A. Structure of Lignosulphonates from Acidic Magnesium-Based Sulphite Pulping of *Eucalyptus Globulus*. *J. Wood Chem. Technol.* **2009**, *29*, 337–357. [CrossRef]
60. Lapierre, C. Determining Lignin Structure by Chemical Degradations. In *Lignin and Lignans*; CRC Press: Boca Raton, FL, USA, 2010.
61. Lai, Y.Z. Determination of Phenolic Hydroxyl Groups. In *Methods in Lignin Chemistry*; Lin, S.Y., Dence, C.W., Eds.; Springer: Berlin/Heidelberg, Germany, 1992.
62. Stücker, A.; Podschun, J.; Saake, B.; Lehnen, R. A Novel Quantitative ^{31} P NMR Spectroscopic Analysis of Hydroxyl Groups in Lignosulfonic Acids. *Anal. Methods* **2018**, *10*, 3281–3488. [CrossRef]
63. Alonso, M.V.; Oliet, M.; Rodríguez, F.; Astarloa, G.; Echeverría, J.M. Use of a Methylolated Softwood Ammonium Lignosulfonate as Partial Substitute of Phenol in Resol Resins Manufacture. *J. Appl. Polym. Sci.* **2004**, *94*, 643–650. [CrossRef]
64. Goldschmid, O. Determination of Phenolic Hydroxyl Content of Lignin Preparations by Ultraviolet Spectrophotometry. *Anal. Chem.* **1954**, *9*, 1421–1423. [CrossRef]
65. Brunow, G.; Lundquist, K. Functional Groups and Bonding Patterns in Lignin (Including the Lignin-Carbohydrate Complexes). In *Lignin and Lignans*; CRC Press: Boca Raton, FL, USA, 2010.
66. Chen, C.-L. Determination of Methoxyl Groups. In *Methods in Lignin Chemistry*; Lin, S.Y., Dence, C.W., Eds.; Springer: Berlin/Heidelberg, Germany, 1992.
67. Hatakeyama, H. Thermal Analysis. In *Methods in Lignin Chemistry*; Lin, S.Y., Dence, C.W., Eds.; Springer: Berlin/Heidelberg, Germany, 1992.
68. Younesi-Kordkheili, H.; Pizzi, A.; Honarbakhsh-Raouf, A.; Nemati, F. The Effect of Soda Bagasse Lignin Modified by Ionic Liquids on Properties of the Urea–Formaldehyde Resin as a Wood Adhesive. *J. Adhes.* **2017**, *93*, 914–925. [CrossRef]
69. Lourençon, T.V.; Alakurtti, S.; Virtanen, T.; Jääskeläinen, A.-S.; Liitiä, T.; Hughes, M.; Magalhães, W.L.E.; Muniz, G.I.B.; Tamminen, T. Phenol-Formaldehyde Resins with Suitable Bonding Strength Synthesized from "Less-Reactive" Hardwood Lignin Fractions. *Holzforschung* **2020**, *74*, 175–183. [CrossRef]
70. Hemmilä, V.; Adamopoulos, S.; Hosseinpourpia, R.; Ahmed, S.A. Ammonium Lignosulfonate Adhesives for Particleboards with PMDI and Furfuryl Alcohol as Crosslinkers. *Polymers* **2019**, *11*, 1633. [CrossRef]
71. Londoño Zuluaga, C.; Du, J.; Chang, H.-M.; Jameel, H.; Gonzalez, R.W. Lignin Modifications and Perspectives towards Applications of Phenolic Foams: A Review. *BioResources* **2018**, *13*, 9157–9179. [CrossRef]
72. Hu, L.; Pan, H.; Zhou, Y.; Zhang, M. Methods to Improve Lignin's Reactivity as a Phenol Substitute and as Replacement for Other Phenolic Compounds: A Brief Review. *BioResources* **2011**, *6*, 3515–3525. [CrossRef]
73. Alonso, M.V.; Oliet, M.; Rodríguez, F.; García, J.; Gilarranz, M.A.; Rodríguez, J.J. Modification of Ammonium Lignosulfonate by Phenolation for Use in Phenolic Resins. *Bioresour. Technol.* **2005**, *96*, 1013–1018. [CrossRef] [PubMed]
74. Hu, L.; Zhou, Y.; Zhang, M.; Liu, R. Characterization and Properties of a Lignosulfonate-Based Phenolic Foam. *BioResources* **2012**, *7*, 554–564.
75. Gilca, I.A.; Ghitescu, R.E.; Puitel, A.C.; Popa, V.I. Preparation of Lignin Nanoparticles by Chemical Modification. *Iran. Polym. J.* **2014**, *23*, 355–363. [CrossRef]
76. Pizzi, A. Lignin-Based Wood Adhesives. In *Advanced Wood Adhesives Technology*; Marcel Dekker: New York, NY, USA, 1994; pp. 219–242.
77. Pang, Y.-X.; Qiu, X.-Q.; Yang, D.-J.; Lou, H.-M. Influence of Oxidation, Hydroxymethylation and Sulfomethylation on the Physicochemical Properties of Calcium Lignosulfonate. *Colloids Surf. A Physicochem. Eng. Asp.* **2008**, *312*, 154–159. [CrossRef]
78. el Mansouri, N.; Pizzi, A.; Salvadó, J. Lignin-Based Wood Panel Adhesives without Formaldehyde. *Holz Als Roh Werkst.* **2007**, *65*, 65–70. [CrossRef]

79. Kielhorn, J.; Pohlenz-Michel, C.; Schmidt, S.; Mangelsdorf, I. Glyoxal. Available online: http://apps.who.int/iris/bitstream/handle/10665/42867/924153057X.pdf;jsessionid=32EC7F5159659F1A18158AB726C50A3A?sequence=1 (accessed on 12 October 2021).
80. Liteplo, R.; Beauchamp, R.; Meek, M.; Chénier, R. Formaldehyde. Available online: https://www.who.int/ipcs/publications/cicad/en/cicad40.pdf (accessed on 12 October 2021).
81. Vangeel, T.; Schutyser, W.; Renders, T.; Sels, B.F. Perspective on Lignin Oxidation: Advances, Challenges, and Future Directions. *Top. Curr. Chem.* **2018**, *376*, 30. [CrossRef]
82. Wang, H.; Pu, Y.; Ragauskas, A.; Yang, B. From Lignin to Valuable Products–Strategies, Challenges, and Prospects. *Bioresour. Technol.* **2019**, *271*, 449–461. [CrossRef]
83. Santos, S.G.; Marques, A.P.; Lima, D.L.D.; Evtuguin, D.V.; Esteves, V.I. Kinetics of Eucalypt Lignosulfonate Oxidation to Aromatic Aldehydes by Oxygen in Alkaline Medium. *Ind. Eng. Chem. Res.* **2011**, *50*, 291–298. [CrossRef]
84. Yuan, Y.; Guo, M.H.; Liu, F.Y. Preparation and Evaluation of Green Composites Using Modified Ammonium Lignosulfonate and Polyethylenimine as a Binder. *BioResources* **2013**, *9*, 836–848. [CrossRef]
85. Hu, L.; Zhou, Y.; Liu, R.; Zhang, M.; Yang, X. Synthesis of Foaming Resol Resin Modified with Oxidatively Degraded Lignosulfonate. *Ind. Crop. Prod.* **2013**, *44*, 364–366. [CrossRef]
86. el Mansouri, N.-E.; Farriol, X.; Salvadó, J. Structural Modification and Characterization of Lignosulfonate by a Reaction in an Alkaline Medium for Its Incorporation into Phenolic Resins. *J. Appl. Polym. Sci.* **2006**, *102*, 3286–3292. [CrossRef]
87. Szalaty, T.J.; Klapiszewski, Ł.; Jesionowski, T. Recent Developments in Modification of Lignin Using Ionic Liquids for the Fabrication of Advanced Materials–A Review. *J. Mol. Liq.* **2020**, *301*, 112417. [CrossRef]
88. Qu, Y.; Luo, H.; Li, H.; Xu, J. Comparison on Structural Modification of Industrial Lignin by Wet Ball Milling and Ionic Liquid Pretreatment. *Biotechnol. Rep.* **2015**, *6*, 1–7. [CrossRef] [PubMed]
89. Younesi-Kordkheili, H.; Pizzi, A. A Comparison between Lignin Modified by Ionic Liquids and Glyoxalated Lignin as Modifiers of Urea-Formaldehyde Resin. *J. Adhes.* **2017**, *93*, 1120–1130. [CrossRef]
90. Guterman, R.; Molinari, V.; Josef, E. Ionic Liquid Lignosulfonate: Dispersant and Binder for Preparation of Biocomposite Materials. *Angew. Chem. Int. Ed.* **2019**, *58*, 1–8. [CrossRef] [PubMed]
91. Ferreira, A.M.; Pereira, J.; Almeida, M.; Ferra, J.; Paiva, N.; Martins, J.; Magalhães, F.D.; Carvalho, L.H. Low-Cost Natural Binder for Particleboards Production: Study of Manufacture Conditions and Stability. *Int. J. Adhes. Adhes.* **2019**, *93*, 59–63. [CrossRef]
92. Ferreira, A.M.; Pereira, J.; Almeida, M.; Ferra, J.; Paiva, N.; Martins, J.; Carvalho, L.H.; Magalhães, F.D. Effect of Spent Sulfite Liquor on Urea-Formaldehyde Resin Performance. *J. Appl. Polym. Sci.* **2019**, *136*, 47389. [CrossRef]
93. Ji, X.; Guo, M. Preparation and Properties of a Chitosan-Lignin Wood Adhesive. *Int. J. Adhes. Adhes.* **2018**, *82*, 8–13. [CrossRef]
94. Rhazi, N.; Oumam, M.; Sesbou, A.; Hannache, H.; Charrier-El Bouhtoury, F. Physico-Mechanical Properties of Plywood Bonded with Ecological Adhesives from Acacia Mollissima Tannins and Lignosulfonates. *Eur. Phys. J. Appl. Phys.* **2017**, *78*, 34813. [CrossRef]
95. Ferreira, A.; Pereira, J.; Almeida, M.; Ferra, J.; Paiva, N.; Martins, J.; Magalhães, F.; Carvalho, L. Biosourced Binder for Wood Particleboards Based on Spent Sulfite Liquor and Wheat Flour. *Polymers* **2018**, *10*, 1070. [CrossRef]
96. Schmitt, L.; Hollis, J. Non-Toxic, Stable Lignosulfonate-Urea-Formaldehyde Composition and Method of Preparation Thereof. U.S. Patent 5,075,402, 21 November 1991.
97. Bekhta, P.; Noshchenko, G.; Réh, R.; Kristak, L.; Sedliačik, J.; Antov, P.; Mirski, R.; Savov, V. Properties of Eco-Friendly Particleboards Bonded with Lignosulfonate-Urea-Formaldehyde Adhesives and PMDI as a Crosslinker. *Materials* **2021**, *14*, 4875. [CrossRef] [PubMed]
98. Antov, P.; Savov, V.; Krišťák, Ľ.; Réh, R.; Mantanis, G.I. Eco-Friendly, High-Density Fiberboards Bonded with Urea-Formaldehyde and Ammonium Lignosulfonate. *Polymers* **2021**, *13*, 220. [CrossRef] [PubMed]
99. Gao, S.; Cheng, Z.; Zhou, X.; Liu, Y.; Chen, R.; Wang, J.; Wang, C.; Chu, F.; Xu, F.; Zhang, D. Unexpected Role of Amphiphilic Lignosulfonate to Improve the Storage Stability of Urea Formaldehyde Resin and Its Application as Adhesives. *Int. J. Biol. Macromol.* **2020**, *161*, 755–762. [CrossRef] [PubMed]
100. Ghorbani, M.; Konnerth, J.; van Herwijnen, H.W.G.; Zinovyev, G.; Budjav, E.; Requejo Silva, A.; Liebner, F. Commercial Lignosulfonates from Different Sulfite Processes as Partial Phenol Replacement in PF Resole Resins. *J. Appl. Polym. Sci.* **2018**, *135*, 45893. [CrossRef]
101. Antov, P.; Mantanis, G.I.; Savov, V. Development of Wood Composites from Recycled Fibres Bonded with Magnesium Lignosulfonate. *Forests* **2020**, *11*, 613. [CrossRef]
102. el Mansouri, N.-E.; Salvadó, J. Structural Characterization of Technical Lignins for the Production of Adhesives: Application to Lignosulfonate, Kraft, Soda-Anthraquinone, Organosolv and Ethanol Process Lignins. *Ind. Crop. Prod.* **2006**, *24*, 8–16. [CrossRef]
103. Domínguez, J.C.; Oliet, M.; Alonso, M.V.; Rojo, E.; Rodríguez, F. Structural, Thermal and Rheological Behavior of a Bio-Based Phenolic Resin in Relation to a Commercial Resol Resin. *Ind. Crop. Prod.* **2013**, *42*, 308–314. [CrossRef]
104. Antov, P.; Savov, V.; Mantanis, G.I.; Neykov, N. Medium-Density Fibreboards Bonded with Phenol-Formaldehyde Resin and Calcium Lignosulfonate as an Eco-Friendly Additive. *Wood Mater. Sci. Eng.* **2021**, *16*, 1–7. [CrossRef]

polymers
MDPI

Article

Properties Enhancement Nano Coconut Shell Filled in Packaging Plastic Waste Bionanocomposite

Ismail Ismail [1,*], Quratul Aini [1], Zulkarnain Jalil [1], Niyi Gideon Olaiya [2], Mursal Mursal [1], C.K. Abdullah [2] and Abdul Khalil H.P.S. [2,*]

[1] Physics Department, Mathematics and Natural Sciences Faculty, Universitas Syiah Kuala, Banda Aceh 23111, Indonesia; quratulainifisika@gmail.com (Q.A.); zjalil@unsyiah.ac.id (Z.J.); mursal@unsyiah.ac.id (M.M.)
[2] School of Industrial Technology, Universiti Sains Malaysia, Penang 11800, Malaysia; ngolaiya@futa.edu.ng (N.G.O.); ck_abdullah@usm.my (C.K.A.)
* Correspondence: ismailab@unsyiah.ac.id (I.I.); akhalilhps@gmail.com (A.K.H.P.S.)

Abstract: Plastic waste recycling has been proposed as a long-term solution to eliminate land and marine deposit. This study proposed a new approach to fabricate biocomposites of nano-sized fillers and low matrix compositions with a great performance by using plastic packaging waste different from the conventional biocomposite. Coconut shell, an agricultural waste, was bonden with waste plastic to form a biocomposite with a coupling agent. The optimum percentage composition and the effect of coconut shell ball milling time on the properties of the biocomposite were studied with density, thickness swelling, porosity flexural strength, flexural modulus, compressive strength, thermogravimetric analysis, differential scanning calorimetry, scanning electron microscope (SEM), and atomic force microscopy (AFM). The results showed that the optimum performance of biocomposite was obtained at 30/70 (wt.%) plastic waste to coconut shell ratio, where 70 wt.% was the highest coconut shell composition that can be achieved. Furthermore, for 30 wt.% of polypropylene (low matrix), the performance of biocomposite improved significantly with milling time due to enhanced interaction between filler and matrix. As the milling time was increased from 0 to 40 h, the density increased from 0.9 to 1.02 g/cm^3; thickness swelling decreased from 3.4 to 1.8%; porosity decreased from 7.0 to 3.0%; flexural strength increased from 8.19 to 12.26 MPa; flexural modulus increased from 1.67 to 2.87 GPa, and compressive strength increased from 16.00 to 27.20 MPa. The degradation temperature of biocomposite also increased as the milling duration increased from 0 to 40 h. The melting temperature increased significantly from 160 to 170 °C as the milling duration increased from 0 to 40 h. The depolymerisation occurred at 350 °C, which also increased with milling duration. This study revealed that the performance of biocomposite improved significantly with a lower percentage matrix and fillernanoparticle rather than increasing the percentage of the matrix. The nanocomposite can be used as a panelboard in industrial applications.

Keywords: biocomposite; mechanical properties; physical properties; polypropylene; thermal properties

Citation: Ismail, I.; Aini, Q.; Jalil, Z.; Olaiya, N.G.; Mursal, M.; Abdullah, C.K.; H.P.S., A.K. Properties Enhancement Nano Coconut Shell Filled in Packaging Plastic Waste Bionanocomposite. *Polymers* **2022**, *14*, 772. https://doi.org/10.3390/polym14040772

Academic Editor: Pavlo Bekhta

Received: 30 December 2021
Accepted: 11 February 2022
Published: 16 February 2022

1. Introduction

The use of composite materials has been increased significantly. Many composites applications include furniture, household appliance, electronic device, automotive, aircraft, etc. [1]. Consequently, composite materials have gained interest among researchers and industries [2,3]. As it is known, composite materials are new materials resulting from combining two or more compatible materials. One material is called the dispersed phase (filler), and the other is the matrix phase (adhesive). Conventionally, composite fillers are synthetic materials, such as fibreglass, carbon, silica carbide, aramid [1]. Synthetic composite fillers have a high performance, such as strong but light-weight. However, they are not good for the environment, are not eco-friendly, cause microplastic pollution, and are not sustainable. Because of the environmental concern, scientists are developing composites by using sustainable fillers that are derived from natural fibres (flax, sisal, kenaf, bamboo), agro or forestry residues (rice straw, coconut shell, coconut coir, wheat, corn), recycled fillers (carpet, cardboard, carbon), and industrial co-products (baggase, grape pomade, lignin, etc.) [4,5]. Those sustainable fillers can be used to reinforce polymer matrices (such as epoxy, polypropylene, polyethene) to form composites, called biocomposites [6,7]. Biocomposite research is in great demand today to address environmental issues [8,9]. Recently, biodegradable spoons were successfully fabricated from the mixture of grape seed, wheat, millet, and xanthan [10]. To produce a high-quality biocomposite from natural fibres is a big challenge because of the compatibility between biofibres and polymers [11]. Biofibre (filler) is a hydrophilic material, while polymer (matrix) is hydrophobic, which causes poor bonding between filler and matrix. An alternative solution was to use a coupling agent (chemical treatment) to improve the compatibility between the filler and matrix, which may improve the performance of biocomposites [12]. Nonetheless, it is very challenging to have good properties of biocomposite for the high load of fillers, less composition of the matrix.

One of the abundant agricultural waste materials (biofibres) in tropical countries is coconut shells. Science agriculture reported that the production of coconut worldwide was about 61 million tons in 2019 [13]. Coconut shell contains 34% cellulose, 21% hemicellulose, and 27% lignin [14]. Previous studies showed that the coconut shell was potentially a filler to form a composite [14–16]. Singh formed a coconut shell composite using epoxy resin. The size of particles was 212–850 μm. The compositions of coconut shell particles were 20, 30, 40 wt.%. The mechanical properties of the composite decreased as for the composition 40 wt.% of fillers [15]. Bhaskar studied composite made of coconut shell particles (200–800 μm) and epoxy resin. The compositions of coconut shells were 20, 25, 30, and 35 wt.%. They reported that the ultimate strength and modulus of elasticity of composites decreased as the compositions of coconut shells were increased [16]. There were a number of other studies on coconut shell composites [17–22]. However, the compositions of fillers for most biocomposites were below 50% and micrometre of filler sizes. On the other hand, it is expected to have a good performance of biocomposite for high loading of fillers, less matrix to use more sustainable resources (biofibres).

Plastic pollution is one of the major environmental issues. The amount of plastics used in the world is increasing every year. In 1950, the total plastic production in the world was about 2.3 million tonnes. It increased to about 448 million tons in 2015. Plastic production worldwide will be around 900 million tons in 2050 [23]. According to MacArthur, about 8 million tonnes of plastic wastes enter the ocean every year [24]. This number will increase following the increase in plastic production in the world. There are several types of plastic waste. One of them is polypropylene (code 05). Polypropylene is a thermoplastic polymer widely used in various applications, such as product packaging [25].

Therefore, waste polypropylene around us should be utilised to overcome the environmental issue. The polypropylene from plastic waste can be recycled as a matrix to mix with sustainable fillers in forming the biocomposite. However, few studies on biocomposites use waste plastic (polypropylene) as an adhesive. Chun et al. prepared a composite from coconut shell particles and recycled polypropylene. The size of the filler was 70 mesh, and the composition of the coconut shell was 0–40% [26]. It was found that the composition of fillers significantly affected the mechanical and thermal properties of the composite. However, the polypropylene composition from those studies was large, i.e., 60–100%. The physical properties of the composite were not reported [26]. Agunsoye also studied the recycled polypropylene reinforced with the coconut shell using coconut shell particles (100 and 200 mesh). However, the filler composition (coconut shell particles) was only 5–25%, i.e., 75–95% polypropylene. The impact energy decreased as the composition of coconut shell particles was above 15%. However, they observed that the impact energy for small particles (200 mesh) was higher than 100 mesh for certain compositions [19]. Thus, it is interesting to study the effect of particle (filler) sizes on the properties of biocomposites.

Several previous studies have shown that nanometer-sized fillers can improve the performance of composites [20,27–30]. Ball milling can be applied to produce nanoparticles or nanopowders [31]. Duration of milling or milling time is one of the important parameters in producing nanoparticles [32]. The milling duration significantly affected the composites' properties [32–34]. A recent study showed that nanoparticles of coconut shell mixed with epoxy resin improved the biocomposite performance [32]. There is no study about nanocomposite or nano board made of coconut shell and the waste plastic packaging as matrix found in the literature. Meanwhile, it is very important to utilise plastic waste to overcome the environmental issue. Other than that, nano board or nanocomposite has a very high prospect of developing biocomposites.

The present work is to prepare a biocomposite of coconut shell particles with a new approach. The coconut shell particles were added into the polypropylene matrix from the packing-plastic waste to form the biocomposite. Benzoyl peroxide, xylene, methanol, and anhydride maleic were used as the coupling agent to improve the compatibility between coconut shell particles and polypropylene. The size of coconut shell particles was reduced to nano-size to improve the performance of the biocomposite. The physical, mechanical, thermal, and morphological properties of biocomposite were examined.

2. Materials and Methods

2.1. Materials

The waste plastic bottle packaging for mineral water (polypropylene, code 5, PP) was collected from the recycled plastic place in Banda Aceh, Indonesia. The waste plastic was cleaned and cut by about 1 cm × 1 cm. PT Indratma Sahitaguna, Indonesia, supplied the coconut shell powder having 200 mesh sizes. Benzoyl peroxide (luperox A70S, 632651), xylene (reagent grade, 214736), methanol (gradient grade, 34885), maleic anhydride (99%, M188) were purchased from Sigma Aldrich, Indonesia.

2.2. Sample Preparation

The coconut shell particles with particle size 200 mesh, waste plastic packaging (recycled polypropylene dissolved in xylene as solvent), and coupling agent (benzoyl peroxide (initiator), methanol (reaction solvent), anhydride maleic (modifier)) were mixed using a rheomixer (manufactured by Polytechnic, Medan, Indonesia) at 60 °C for 60 min [35,36]. The chamber volume of the mixer was 625 cm^3, equipped with electrical heating with atemperature up to 350 °C. Figure 1 shows the possible coupling agent reaction in the biocomposite.

The sample mixture was transferred into the extrusion (manufactured by Polytechnic, Medan, Indonesia) heated at 180 °C with 30 rpm of the rotor speed. After that, the mixture was poured into the compression moulding machine and pressed with 3 tons of load for 1 h at 175 °C to obtain a biocomposite sample. The mould was made of steel with 200 mm × 150 mm. The moulding machine (manufactured by Polytechnic, Medan, Indonesia) was equipped with an electric heater a hydraulic press with a load of up to 4 tons. The temperature can be controlled automatically, up to 350 °C.

The composition of biocomposite samples for 200 mesh of coconut shell particles (CSP) is listed in Table 1. The composition of polypropylene (PP) from packaging plastic waste was varied from 30 to 100 wt.%. The size of the biocomposite sample fabricated was 200 mm × 150 mm × 10 mm. The samples were stored in a dried place to prevent moisture.

To study the effect of particle size and milling time on the properties of the biocomposite, the coconut shell powder (200 mesh) was milled at room temperature by a ball mill to produce nanoparticles. The ball mill used in this study was a planetary type, Planetary Mono Mill Pulverisette 6 manufactured by Fritsch Germany, consisting of one working station. The grinding bowl size was 250 mL with a grinding ball diameter of 15 mm. The ratio of ball to powder was 10:1 (wt.). The milling was conducted at room temperature with a rotational speed of 350 rpm (constant speed) and a dry process. The milling duration was varied from 0, 10, 20, 30 to 40 h. The resulting crystallite nanoparticle of the coconut shell was confirmed with X-ray diffraction and Transmission Electron Microscopy (TEM) [32]. The coconut shell nanoparticle was used in biocomposite with propylene waste at an optimum percentage obtained from the composition variation and studied properties.

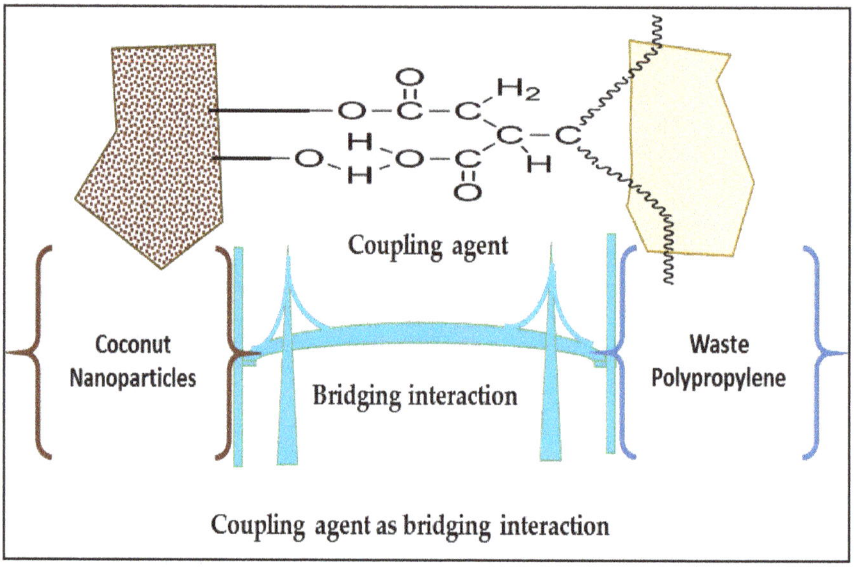

Figure 1. Schematic of possible coupling agent bridging interaction in the biocomposite.

Table 1. The composition of biocomposite sample for 200 mesh coconut particle size.

Sample No.	Particle Size (Mesh)	CSP (wt.%)	PP (wt.%)
A1	200	70 (210 g)	30 (90 g)
A2	200	60 (180 g)	40 (120 g)
A3	200	50 (150 g)	50 (150 g)
A4	200	40 (120 g)	60 (180 g)
A5	200	0 (0 g)	100 (300 g)

2.3. Characterisation of the Biocomposite

The testing of the physical properties of samples was conducted according to the Indonesian National Standard for particleboard [37]. The density of the composite sample (ρ) was obtained using the Equation (1).

$$\rho = \frac{m}{V} \tag{1}$$

where V is the volume of the sample; m is the mass of the sample. The porosity of the sample (PR) was determined using Equation (2).

$$PR = \frac{m_{wet} - m_{dry}}{V_{bulk}} \times \frac{1}{\rho_{water}} \times 100\% \tag{2}$$

where V_{bulk} is sample volume; ρ_{water} is the density of water; m_{dry} and m_{wet} are the mass of the sample before and after the sample was immersed in the water for 24 h. The sample's thickness swelling (TSW) percentage was calculated using Equation (3).

$$TSW = \frac{T_{wet} - T_{dry}}{T_{dry}} \times 100\% \quad (3)$$

where T_{dry} is the sample thickness before immersing in the water; T_{wet} is the thickness of the sample after the sample was immersed in the water for 24 h.

The mechanical properties of biocomposite were determined according to the Indonesian National Standard for particleboard [37]. The measurement was conducted using the universal testing machine manufactured by Hung Ta Company (Taiwan). The flexural strength (*FS*) was obtained using Equation (4).

$$FS = \frac{3 \times B \times S}{2 \times L \times T^2} \quad (4)$$

where B is the maximum load; S is the span; T is the sample thickness; L is the sample width. The flexural modulus (*FM*) was calculated using Equation (5).

$$FM = (\frac{\Delta B}{\Delta D})\frac{S^3}{4 \times L \times T^3} \quad (5)$$

where S, L, and T are the same as in Equation (4); $(\Delta B/\Delta D)$ is the slope of the force to deformation. The compressive strength (*CS*) was determined using Equation (6).

$$CS = \frac{F_{\max}}{A} \quad (6)$$

where $F_{\max}$ is the load at the point of failure, and A is the cross-sectional area of the sample.

The thermogravimetric analysis (TGA) and differential scanning calorimetry (DSC) was utilised to determine the thermal properties of the biocomposite samples. The TGA equipment was produced by Shimadzu, type DTG–60 (Shimadzu, Kyoto, Japan). The DSC equipment was manufactured by Shimadzu, type DSC–60 (Shimadzu, Kyoto, Japan). The heating rate of the sample for both DSC and TGA measurements was 10 °C per minute.

Atomic force microscopy (AFM) (Nanosurf, Liestal, Switzerland) and a scanning electron microscope (SEM) (Thermo Fisher Scientific, Waltham, MA, USA) have been used to evaluate the morphological properties of composite samples. The sample size was 1 cm × 1 cm with 1 cm thickness. The sample was put on the carbon conductive double tape where the tape was stuck on the stub. Then, the samples were gold coated, and the stub with the sample on it was inserted into the chamber. The SEM chamber was operated at the high vacuum mode.

2.4. Statistical Analysis

One-way analysis of variance (ANOVA) with Duncan's multiple range tests was used to test the homogeneity of variance and determine the effect of PP composition and duration of milling on the physical (Tables A1 and A3) and mechanical (Tables A2 and A4) properties of biocomposite. The statistical significance used in the analysis was 0.05 (5%). The calculation was conducted by using the SPSS software version 16.0 (IBM, Chicago, IL, USA).

3. Results and Discussion

3.1. Properties of Biocomposite with Varying Composition

The biocomposite from coconut shell particles using the plastic waste (recycled polypropylene) as the matrix has been successfully prepared. The possible schematic bonding between the biocomposite materials is shown in Figure 1. The coupling agent formed a bridging effect between the coconut shell and the propylene. This enhances the binding efficiency of propylene in the biocomposite. Previous studies reported the bridging effect of the coupling agent used in this study [36,38–40].

3.1.1. Physical Properties

The density, thickness, swelling, and porosity of biocomposite with coconut shell for various compositions of polypropylene (PP) are shown in Figure 2. For the composition of PP 30 wt.%, composite density was 0.90 g/cm^3. The density increased to 0.97 g/cm^3 for PP 50 wt.%. The increasing density of composite from 30 to 50 wt.% of PP (see filled circles in Figure 2) could be related to the well-blended between PP and coconut shell particle at the ratio composition of 50/50 wt.%. However, the density decreased to 0.95 and 0.85 g/cm^3 for 60 and 100 wt.%, respectively. The decreasing density of composite for the composition of PP above 50 wt.% is because the density of PP is low; i.e., the density of virgin PP is 0.855 g/cm^3. The highest composite density from the present study was 0.97 g/cm^3, which occurred at 50 wt.% coconut shell and 50 wt.% PP. The density of composite for 75 wt.% HDPE (25 wt.% coconut shell particle) is about 0.91 g/cm^3 [41], almost the same density as this study.

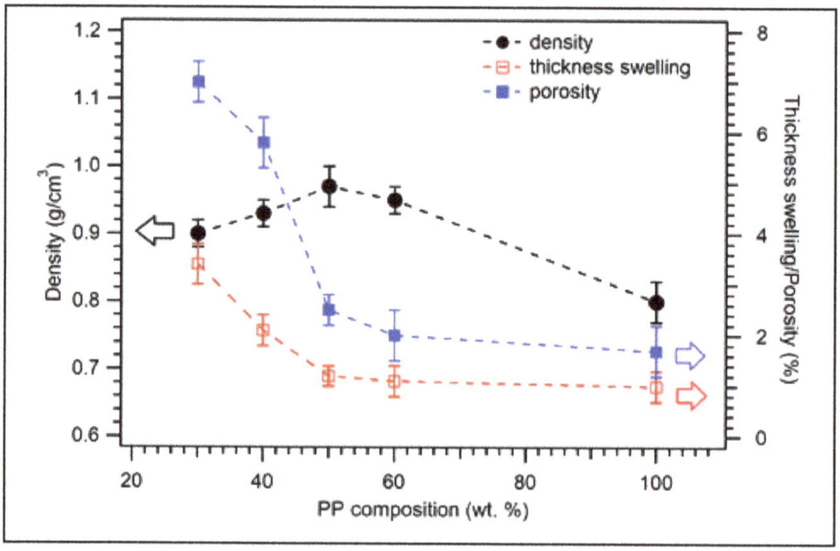

Figure 2. Physical properties of coconut shell biocomposite for various PP compositions with 200 mesh of coconut shell particles.

The thickness swelling of biocomposite for various PP compositions is displayed in Figure 2 (unfilled squares). For 30 wt.% of PP, the thickness swelling of the composite was 3.4%. The thickness swelling of the sample decreased to 1.1% and 1.0% for the composition of PP 60 and 100 wt.%, respectively. The thickness swelling of the composite decreased significantly as the composition of PP was increased because PP is hydrophobic (repelling water). Figure 2 (filled squares) shows the porosity of biocomposite samples for various PP compositions with 200 mesh particle sizes. The porosity of the sample for 30 wt.% of PP is 7%. For 40 wt.% of PP, the porosity decreased to 5.8%. The porosity of the composite continued decreasing as the composition of PP was increased. The porosity was 1.7% for 100 wt.% of PP. The trend of porosity is the same as that of thickness swelling. In general, the physical properties (thickness swelling and porosity) of coconut shell biocomposite improved as the composition of PP was increased. Nonetheless, it is expected that the composition of PP should be minimised (low matrix) for biocomposites.

The results of statistical analysis of physical properties of biocomposite with various PP compositions are listed in Table A1. The statistically significant values of density, thickness swelling, and porosity measurements are 0.740, 0.434, and 0.483, respectively. These numbers are larger than 0.05, indicating that the variances of density, thickness swelling, and porosity data are homogenous. The F values from the calculation for density, thickness swelling, and porosity are 21.800, 54.138, and 155.353, respectively. The value of F theoretical (5%) for df 4/20 is 2.87. The value of F calculation is larger than the values of F theoretical, which indicates the physical properties of biocomposite are significant differences for different PP compositions. This finding confirms a significant effect of PP composition on the physical properties of biocomposite.

3.1.2. Mechanical Properties

The flexural strength, flexural modulus, and compressive strength of coconut shell biocomposite weremeasured. Figure 3 displays the flexural strength, flexural modulus, and compressive strength of biocomposite samples for 200 mesh particle size with various PP compositions. As the composition of PP was increased, it was found that the flexural strength (FS) of the composite increased from 8.19 MPa for 30 wt.% of PPto 11.00 MPa for 60 wt.% of PP, see filled circles in Figure 3. For 100 wt.% of PP, the flexural strength increased to 12.80 MPa. The FS for 100 wt.% PP is significantly lower than the value of FS for virgin PP is 32 MPa [42].

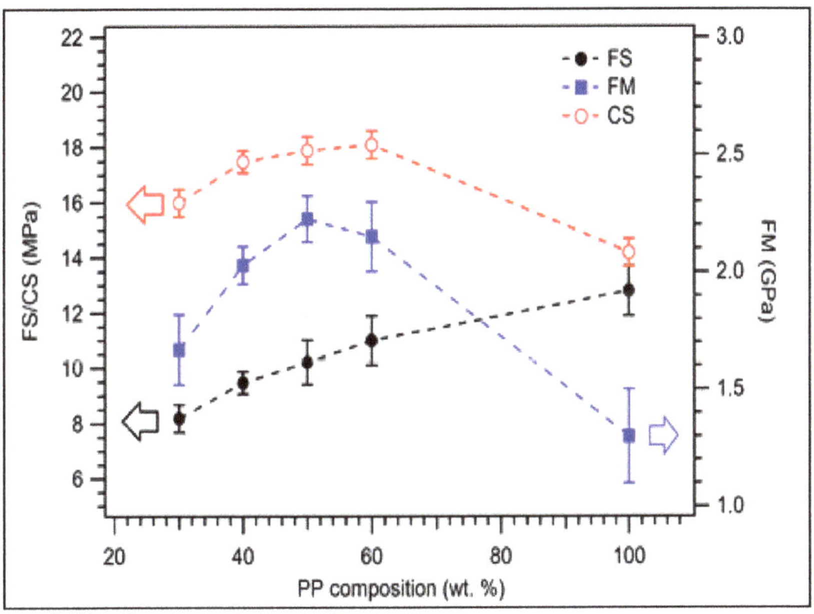

Figure 3. Flexural properties of coconut shell biocomposite for various PP compositions with 200 mesh of coconut shell particles.

The flexural modulus of composite was 1.67 GPa for 30 wt.% of PP. Its value increased to 2.23 GPa for 50 wt.% of PP. However, the flexural modulus slightly decreased to 2.15 GPa for 60 wt.% compositions of PP. The flexural modulus was 1.30 GPa for 100 wt.% PP. The highest flexural modulus of biocomposite was found at 50 wt.% PP, see filled square in Figure 3. For the composition of PP greater than 50 wt.%, the flexural modulus decreases because the flexural modulus of virgin PP is about 1.45 GPa [42]. The measured compressive strength is displayed in unfilled circles in Figure 3. Its value was 16.00 MPa for 30 wt.%

of PP and increased to 18.10 MPa for 60 wt.% of PP. The compressive strength decreased to 14.20 MPa for 100 wt.% of PP. This value is lower than the value of virgin PP reported previously, which is 34.4 MPa [43]. The discrepancy between the results of this study and previous studies could be due to the purity of the matrix.

In general, this study revealed that the composition of PP significantly influenced the mechanical properties of the biocomposite. This behaviour was observed in the previous study where composite tensile strength and elongation break increased as the matrix (PP) composition increased [26]. The flexural strength, flexural modulus, and compressive strength of rice straw composite also increased as the composition of PP was increased [44]. By considering the best flexural modulus value, the good mechanical properties of coconut shell PP biocomposite were obtained at the 50 wt.% PP composition.

Table A2 displays the statistical analysis of mechanical properties of biocomposite with various PP compositions. The statistically significant values of flexural strength, flexural modulus, and compressive strength data are 0.540, 0.264, and 0.017, respectively, larger than 0.05. This indicates that the variances of flexural strength, flexural modulus, and compressive strength data are homogenous. The F values from the calculation for flexural strength, flexural modulus, and compressive strength are 27.407, 36.774, and 56.935, respectively, larger than the F theoretical value (2.87), which indicate the mechanical properties of biocomposite are significant differences for different PP compositions. This information confirms that PP compositions significantly influence the mechanical properties of biocomposites.

3.1.3. Thermal Properties

The thermal properties of the coconut shell biocomposites have been evaluated. Figure 4 displays the TGA of biocomposite samples with 200 particle size mesh for various PP compositions. The summary of decomposition temperature of biocomposite for various PP compositions is listed in Table 2. For 30 wt.% of PP, the decomposition temperature was 199 °C for 95% of the weight. At 242 °C, the remaining sample was 90%. The weight of the sample was still 80% at 286 °C. Above that temperature, decomposition significantly occurred. At the temperature of 356 °C, only 50% of thesample was left. The weight of the sample was 20% at 443 °C. The result obtained is almost identical to the previous study, where coconut shell powder's degradation temperature is 250–450 °C [45]. As the composition of PP was increased, the decomposition of temperature increased, as shown in Figure 4 and Table 2.

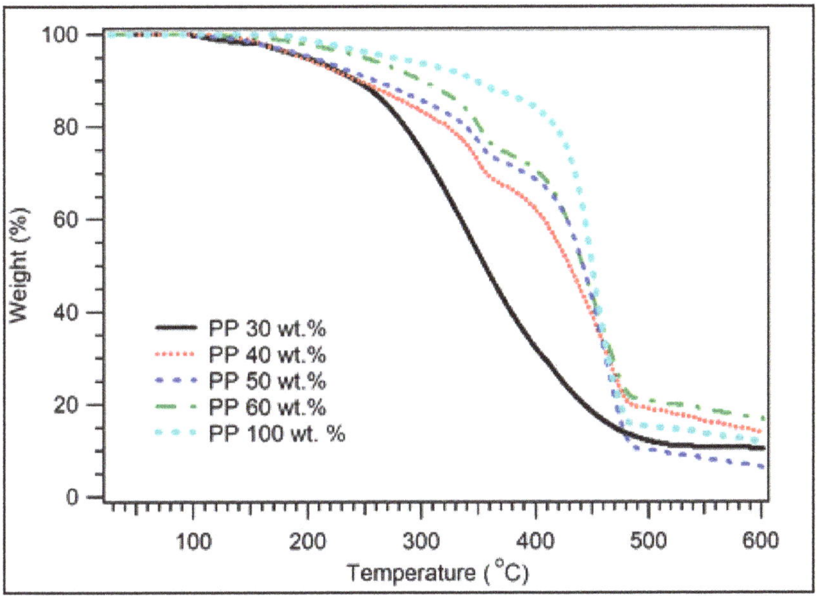

Figure 4. TGA of coconut shell biocomposite for various PP compositions with 200 mesh of coconut shell particles.

Table 2. Decomposition temperature of coconut shell biocomposite for various PP compositions.

Sample Weight (%)	Decomposition Temperature (°C)				
	PP 30 wt.%	**PP 40 wt.%**	**PP 50 wt.%**	**PP 60 wt.%**	**PP 100 wt.%**
95	199	195	202	249	275
90	242	245	260	300	346
80	286	323	339	351	417
50	356	429	441	442	450
20	443	480	471	500	474

The DSC curves for the coconut shell biocomposites for various PP compositions are displayed in Figure 5. There was a broad peak observed at 60 °C. This peak was associated with coconut shell biocomposite's glass transition temperature (Tg). As the composition of PP increased, the peak widened and disappeared at 100 wt.% of PP. The disappearance of the Tg peak at the 100% PP composition is due to the Tg of PP at a low temperature of about −10 °C. The observed Tg for this study was about the same as Chun et al. for coconut shell particle polylactic acid composite 63 °C [26]. The melting temperature (Tm) was observed around 162 °C. As the composition of PP increased, the Tm of biocomposite improved. Another peak was observed at 350 °C, which was related to depolymerisation. As shown in the TGA results, decomposition occurs drastically at 350 °C. The summary of DSC data of coconut shell biocomposite for various PP compositions is listed in Table 3. As the composition of PP increased, the Tg and Td of biocomposite decreased. However, its Tm increased from 160 to 164 °C.

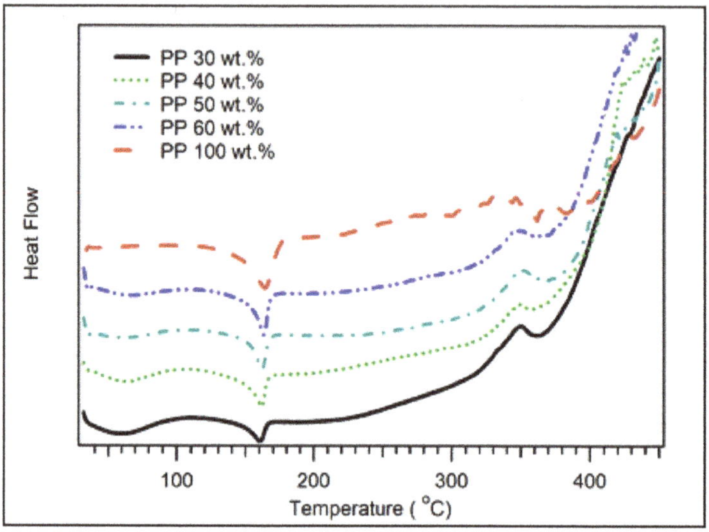

Figure 5. DSC of coconut shell biocomposite for various PP compositions with 200 mesh of coconut shell particles.

Table 3. DSC data of coconut shell biocomposite for various PP compositions.

PP Composition (wt.%)	Tg (°C)	Tm (°C)	Td (°C)
30	60	160	350
40	62	161	350
50	57	161	350
60	57	162	346
100	-	164	340

3.1.4. Morphological Properties

The morphology of biocomposite samples for 200 mesh particle size with the various compositions of PP was examined by SEM and AFM. The results are displayed in Table 4. The surface was rather rough for 30 wt.% compositions of PP (70 wt.% of coconut shell particles). There were some porosities and agglomerations observed. As the PP composition increased 50 wt.%, the coconut shell particles were mixed well with PP, reducing the porosities. The surface corrugation of biocomposite reduced as the composition of PP was increased, as shown by the AFM image in Table 4. The biocomposite becomes denser, so the density becomes greater. The bond between the filler and the adhesive becomes better, increasing the mechanical properties for 100 wt.% PP composition, the surface is homogenous and smooth.

Table 4. SEM and AFM images for the various compositions of PP.

Composition of PP	SEM	AFM
30 wt.%		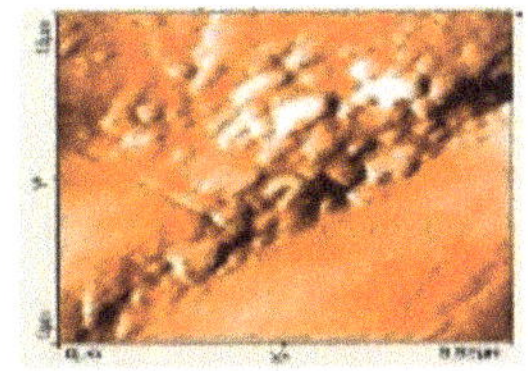
40 wt.%	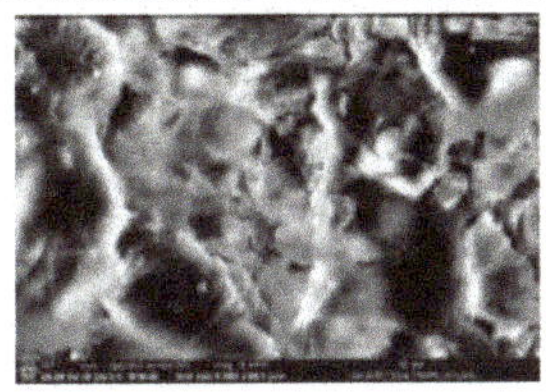	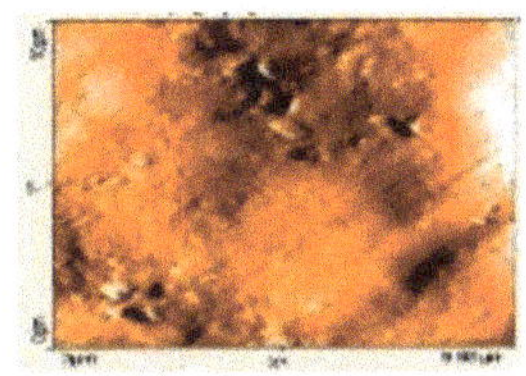
50 wt.%	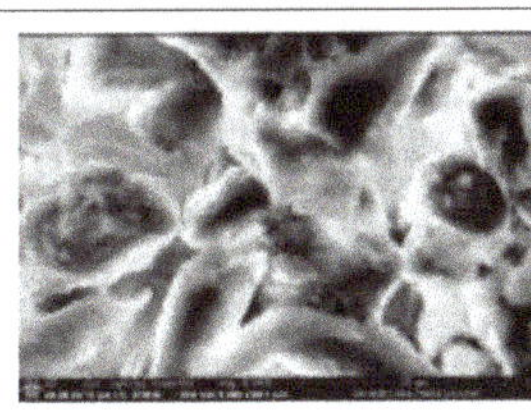	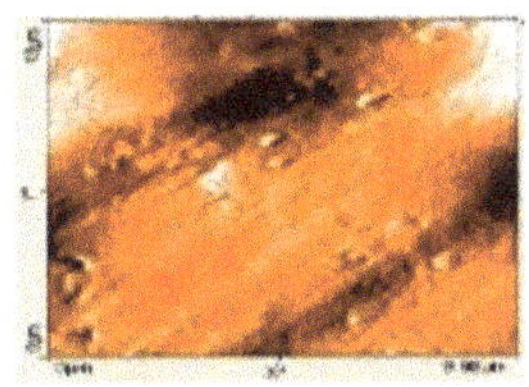
60 wt.%	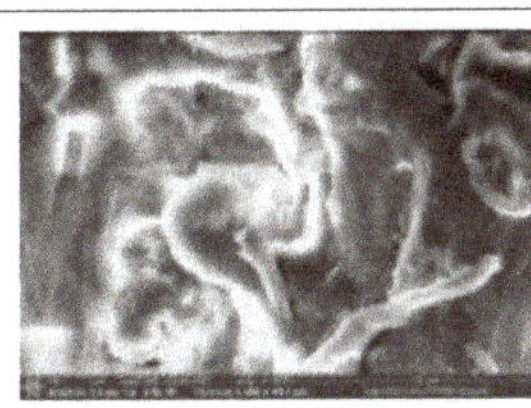	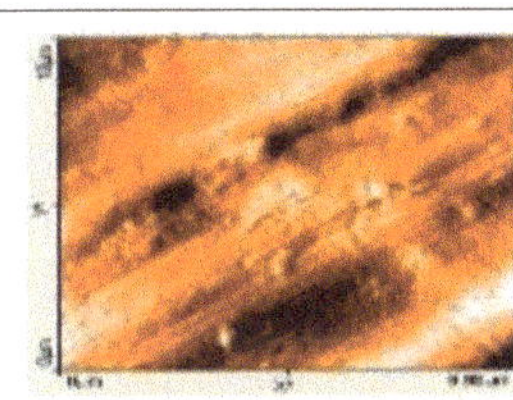
100 wt.%	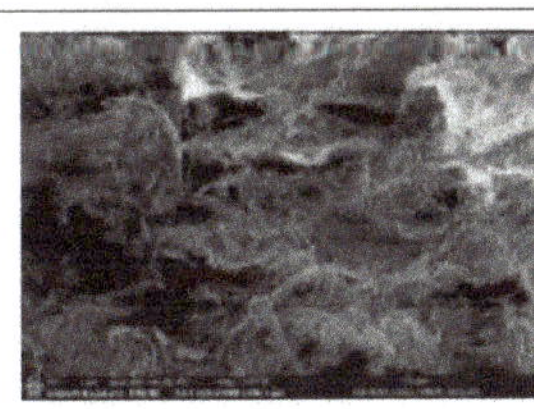	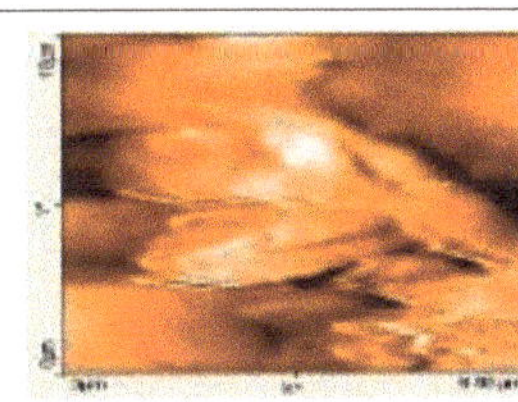

3.2. Properties of Biocomposite with Varying Particle Size

The crystallite size and TEM images of the coconut shell particles for each milling time are shown in Table 5. The TEM images confirmed the production of nanoparticles from the coconut shell milling time. The particle size of the coconut shell reduces with milling time until 40 h. Further milling time has no significant effect on its particle size. The particle size of the coconut shell was found to be 48 nm for 10 h of milling time, which decreases with increasing milling duration, as displayed in filled circles in Table 5. The particle size decreases to 30 nm for a 40 h milling duration. The size of the coconut shell remains constant after 40 h which shows the optimum milling time. The trend of the result was similar to that found in the literature.

Table 5. Crystallite size of coconut shell particles.

Duration of Milling	TEM	XRD Spectra	Crystallite Size (nm)
0 h	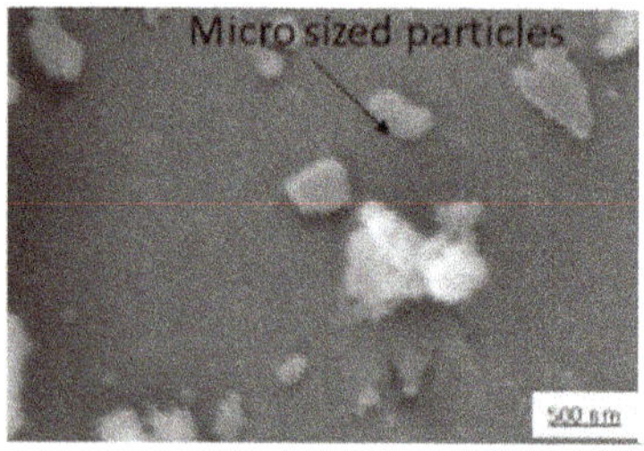	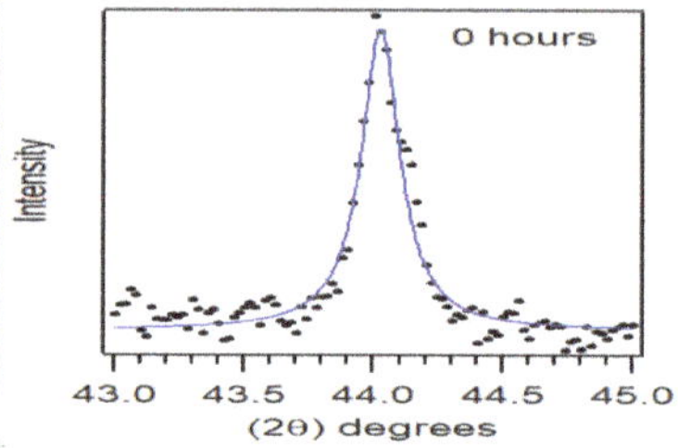	80
10 h	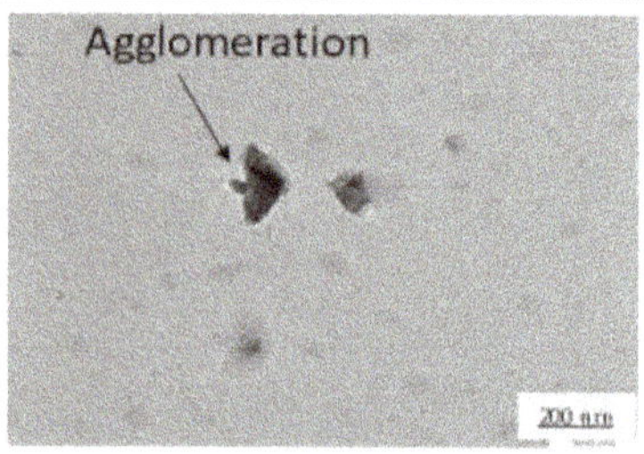	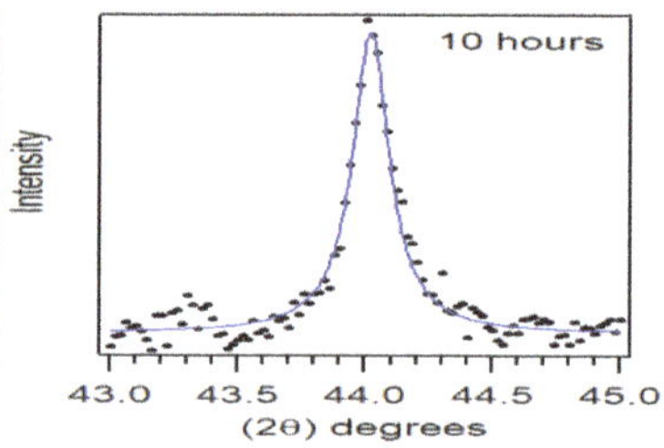	48
20 h	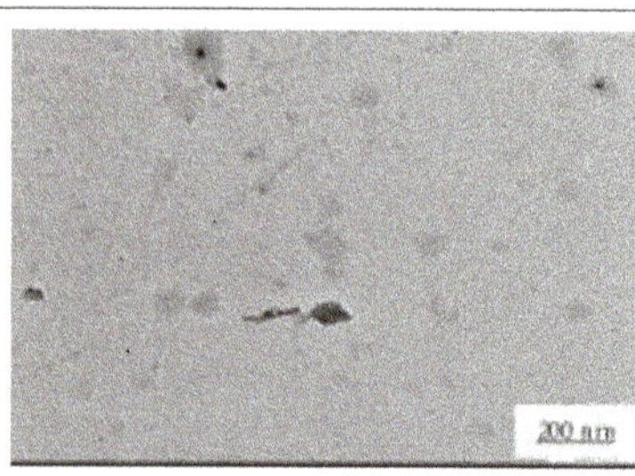	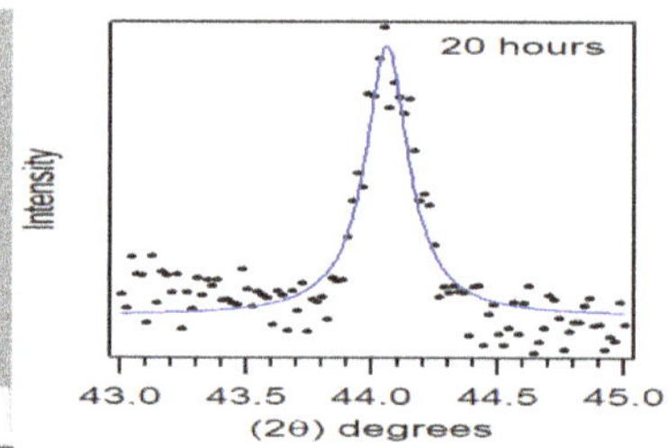	45
30 h	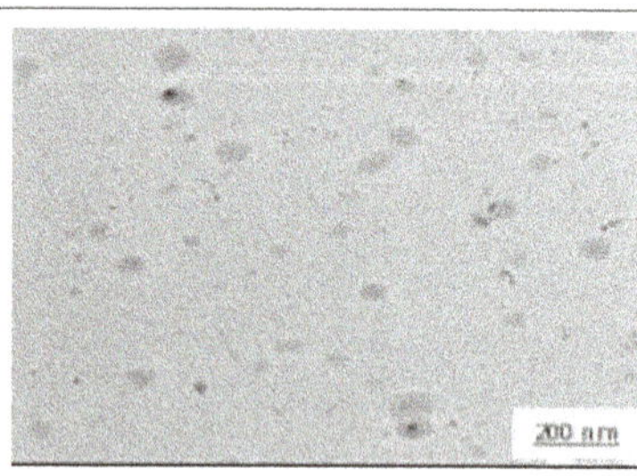	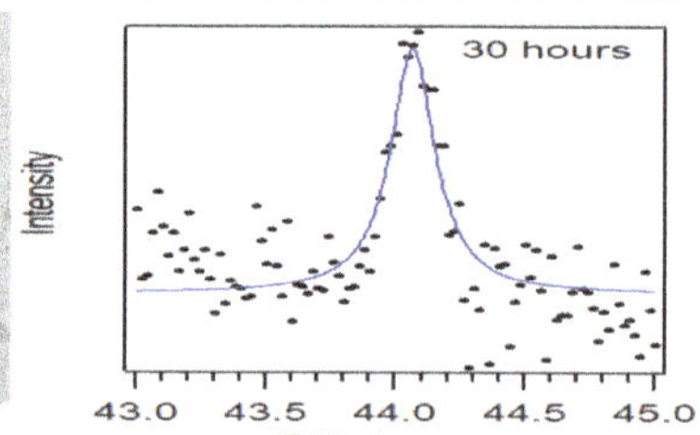	42

Table 5. *Cont.*

Duration of Milling	TEM	XRD Spectra	Crystallite Size (nm)
40 h	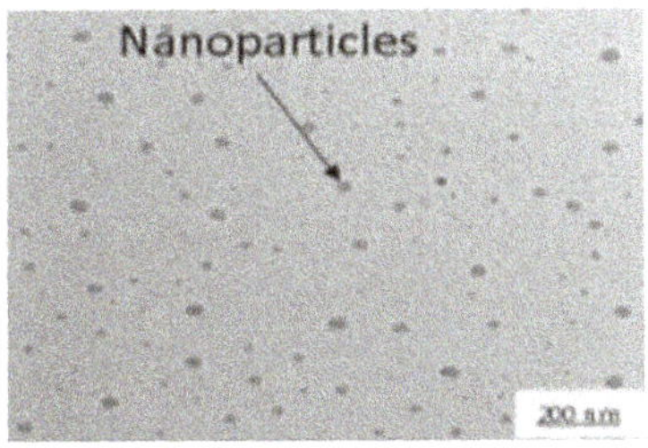	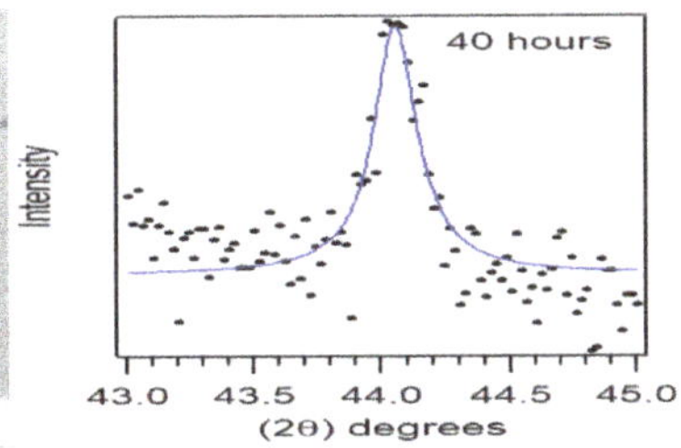	30

3.2.1. Physical Properties

For the composition of samples with 30 wt.% of recycled polypropylene and 70 wt.% of coconut shell particles, the milling duration was varied from 0, 10, 20, 30, to 40 h. The physical properties of nano-biocomposite are shown in Figure 6 for the 30 wt.% compositions of PP (70 wt.% of coconut shell particle). The density of biocomposite was found to be 0.91 g/cm^3 for 10 h of milling time. Its density increased with increasing milling duration, as displayed in the filled circles in Figure 6. The density was 1.02 g/cm^3 for a 40 h milling duration. The density of biocomposite found in this study is lower than that of coconut shell nanoparticles with epoxy resin (1.03–1.19 g/cm^3). However, its trend is the same [32]. Figure 6 (unfilled squares) displays the thickness swelling of biocomposite for several milling durations. For 0 h duration of milling, the thickness swelling of biocomposite was 3.4%. As the milling time was increased, the thickness swelling decreased. For 40 h duration of milling, the thickness swelling reduced to 1.8%. Filled squares depict the porosity of biocomposites in Figure 6. For 10 h of milling time, the porosity of the composite sample was 6.6%. The porosity of the sample was reduced to 3.0% for 40 h. The porosity of the composite decreased with the increase of milling times which is the same trend for thickness swelling. The previous work also observed this behavior [32].

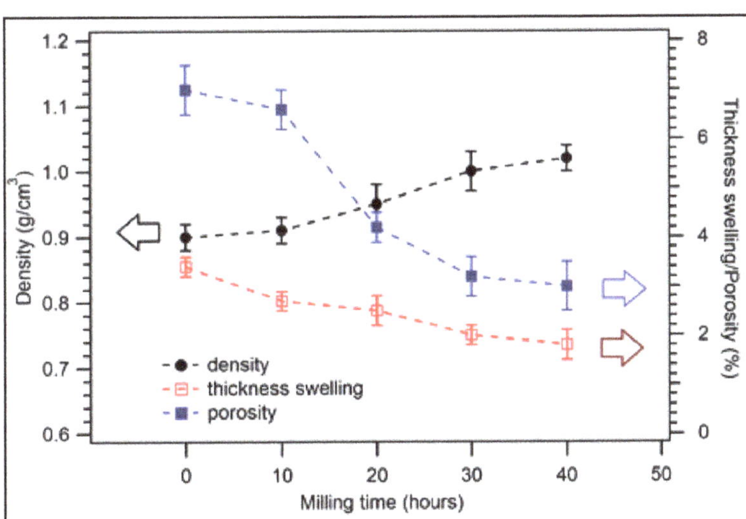

Figure 6. Physical properties of coconut shell nano-biocomposite for various milling times with 30 wt.% of PP composition.

In general, the physical properties of the biocomposite improve significantly as the milling times increase or reduce the particle sizes. This property occurs due to the nano-sized of coconut shell [32]. Nano-sized filler can blend well with the matrix during the formation of composite, which causes the composite to be dense. As a result, the porosity of the composite decreases, which improves the density of the biocomposite. This study indicates that it is unnecessary to increase an adhesive or matrix composition to obtain good physical properties of biocomposite. Still, it can be achieved by reducing the particle size of filler from micro to nanometer.

The statistical analysis of the physical properties of biocomposite with various durations of milling is listed in Table A3. The significant statistical values of density, thickness swelling, and porosity data are 0.170, 0.274, and 0.361, respectively, are larger than 0.05. This shows that the variances of density, thickness swelling, and porosity data for various milling times are homogenous. The calculated F values for density, thickness swelling, and porosity are 21.938, 28.571, and 117.353, respectively. The value of F theoretical (5%) for df 4/20 is 2.87. The calculated F values are larger than the values of F theoretical, which means the physical properties of biocomposite are significant differences for different duration of millings. This confirms that the milling duration affects the physical properties of biocomposite significantly.

3.2.2. Mechanical Properties

Figure 7 displays the flexural strength, flexural modulus, and compressive strength of the biocomposite for various milling times with 30 wt.% of PP composition. For the 10 h duration of milling, the flexural strength was 9.81 MPa which was larger than the value for 0 h of milling (8.19 MPa). The flexural strength of biocomposite increased significantly with the increasing milling times, as depicted by the filled circles in Figure 7. For 40 h milling duration, the flexural strength of the composite was found to be 12.26 MPa which was larger than the value for 60 wt.% of PP. This finding shows that the flexural strength of coconut shell bio-nanocomposite can be improved significantly by increasing the duration of milling time without increasing the composition of PP.

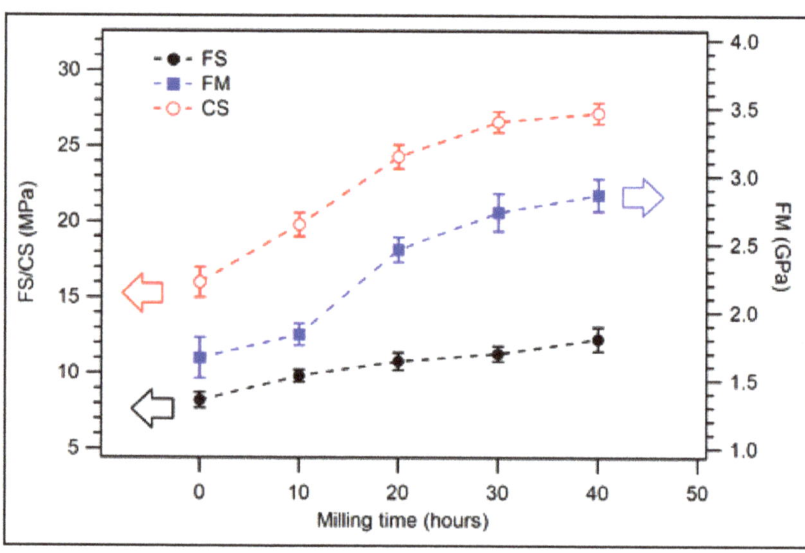

Figure 7. Mechanical properties of coconut shell nano-biocomposite for various milling times with 30 wt.% of PP composition.

Figure 7 shows the flexural modulus of coconut shell bio-nanocomposite samples for 30 wt.% of PP with various milling duration. For 10 h of milling, the flexural modulus of coconut shell bio-nanocomposite was 1.842 GPa. The flexural modulus value increases as the milling duration increases, as displayed in filled squares in Figure 7. For the duration of milling 40 h, the flexural modulus of coconut shell bio-nanocomposite was 2.767 GPa. Its compressive strength was also affected significantly by the duration of milling (see unfilled circles in Figure 7). The compressive strength increased from 16.0 MPa for 0 h of milling to 27.2 MPa for 40 h of milling.

Similar to the physical properties, the mechanical properties of biocomposite improve as the duration of milling increases. This implies that particle sizes (nanoparticles) play an important role in the properties of a composite [32]. As the coconut shell particles become smaller, the contact surface areas between the fillers and matrix (PP) increase. Consequently, the bonding between matrix and fillers increases [30]. Moreover, nanoparticles are well blended with the matrix. As a result, the physical and mechanical properties of the composite improve significantly. This behaviour is similar to that observed in the previous study (coconut shell–epoxy resin composite) [32].

This study's highest flexural strength and modulus values are obtained at the 40 h milling time with 30 wt.% of PP 12.260 MPa and 2.767 GPa, respectively. The values are met the Indonesian National Standard for particleboard. The results obtained from this study are almost the same as the results from the previous study, a rice straw composite using PP as the matrix [14,44].

Table A4 shows the statistical analysis of mechanical properties of biocomposite with various milling times. The significant statistical values of flexural strength, flexural modulus, and compressive strength data are 0.305, 0.653, and 0.715, respectively, larger than 0.05. This means that the variances of flexural strength, flexural modulus, and compressive strength data for various milling durations are homogenous. The calculated F values for flexural strength, flexural modulus, and compressive strength with various milling times are 39.028, 102.017, and 228.771, respectively, larger than the F theoretical value (2.87). This finding indicates the mechanical properties of biocomposite are significant differences for different durations of milling (0, 10, 20, 30, and 40 h). The statistical results confirm that the mechanical properties of biocomposite are significantly influenced by the duration of milling or particle sizes.

3.2.3. Thermal Properties

Figure 8 shows the TGA of coconut shell bio-nanocomposites for 30 wt.% of PP with various durations of milling times. The composite samples started to decompose at a temperature of 150 °C. For 10 and 40 h of milling time, the sample weights were 99.0% and 99.5%, respectively. At a temperature of 200 °C, the sample weight for 0 h milling time remained 94.8%, while it remained 97.8% for 40 h milling time. For sample weight 95%, the decomposition temperature was 199 °C for 0 h milling time. As the milling duration was increased to 10 h, the decomposition temperature increased to 225 °C. It increased to 243 °C for 40 h of milling time. In general, as the duration of milling time was increased, the decomposition of temperature increased, as shown in Figure 8 and Table 6. For 20, 30, and 40 h of milling time, there was a second peak observed at 360 °C. This peak was related to the decomposition of cellulose. At 500 °C, the residual samples were 10–30% of the weight. As shown in Figure 8, as milling time increased (reduced the particle size), the decomposition temperature increased significantly, especially from 0 to 10 h of milling time. The decomposition temperature of biocomposite increased significantly as the milling duration increased because the bond between the filler and adhesive increased with decreasing particle size (increasing milling duration).

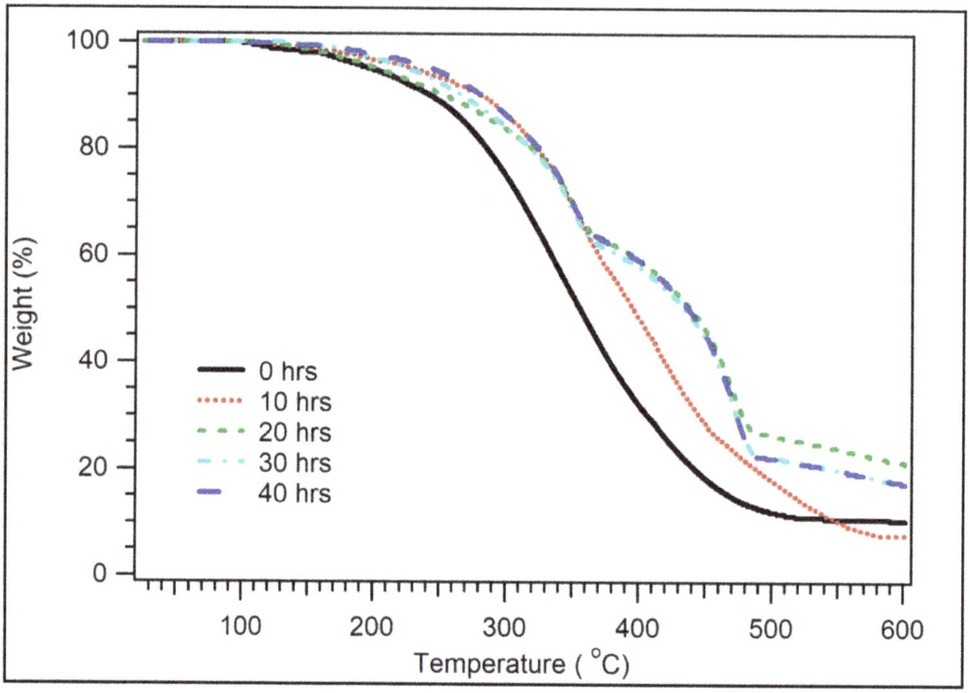

Figure 8. TGA of coconut shell nano-biocomposite for various milling times with 30 wt.% of PP composition.

Table 6. Decomposition temperature of coconut shell biocomposite for various milling times.

Sample Weight (%)	Decomposition Temperature (°C)				
	0 h	**10 h**	**20 h**	**30 h**	**40 h**
95	199	225	205	226	243
90	242	275	253	267	281
80	286	323	321	318	325
50	356	396	439	435	438
20	443	489	600	544	547

Figure 9 displays the DSC curves for various milling times with 30 wt.% of PP. The glass transition temperature (Tg) of biocomposite was observed at 60 °C, which increased as the milling times were increased. The melting temperature (Tm) of biocomposite also increased with the milling duration from 0 to 40 h. The depolymerisation temperature (Td) was observed at 350 °C, which increased with the milling times. The summary of the DSC data of coconut shell biocomposite for various milling duration is listed in Table 7. The Tg of biocomposite increased from 60 °C to 75 °C as the milling duration increased from 0 to 40 h. The Tm also increased from 160 °C to 172 °C. The increasing value of glass transition, melting, and depolymerisation temperatures were due to improved bond strength between filler and matrix for nanocomposites. Consequently, the thermal stability of biocomposite improved.

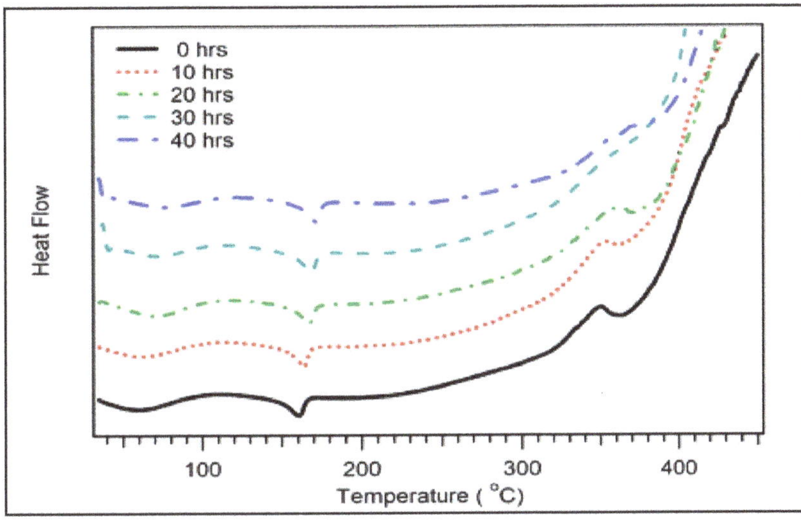

Figure 9. DSC of coconut shell nano-biocomposite for various milling times with 30 wt.% of PP composition.

Table 7. DSC data of coconut shell biocomposite for various milling times.

Milling Time	Tg (°C)	Tm (°C)	Td (°C)
0 h	60	160	350
10 h	65	162	352
20 h	70	165	359
30 h	70	167	-
40 h	75	172	-

3.2.4. Morphological Properties

The SEM and AFM images of biocomposite for various milling times with 30 wt.% of PP composition are displayed in Table 8. For 0 h duration of milling, some agglomerations and micro-voids were observed. Some coconut shell particles were detached from the matrix (PP) because of weak bonding between the coconut shell and PP. As the milling duration was increased (10–40 h), the number of agglomerations and micro-voids decreased. The surface became smooth and denser, indicating good miscibility between the coconut shell particles with PP. Consequently, the density of biocomposite increased; the thickness swelling and porosity decreased.

Table 8. SEM and AFM images for various milling times, 30 wt.% of PP.

Duration of Milling	SEM	AFM
0 h	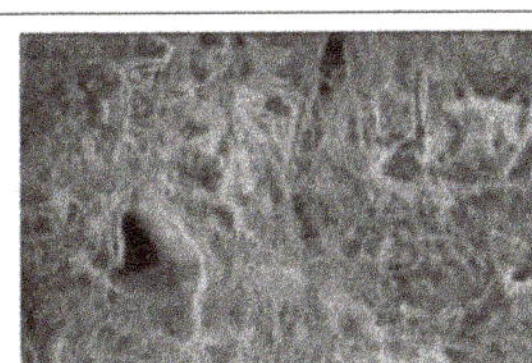	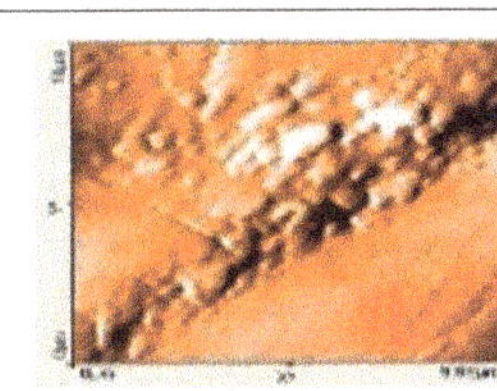

Table 8. *Cont.*

Duration of Milling	SEM	AFM
10 h		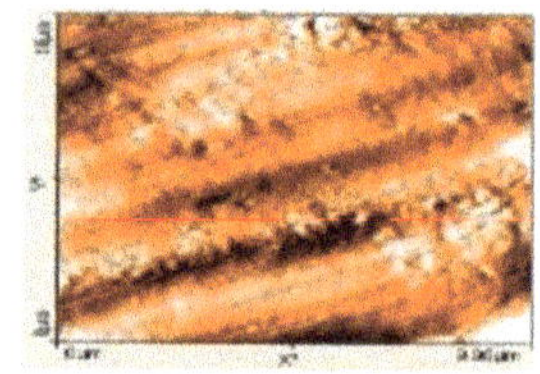
20 h	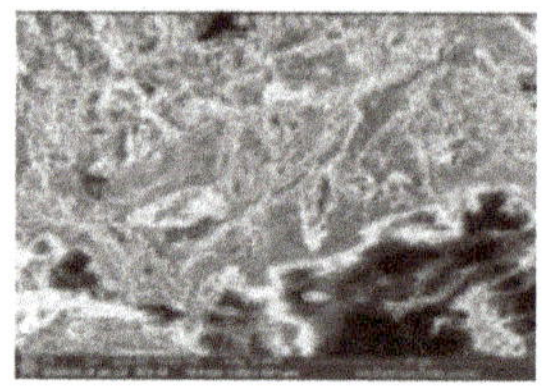	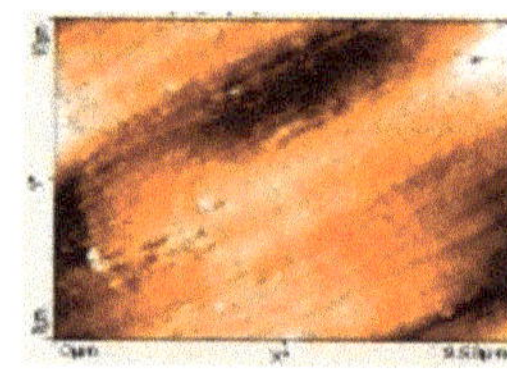
30 h	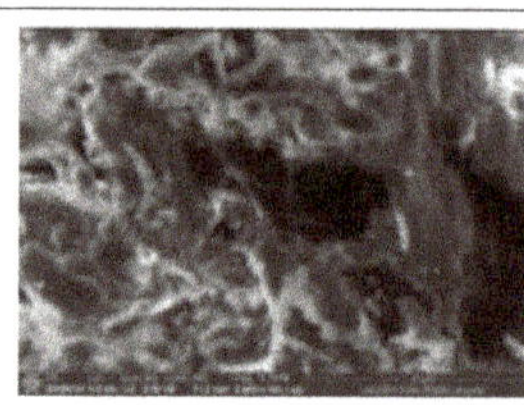	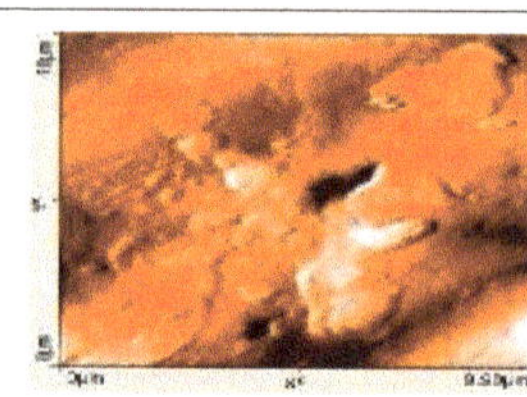
40 h	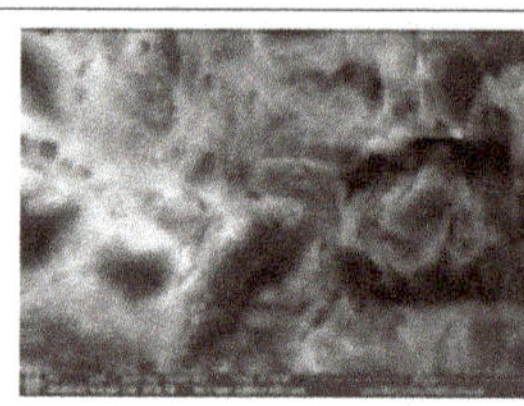	

Reducing the size of the filler will increase the surface area interaction between the filler and matrix. Assume the diameter of filler (particle) is 1 mm. Then, the volume and surface area of this particle is $\pi/6$ mm^3 and π mm^2, respectively. Suppose the particle is ground into nineother particles, as shown in Figure 10. Assuming that the total volume of the milled particles is the same as the previous volume, the diameter of the particles after grinding is obtained at 0.48 mm. Then, the total surface area of nineparticles is found to be 2.1π mm^2. Thus, the surface area interaction between fillers (particles) after grinding becomes twice as before. As a result, the bonding between fillers (particles) and the matrix increases. This simulation can explain why the mechanical and thermal properties of the biocomposite from this study increase with milling time. The bonding between coconut shell particles and PP increased as the milling duration increased. The flexural strength, flexural modulus, and compressive strength of the biocomposite increased. The thermal properties of biocomposite also improved.

This study measured the physical properties, mechanical properties, thermal properties, and morphological properties of coconut shell biocomposite. The results show a close relationship between physical properties and other properties. If the physical properties increase, the mechanical and thermal properties also increase. The morphology of the biocomposite is getting better. All of this has a lot to do with particle size. In other words, particle size becomes important in producing an excellent composite in the future.

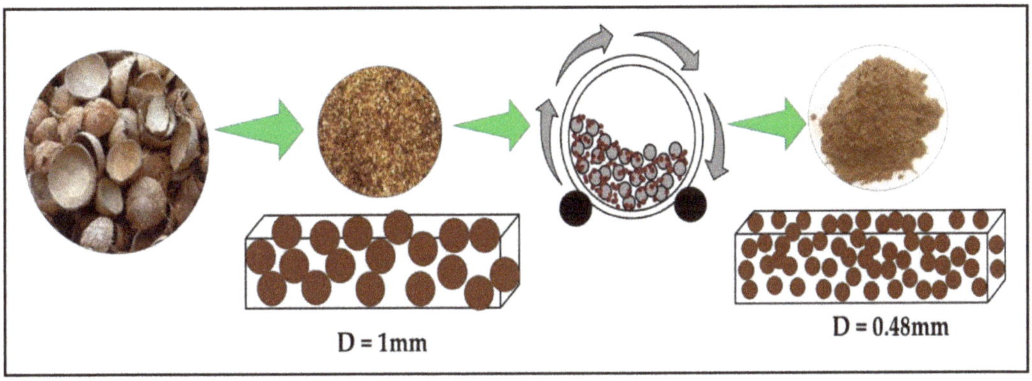

Figure 10. Schematic of reducing the size of filler (particle) from one to nine particles.

4. Conclusions

The coconut shell nano-biocomposites using waste polypropylene plastic packaging as a matrix have been successfully prepared and characterised. The physical, mechanical, and thermal properties of biocomposite were dependent on the composition of polypropylene. Instead of increasing the PP composition, the properties of the biocomposite can be improved by reducing the particle size of the coconut shell (increasing the duration of milling). The physical and mechanical properties improved from 0 to 40 h of milling times (density increased from 0.9 to 1.02 g/cm^3; thickness swelling decreased from 3.4 to 1.8%; porosity decreased from 7.0 to 3.0%; flexural strength increased from 8.19 to 12.26 MPa; flexural modulus increased from 1.67 to 2.87 GPa, and compressive strength increased from 16.00 to 27.20 MPa). Similarly, the thermal properties of biocomposite also improved as the particle size reduced. The degradation temperature increased as the milling duration increased from 0 to 40 h. The glass transition temperature of biocomposite increased slightly. The melting temperature increased significantly from 160 to 170 °C as the milling duration increased from 0 to 40 h. The depolymerisation occurred at 350 °C, which also increased with milling duration. The improvement properties of the biocomposite were due to the increase of bond strength between filler and matrix. This finding indicates that nanoparticles play an important role in biocomposite properties. Two important things can be drawn from this study. First, waste plastic packaging can be utilised to fabricate a high-quality biocomposite. Second, the properties of the biocomposite can be improved by reducing the particle size of filler to nanometres without having to increase the adhesive composition.

Author Contributions: Conceptualisation, I.I. and A.K.H.P.S.; methodology, I.I.; investigation, I.I., Q.A., M.M. and Z.J.; formal analysis, I.I., C.K.A. and A.K.H.P.S.; writing—original draft preparation, I.I.; writing—review and editing, I.I., C.K.A., N.G.O. and A.K.H.P.S.; project administration, I.I. and Z.J.; resources, I.I. and A.K.H.P.S. All authors have read and agreed to the published version of the manuscript.

Funding: This study was supported by Research Grant (WCR scheme) from the Indonesian Government, Ristekdikti under contract number No.: 215/SP2H/LT/DPRM/2020 dated 20th of May 2020.

Institutional Review Board Statement: Not applicable.

Informed Consent Statement: Not applicable.

Data Availability Statement: Not applicable.

Acknowledgments: The authors thank the Indonesian government for providing a research grant and Universiti Sains Malaysia, Penang, Malaysia for the collaboration. The author would like to thank the RUI Grant No (1001/PTEKIND/8014119) that has made this work possible. This work is a research collaboration between the Universitas Syiah Kuala, Banda Aceh, Indonesia and Universiti Sains Malaysia, Penang, Malaysia.

Conflicts of Interest: The authors declare no conflict of interest.

Appendix A

Table A1. Analysis of variance for physical properties with various PP compositions.

Physical Properties			Sum of Squares	df	Mean Square	F	Sig.
Density	Levene Statistic	0.495					
	df1	4					
	df2	20					
	Sig.	0.740					
	Between Groups		0.044	4	0.011	21.800	0.000
	Within Groups		0.010	20	0.000		
	Total		0.054	24			
Thickness Swelling	Levene Statistic	0.993					
	df1	4					
	df2	20					
	Sig.	0.434					
	Between Groups		21.006	4	5.251	54.138	0.000
	Within Groups		1.940	20	0.097		
	Total		22.946	24			
Porosity	Levene Statistic	0.899					
	df1	4					
	df2	20					
	Sig.	0.483					
	Between Groups		117.322	4	29.331	155.353	0.000
	Within Groups		3.776	20	0.189		
	Total		121.098	24			

Table A2. Analysis of variance for mechanical properties with various PP compositions.

Mechanical Properties			Sum of Squares	df	Mean Square	F	Sig.
Flexural Strength	Levene Statistic	0.799					
	df1	4					
	df2	20					
	Sig.	0.540					
	Between Groups		59.230	4	14.808	27.407	0.000
	Within Groups		10.806	20	0.540		
	Total		70.036	24			
Flexural Modulus	Levene Statistic	1.418					
	df1	4					
	df2	20					
	Sig.	0.264					
	Between Groups		2.975	4	0.744	36.774	0.000
	Within Groups		0.405	20	0.020		
	Total		3.380	24			
Compressive Strength	Levene Statistic	0.233					
	df1	4					
	df2	20					
	Sig.	0.917					
	Between Groups		53.382	4	13.346	56.935	0.000
	Within Groups		4.688	20	0.234		
	Total		58.070	24			

Table A3. Analysis of variance for physical properties with various milling times.

	Physical Properties		Sum of Squares	df	Mean Square	F	Sig.
Density	Levene Statistic	1.793					
	df1	4					
	df2	20					
	Sig.	0.170					
	Between Groups		0.054	4	0.014	21.938	0.000
	Within Groups		0.012	20	0.001		
	Total		0.066	24			
Thickness Swelling	Levene Statistic	1.388					
	df1	4					
	df2	20					
	Sig.	0.274					
	Between Groups		8.046	4	2.011	28.571	0.000
	Within Groups		1.408	20	0.070		
	Total		9.454	24			
Porosity	Levene Statistic	1.152					
	df1	4					
	df2	20					
	Sig.	0.361					
	Between Groups		72.290	4	18.072	117.353	0.000
	Within Groups		3.080	20	0.154		
	Total		75.370	24			

Table A4. Analysis of variance for mechanical properties with various milling times.

	Mechanical Properties		Sum of Squares	df	Mean Square	F	Sig.
Flexural Strength	Levene Statistic	1.298					
	df1	4					
	df2	20					
	Sig.	0.305					
	Between Groups		48.019	4	12.005	39.028	0.000
	Within Groups		6.152	20	0.308		
	Total		54.171	24			
Flexural Modulus	Levene Statistic	0.621					
	df1	4					
	df2	20					
	Sig.	0.653					
	Between Groups		5.723	4	1.431	102.017	0.000
	Within Groups		0.280	20	0.014		
	Total		6.004	24			
Compressive Strength	Levene Statistic	0.530					
	df1	4					
	df2	20					
	Sig.	0.715					
	Between Groups		456.262	4	114.065	228.771	0.000
	Within Groups		9.972	20	0.499		
	Total		466.234	24			

References

1. Chen, Y.; Zheng, K.; Niu, L.; Zhang, Y.; Liu, Y.; Wang, C.; Chu, F. Highly mechanical properties nanocomposite hydrogels with biorenewable lignin nanoparticles. *Int. J. Biol. Macromol.* **2019**, *128*, 414–420. [CrossRef] [PubMed]
2. Zheng, D.; Zhang, Y.; Guo, Y.; Yue, J. Isolation and characterization of nanocellulose with a novel shape from walnut (*Juglans Regia* L.) shell agricultural waste. *Polymers* **2019**, *11*, 1130. [CrossRef] [PubMed]
3. Kai, D.; Chong, H.M.; Chow, L.P.; Jiang, L.; Lin, Q.; Zhang, K.; Zhang, H.; Zhang, Z.; Loh, X.J. Strong and biocompatible lignin/poly (3-hydroxybutyrate) composite nanofibers. *Compos. Sci. Technol.* **2018**, *158*, 26–33. [CrossRef]

4. Mohanty, A.K.; Vivekanandhan, S.; Pin, J.-M.; Misra, M. Composites from renewable and sustainable resources: Challenges and innovations. *Science* **2018**, *362*, 536–542. [CrossRef] [PubMed]
5. Mu, L.; Shi, Y.; Hua, J.; Zhuang, W.; Zhu, J. Engineering hydrogen bonding interaction and charge separation in bio-polymers for green lubrication. *J. Phys. Chem. B* **2017**, *121*, 5669–5678. [CrossRef] [PubMed]
6. Faruk, O.; Bledzki, A.K.; Fink, H.-P.; Sain, M. Biocomposites reinforced with natural fibers: 2000–2010. *Prog. Polym. Sci.* **2012**, *37*, 1552–1596. [CrossRef]
7. Nagalakshmaiah, M.; Afrin, S.; Malladi, R.P.; Elkoun, S.; Robert, M.; Ansari, M.A.; Svedberg, A.; Karim, Z. Biocomposites: Present trends and challenges for the future. In *Green Composites for Automotive Applications*; Elsevier: Amsterdam, The Netherlands, 2019; pp. 197–215.
8. Motaleb, K.; Ahad, A.; Laureckiene, G.; Milasius, R. Innovative Banana Fiber Nonwoven Reinforced Polymer Composites: Pre-and Post-Treatment Effects on Physical and Mechanical Properties. *Polymers* **2021**, *13*, 3744. [CrossRef]
9. Sherwani, S.; Zainudin, E.; Sapuan, S.; Leman, Z.; Abdan, K. Mechanical properties of sugar palm (Arenga pinnata Wurmb. Merr)/glass fiber-reinforced poly (lactic acid) hybrid composites for potential use in motorcycle components. *Polymers* **2021**, *13*, 3061. [CrossRef]
10. Dordevic, D.; Necasova, L.; Antonic, B.; Jancikova, S.; Tremlová, B. Plastic cutlery alternative: Case study with biodegradable spoons. *Foods* **2021**, *10*, 1612. [CrossRef]
11. Hasan, K.; Horváth, P.G.; Alpár, T. Potential natural fiber polymeric nanobiocomposites: A review. *Polymers* **2020**, *12*, 1072. [CrossRef]
12. Alsewailem, F.D.; Binkhder, Y.A. Effect of Coupling Agent on the Properties of Polymer/Date Pits Composites. *J. Compos.* **2014**, *2014*, 412432. [CrossRef]
13. The Science Agriculture. Available online: https://scienceagri.com/10-worlds-largest-coconut-producing-countries (accessed on 8 November 2021).
14. Bledzki, A.K.; Mamun, A.A.; Volk, J. Barley husk and coconut shell reinforced polypropylene composites: The effect of fibre physical, chemical and surface properties. *Compos. Sci. Technol.* **2010**, *70*, 840–846. [CrossRef]
15. Singh, A.; Singh, S.; Kumar, A. Study of mechanical properties and absorption behaviour of coconut shell powder-epoxy composites. *Int. J. Mater. Sci. Appl.* **2013**, *2*, 157–161. [CrossRef]
16. Bhaskar, J.; Singh, V. Physical and mechanical properties of coconut shell particle reinforced-epoxy composite. *J. Mater. Environ. Sci.* **2013**, *4*, 227–232.
17. Kumar, R.; Singh, T.; Singh, H. Solid waste-based hybrid natural fiber polymeric composites. *J. Reinf. Plast. Compos.* **2015**, *34*, 1979–1985. [CrossRef]
18. Vasu, A.; Reddy, C.; Danaboyina, S.; Manchala, G.; Chavali, M. The Improvement in mechanical properties of coconut shell powder as filter in HDPE composites. *J. Polym. Sci. Appl.* **2017**, *2*, 2.
19. Agunsoye, J.O.; Bello, S.A.; Azeez, S.O.; Yekinni, A.A.; Adeyemo, R.G. Recycled Polypropylene Reinforced Coconut Shell Composite: Surface Treatment Morphological, Mechanical and Thermal Studies. *Int. J. Compos. Mater.* **2014**, *4*, 168–178.
20. Bello, S.A.; Agunsoye, J.O.; Hassan, S.B. Synthesis of coconut shell nanoparticles via a top down approach: Assessment of milling duration on the particle sizes and morphologies of coconut shell nanoparticles. *Mater. Lett.* **2015**, *159*, 514–519. [CrossRef]
21. Obasi, H.C.; Mark, U.C.; Mark, U. Improving the mechanical properties of polypropylene composites with coconut shell particles. *Compos. Adv. Mater.* **2021**, *30*, 26349833211007497. [CrossRef]
22. Kirby, M.; Lewis, B.; Peterson, B.; Anggono, J.; Bradley, W. The Effect of Coconut Shell Powder as Functional Filler in Polypropylene during Compounding and Subsequent Molding. *E3S Web. Conf.* **2019**, *130*, 01021. [CrossRef]
23. Borrelle, S.B.; Ringma, J.; Law, K.L.; Monnahan, C.C.; Lebreton, L.; McGivern, A.; Murphy, E.; Jambeck, J.; Leonard, G.H.; Hilleary, M.A. Predicted growth in plastic waste exceeds efforts to mitigate plastic pollution. *Science* **2020**, *369*, 1515–1518. [CrossRef] [PubMed]
24. MacArthur, E. Beyond plastic waste. *Science* **2017**, *358*, 843. [CrossRef] [PubMed]
25. Alsabri, A.; Tahir, F.; Al-Ghamdi, S.G. Environmental impacts of polypropylene (PP) production and prospects of its recycling in the GCC region. *Mater. Today Proc.* 2021; *in press*. [CrossRef]
26. Chun, K.S.; Husseinsyah, S.; Azizi, F.N. Characterization and properties of recycled polypropylene/coconut shell powder composites: Effect of sodium dodecyl sulfate modification. *Polym. Plast. Technol. Eng.* **2013**, *52*, 287–294. [CrossRef]
27. Muller, J.; González-Martínez, C.; Chiralt, A. Combination of poly (lactic) acid and starch for biodegradable food packaging. *Materials* **2017**, *10*, 952. [CrossRef]
28. Alwani, M.S.; Abdul Khalil, H.P.S.; Asniza, M.; Suhaily, S.; Amiranajwa, A.N.; Jawaid, M. Agricultural biomass raw materials: The current state and future potentialities. In *Biomass Bioenergy*; Springer: Cham, Switzerland, 2014; pp. 77–100.
29. Atiqah, M.; Gopakumar, D.A.; Owolabi, F.A.T.; Pottathara, Y.B.; Rizal, S.; Aprilia, N.; Hermawan, D.; Paridah, M.; Thomas, S.; Abdul Khalil, H.P.S. Extraction of Cellulose Nanofibers via Eco-friendly Supercritical Carbon Dioxide Treatment Followed by Mild Acid Hydrolysis and the Fabrication of Cellulose Nanopapers. *Polymers* **2019**, *11*, 1813. [CrossRef]
30. Abdul Khalil, H.P.S.; Adnan, A.; Yahya, E.B.; Olaiya, N.; Safrida, S.; Hossain, M.; Balakrishnan, V.; Gopakumar, D.A.; Abdullah, C.; Oyekanmi, A. A Review on plant cellulose nanofibre-based aerogels for biomedical applications. *Polymers* **2020**, *12*, 1759. [CrossRef]

31. Abdullah, C.; Ismail, I.; Nurul Fazita, M.; Olaiya, N.; Nasution, H.; Oyekanmi, A.; Nuryawan, A.; Abdul Khalil, H.P.S. Properties and Interfacial Bonding Enhancement of Oil Palm Bio-Ash Nanoparticles Biocomposites. *Polymers* **2021**, *13*, 1615. [CrossRef]
32. Ismail, I.; Arliyani; Jalil, Z.; Olaiya, N.; Abdullah, C.; Fazita, M.; Abdul Khalil, H.P.S. Properties and Characterization of New Approach Organic Nanoparticle-Based Biocomposite Board. *Polymers* **2020**, *12*, 2236. [CrossRef]
33. Hewitt, S.A.; Kibble, K.A. Effects of ball milling time on the synthesis and consolidation of nanostructured W.C.–Co composites. *Int. J. Refract. Met. Hard Mater.* **2009**, *27*, 937–948. [CrossRef]
34. Singh, P.; Abhash, A.; Yadav, B.; Shafeeq, M.; Singh, I.; Mondal, D. Effect of milling time on powder characteristics and mechanical performance of Ti4wt% Al alloy. *Powder Technol.* **2019**, *342*, 275–287. [CrossRef]
35. Sitorus, R.; Wirjosentono, B.; Tamrin; Siregar, A.H.; Nasution, D.A. Characteristics of maleic anhydride-modified polystyrene and natural rubber blends containing "Talang" bamboo powder as sound damping material. *Proc. AIP Conf. Proc.* **2020**, *2267*, 020050.
36. Bonilla-Cruz, J.; Hernández-Mireles, B.; Mendoza-Carrizales, R.; Ramírez-Leal, L.A.; Torres-Lubián, R.; RamosdeValle, L.F.; Paul, D.R.; Saldívar-Guerra, E. Chemical modification of butyl rubber with maleic anhydride via nitroxide chemistry and its application in polymer blends. *Polymers* **2017**, *9*, 63. [CrossRef] [PubMed]
37. Lubis, M.R.; Maimun, T.; Kardi, J.; Masra, R.B. Characterizing particle board made of oil palm empty fruit bunch using central composite design. *Makara J. Sci.* **2018**, *22*, 3. [CrossRef]
38. Tai, Y.; Qian, J.; Zhang, Y.; Huang, J. Study of surface modification of nano-SiO_2 with macromolecular coupling agent (LMPB-g-MAH). *Biochem. Eng. J.* **2008**, *141*, 354–361. [CrossRef]
39. González-López, M.; Robledo-Ortíz, J.; Manríquez-González, R.; Silva-Guzmán, J.; Pérez-Fonseca, A. Polylactic acid functionalization with maleic anhydride and its use as coupling agent in natural fiber biocomposites: A review. *Compos. Interfaces* **2018**, *25*, 515–538. [CrossRef]
40. Severini, F.; Pegoraro, M.; Yuan, L.; Ricca, G.; Fanti, N. Free radical grafting of maleic anhydride in vapour phase on polypropylene film. *Polymer* **1999**, *40*, 7059–7064. [CrossRef]
41. Obiukwu, O.; Uchechukwu, M.; Nwaogwugwu, M. Study on the properties of coconut shell powder reinforced high-density polyethylene composite. *Futo J. Ser* **2016**, *2*, 43–55.
42. Kada, D.; Migneault, S.; Tabak, G.; Koubaa, A. Physical and mechanical properties of polypropylene-wood-carbon fiber hybrid composites. *BioResources* **2016**, *11*, 1393–1406. [CrossRef]
43. Shayuti, M.S.M.; Abdullah, M.Z.; Yusoff, P. Compressive properties and morphology of polypropylene/polycarbonate blends. In Proceedings of the International Conference on Environment and Industrial Innovation (ICEII 2011), Kuala Lumpur, Malaysia, 4–5 June 2011.
44. Ismail, I.; Fitri, N.; Zulfalina; Fadzullah, S.H.S.M. Evaluation possibilities to utilize rice straw and plastic waste for particleboard. *IOP Conf. Series J. Phys.* **2018**, *1120*, 012015. [CrossRef]
45. Liyanage, C.D.; Pieris, M. A physico-chemical analysis of coconut shell powder. *Procedia Chem.* **2015**, *16*, 222–228. [CrossRef]

polymers

MDPI

Article

Selected Properties of Plywood Bonded with Low-Density Polyethylene Film from Different Wood Species

Pavlo Bekhta [1,*], Orest Chernetskyi [2], Iryna Kusniak [1], Nataliya Bekhta [1] and Olesya Bryn [1]

[1] Department of Wood-Based Composites, Cellulose, and Paper, Ukrainian National Forestry University, 79057 Lviv, Ukraine; kusnyak@nltu.edu.ua (I.K.); n_bekhta@nltu.edu.ua (N.B.); bryn_o@nltu.edu.ua (O.B.)

[2] Shpon Shepetivka LLC, 30400 Shepetivka, Ukraine; Lokiorest@gmail.com

* Correspondence: bekhta@nltu.edu.ua

Abstract: In this work, the effects of wood species and thickness of low-density polyethylene (LDPE) film on the properties of environmentally-friendly plywood were studied. Rotary-cut veneers from four wood species (beech, birch, hornbeam and poplar) and LDPE film of four thicknesses (50, 80, 100 and 150 μm) as an adhesive were used for making plywood samples. The findings of this study demonstrated that plywood samples using all the investigated wood species bonded with LDPE film showed satisfactory physical–mechanical properties. Poplar veneer provided the lowest values for bending strength, modulus of elasticity and thickness swelling of all the plywood samples, but the bonding strength was at the same level as birch and hornbeam veneer. Beech plywood samples had the best mechanical properties. An increase in LDPE film thickness improved the physical–mechanical properties of plastic-bonded plywood.

Keywords: plastic film-bonded plywood; polyethylene film; physical–mechanical properties; bonding strength; wood species

Citation: Bekhta, P.; Chernetskyi, O.; Kusniak, I.; Bekhta, N.; Bryn, O. Selected Properties of Plywood Bonded with Low-Density Polyethylene Film from Different Wood Species. *Polymers* **2022**, *14*, 51. https://doi.org/10.3390/polym14010051

Academic Editor: Hiroshi Yoshihara

Received: 22 November 2021
Accepted: 22 December 2021
Published: 23 December 2021

1. Introduction

Wood-based materials are popular materials that are still used extensively as an alternative to solid wood across many industries. Among wood-based materials, plywood is widely used in diverse fields, including housing construction, vehicles, laser engraving and interior decoration, furniture production and many others [1,2]. According to the latest data released by FAO [3], during the ten year period from 2009 to 2018, plywood global production and consumption volumes increased by 101% and 104%, respectively, which amounted to record values in 2018 of over 163 and 161 million m^3, respectively (Figure 1). However, over the past two years, due to global problems related to the coronavirus, there has been a stagnation in plywood production and consumption. In 2019, global plywood production decreased by 0.19% in comparison with 2018 [3].

Plywood products are manufactured using different formaldehyde-based adhesives [4]. Despite the fact that these adhesives have good adhesive properties [5], the formaldehyde is emitted in the production and use of plywood. Formaldehyde is a human carcinogen [6] and poses a major risk to human health. The impact of the living environment on human health has become increasingly important, and has resulted in a change in attitudes towards issues related to the environment and human health. Therefore, the development of plywood with alternative formaldehyde-free adhesives has attracted increasing attention from both the industrial community and academia in recent years. Demand is growing for plywood that uses new formaldehyde-free adhesive compositions for wood bonding [7]. However, the use of some newly developed adhesives at an industrial scale is limited due to the high cost of modification and poor water-resistance [7].

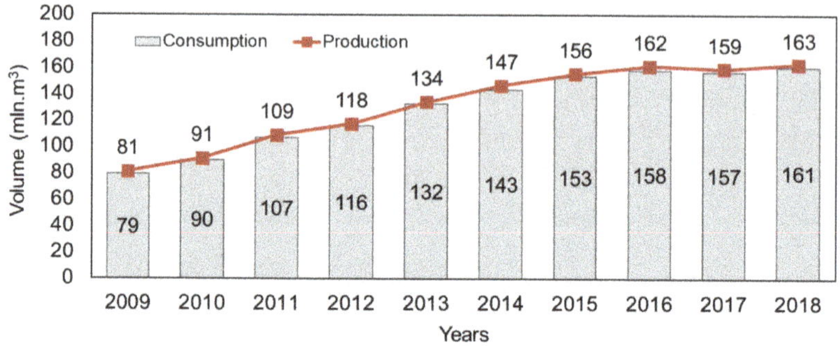

Figure 1. World production and consumption of plywood according to FAO statistics.

Therefore, the application of thermoplastics and their copolymers for bonding plywood is promising [8], and formaldehyde-free plywood has been successfully produced using thermoplastic polymers as adhesives [8–28]. Different types of thermoplastic polymers have been used for wood veneer bonding: low-density polyethylene (LDPE) [9–11], high-density polyethylene (HDPE) [12–21], PS [22,23], PP [8,24,25], PVC [26], poly-b-hydroxybutyrate [27], co-polyamide and co-polyester [11]. Thermoplastic polymers are used in different forms for the bonding of wood veneer [8,11–19,28]. The application of thermoplastic polymers in the form of a film for bonding plywood is the most effective method and greatly simplifies the technology. Some of the advantages and disadvantages of using thermoplastic film for bonding plywood were described in our previous work [11,21]. The great advantage of these polymers is that the amount of formaldehyde emission from the plastic-bonded plywood is almost zero [28]. In addition, the use of plastic film in the production of plywood enables the production of larger-sized products at lower pressure [11,21] compared to the manufacture of conventional UF and PF-bonded plywood or wood–polymer composites (WPC) [29].

Plywood can be produced from different wood species, and the species determines the physical–mechanical properties of the plywood [11,30]. Most of the above-mentioned studies used poplar [13–15,17,18,22,25,28], eucalyptus [19,20,26,27], or birch [8,11] for bonding with thermoplastic polymers. There are also several works on the use of red Seraya [24], spruce [10,11], alder [21], beech [11], Masson pine [12] and Amescla wood [16]. There is limited data in the literature on the use of beech and hornbeam wood veneers. In Ukraine, birch is the most popular wood for plywood production. However, due to the limited stocks of birch raw materials, plywood producers are being forced to find other wood species to replace birch. Beech, hornbeam and poplar wood species are already being considered as alternative raw materials for plywood production.

Therefore, in this study we investigated the possibility of making plywood using beech, hornbeam and poplar, as well as birch for comparison, bonded with thermoplastic film of different thicknesses. This will expand our knowledge of the raw materials that can be used as a base for the manufacture of plastic plywood, as well as expand the applications of such plywood.

2. Materials and Methods

2.1. Materials

Rotary-cut veneers of four hardwood species of poplar (*Populus alba* L.), birch (*Betula verrucosa* Ehrh.), beech (*Fagus sylvatica* L.), and hornbeam (*Carpinus betulus* L.) with a thickness of 0.75 mm, 1.55 mm, 0.45 mm and 1.50 mm, respectively, were used in the experiments. The moisture content of the veneer sheets was about 6 ± 2%. Birch is the wood species most frequently used for the production of peeled veneer and plywood in Ukraine. Poplar, beech and hornbeam were selected as alternative wood species, all of which have significant reserves in Ukraine. Veneer sheets were obtained from a local producer of peeled veneer and plywood panels. Each veneer sheet was visually checked, and sheets without any visible defects were selected. In addition, the veneer was manufactured from one batch of raw materials. This was done to minimize the effects of wood structure and origin on the findings of the experiment.

Low-density polyethylene (LDPE) film (LLC "Planet Plastic", Irpin, Ukraine) with the same dimensions as the veneers and thicknesses of 50 μm, 80 μm, 100 μm and 150 μm, density of 0.92 g/cm^3 and a melting point of 105–110 °C was used for the bonding of the wood veneers.

2.2. Manufacturing of Plywood Samples

Three-layer plywood samples measuring 300 mm × 300 mm were prepared. LDPE film, which mainly works as the adhesive connecting between two adjacent veneers, was used to manufacture the plywood samples. One sheet of film was incorporated between the veneer sheets. The prepared veneer assemblies (Figure 2) were subjected to hot pressing in the lab press at a pressure of 1.4 MPa and temperature of 160 °C for 4.5 min (Table 1). The mass per unit area of LDPE film at thicknesses of 50 μm, 80 μm, 100 μm and 150 μm was approximately 46.0, 73.6, 92.0 and 138 g/m^2, respectively.

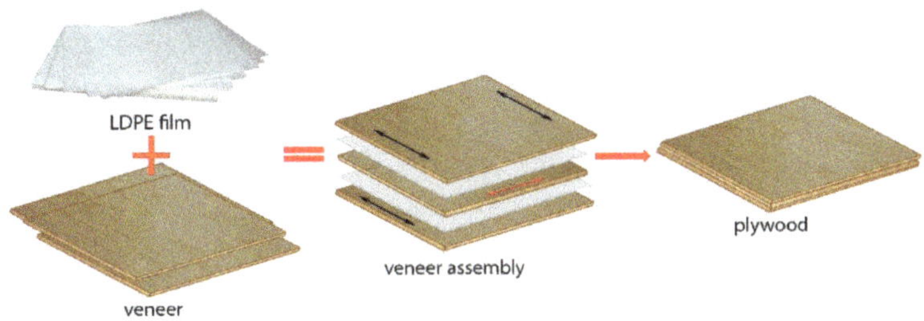

Figure 2. Schematic of plywood samples' production.

Table 1. Manufacturing conditions of plywood samples.

Adhesive	Wood Species	Thickness of Film (μm)	Pressing Temperature (°C)	Pressing Pressure (MPa)	Pressing Time (min)
LDPE film	beech, birch, hornbeam, poplar	50, 80, 100, 150	160	1.4	4.5

After the plywood samples were removed from the hot press, they were subjected to the cold pressing stage at room temperature. This was performed to release internal stresses and reduce the warping of samples. Three plywood samples were prepared at each condition.

2.3. Testing the Plywood Panels

After bonding, the plywood panels were air conditioned at 20 ± 2 °C and 65 ± 5% (RH). The physical properties (density, water absorption (WA) and thickness swelling (TS)) and mechanical properties (bending strength (MOR), modulus of elasticity in bending (MOE) and shear strength) of the LDPE film-bonded plywood samples were determined according to the standards [31–35]. The shear strength was measured after pre-treatment for bonding class 1—dry conditions—plywood test pieces were immersed in water at 20 ± 3 °C for 24 h [33,34]. To determine the WA and TS, the samples were immersed in distilled water for 2 and 24 h. For each variant, at least ten samples were used for the shear strength test and six samples were used to determine MOR, MOE, WA and TS.

Moreover, the measurement of the core temperature inside the veneer package under given wood species and thickness of LDPE film was undertaken (Figure 3) according to the procedure described in our previous work [36].

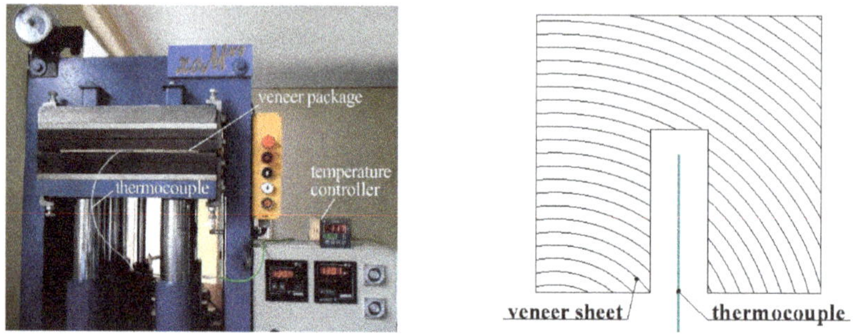

(a) (b)

Figure 3. Measuring the core temperature during pressing of plywood sample (**a**) and positioning the thermocouple in the middle sheet of the plywood (**b**).

2.4. Statistical Analysis

To evaluate the effect of the LDPE film thickness and wood species on the properties of plywood samples, ANOVA analysis was performed using SPSS software program version 22 (IBM Corp., Armonk, NY, USA). The effects were not considered to be statistically significant when the p-value was higher than 0.05 at the 95% confidence level. Duncan's multiple range test was used to determine the significant differences between and among the groups.

3. Results

3.1. Core Layer Temperature Analysis

The melting point of the investigated film was 105–110 °C. However, the plywood samples were pressed at a higher temperature of 160 °C to ensure good fluidity of the molten LDPE film and its good penetration into the cavities of the wood [11,21]. To determine the process for heating the veneer package at a given temperature, as well as the melting and spreading of the LDPE film during pressing, the core layer temperature inside the package was measured for each film thickness and wood species.

The core temperature distribution inside the veneer package during pressing of the plywood panels with different film thicknesses and various wood species is shown in Figure 4. Among the investigated wood species, the samples with of poplar and beech veneers warmed up the fastest (in approximately 17 s) to the melting temperature of LDPE film (110 °C), while the samples with birch and hornbeam veneer warmed up slowly (in approximately 30 s). This can be mainly explained by the influence of the veneer thickness. The beech and poplar veneer were the thinnest, and therefore the samples using these veneers heat up faster. Analysis (ANOVA) also showed that in addition to the thickness of the veneer, the wood species affects the temperature distribution inside the sample, especially at the stage of heating to the melting point of the polymer. When the temperature inside the veneer package reaches about 150 °C (approximately after 100 s of heating), the heating curves coincide for different wood species. The effect of the thickness of the thermoplastic film on the temperature distribution inside the veneer package was insignificant ($p > 0.05$) due to its slight thickness (Figure 4a). The main factors that determine the heating rate of the veneer package with thermoplastic film are the thickness of the veneer and the wood species at the same pressing conditions.

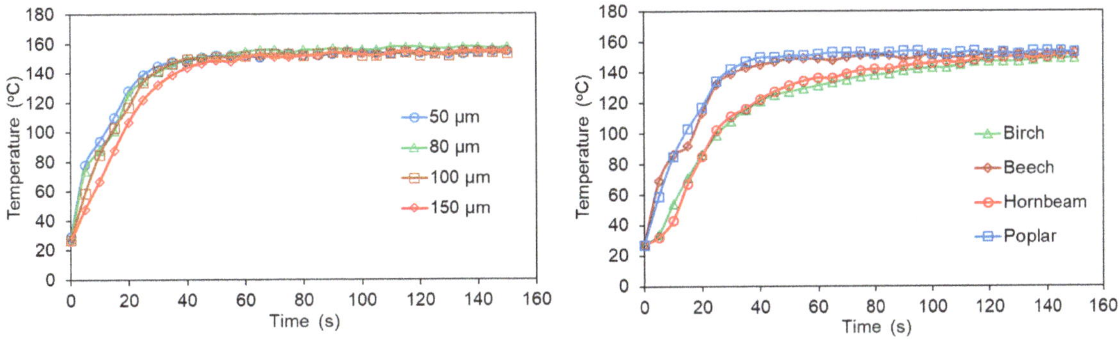

(**a**) (**b**)

Figure 4. Core temperature curves of plywood samples made with (**a**) poplar veneers and different thicknesses of LDPE film, and (**b**) veneers of different wood species and LDPE film of 100 μm thickness.

To improve the bonding properties of plywood, the plastic must penetrate the wood before it is cured [17]. From Figure 4a,b, it can be seen clearly that a temperature of 160 °C and 4.5 min of pressing are sufficient to heat the veneer package, melt the plastic film, allow it to spread on the surface of the veneer and penetrate into the wood before it is cured. Better penetration creates stronger bonds in the sample. A study by Goto et al. [24] also showed that good melting and the penetration of polypropylene into various wooden elements and veneer spaces, generally promoted bonding. Therefore, since wood is a porous material, the most probable gluing mechanism is mechanical locking [8,13,18,19,24,37].

3.2. Density of Plywood Samples

The ANOVA analysis showed the significance of the effect of the wood species and LDPE film thickness on the density of plastic-bonded plywood samples. The greatest influence on the density of plywood samples was the wood species ($F = 1610.956$), while the thickness of the plastic film affects the density to a lesser extent ($F = 23.939$). Based on Duncan's test, it was found that the values for the density of the samples made with beech, birch, hornbeam or poplar veneers differed significantly ($p \leq 0.05$). Duncan's test also confirmed that there was a statistically significant difference in the density values of plywood samples bonded by film with a thickness of 50, 80, 100 or 150 μm ($p \leq 0.05$). Therefore, the density of the LDPE film should be taken into account in the manufacture of plywood panels. However, the differences in the density made with LDPE film of 50, 80 and 100 μm were insignificant ($p > 0.05$). Similar results were obtained by other authors [30,38] who found that thermosetting adhesives had a significant effect on the density of laminated materials.

It was found that the density of plywood samples was significantly affected by the density of the veneers. Naturally, the smallest and highest density values correspond to the samples with poplar (481.8 kg/m^3) and hornbeam (790.8 kg/m^3) veneers, respectively (Figure 5). This is due to the higher density of hornbeam (730 kg/m^3) in comparison with poplar (390 kg/m^3) veneers.

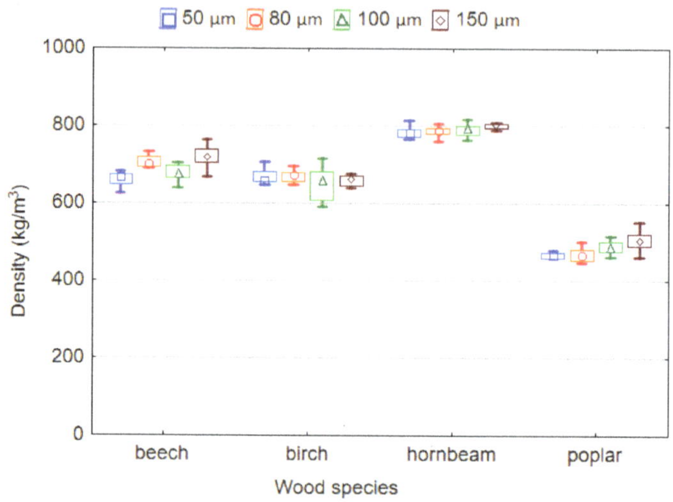

Figure 5. Density of plywood bonded by LDPE film.

3.3. Bending Strength and Modulus of Elasticity of Plywood Samples

The effect of the thickness of LDPE film and wood species on the MOR and MOE of plywood samples is shown in Figures 6 and 7. It was found that both variables significantly affect the MOR and MOE ($p < 0.05$). The ANOVA analysis showed that according to the F-values, the wood species has the greatest effect on the MOR ($F = 122.307$) and MOE ($F = 108.160$) of the plywood samples, while the thickness of the plastic film affected the MOR ($F = 7.488$) and MOE ($F = 5.731$) to a lesser extent. The values of MOR for samples made using film with thicknesses 50, 80 and 100 μm differed insignificantly ($p > 0.05$). Similarly, as for MOR, the difference in the values of MOE for the thicknesses of 80 and 100 μm as well as 100 and 150 μm was also insignificant ($p > 0.05$).

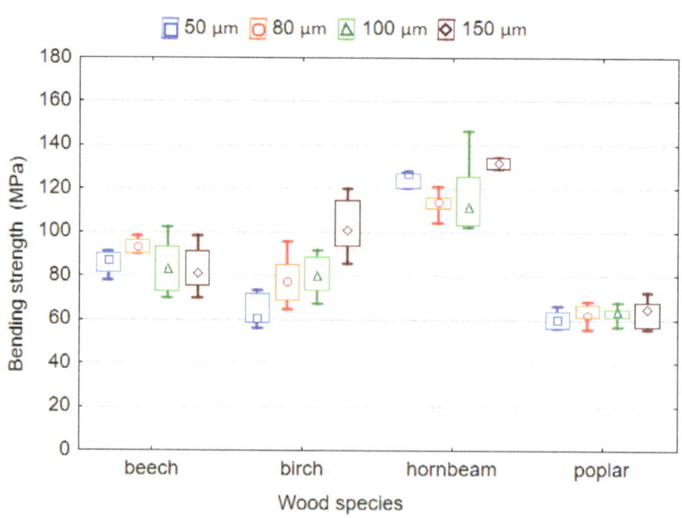

Figure 6. Bending strength of plywood bonded by LDPE film.

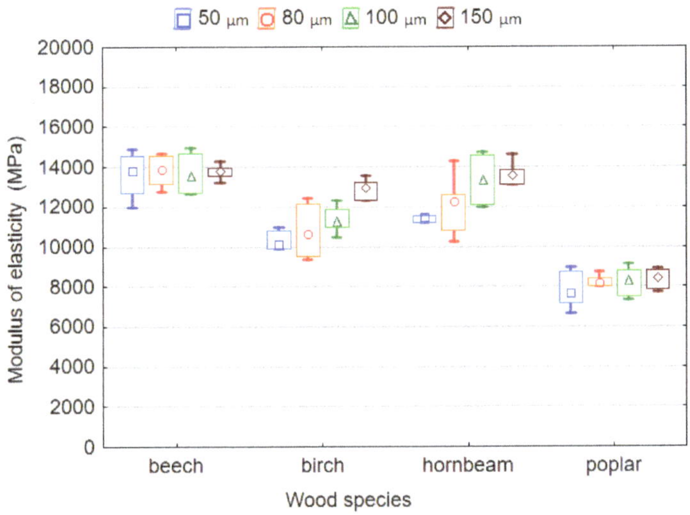

Figure 7. Modulus of elasticity of plywood bonded by LDPE film.

The highest MOR and MOE values of 94.8 MPa and 11,750.0 MPa, respectively, corresponded to samples bonded by LDPE film with a thickness of 150 μm. The smallest MOR and MOE values of 82.6 MPa and 10,490.0 MPa, respectively, corresponded to the samples bonded by LDPE film with a thickness of 50 μm. This was predictable given that previous studies have shown that the properties of plywood bonded by plastic may decrease with low amounts of adhesive [27]. Moreover, it should be noted that both the MOR and MOE of plywood samples increased with an increase in the thickness of the film from 50 to 150 μm. This can be explained by the fact that the consumption of polymer increases with increases in the film thickness, which leads to better and more complete filling of the wood cavities. In turn, this leads to the formation of more adhesive locks. Goto et al. [24] showed that penetration of thermoplastic polymer into various wooden elements and veneer spaces, mainly promoted bonding. Many authors point out that since wood is a porous material, the most probable mechanism for bonding of thermoplastic polymers to wood is mechanical locking [8,13,18,19,24,37].

It is to be expected that the wood species affects the MOR and MOE of plywood samples differently. The highest MOR and MOE values of 120.4 MPa and 13,746.1 MPa, respectively, correspond to the samples made using hornbeam and beech, respectively. The lowest MOR and MOE values of 62.5 MPa and 8156.2 MPa, respectively, correspond to the samples made using poplar veneer. Thus, because hornbeam and beech wood have a higher density and better strength properties than poplar wood, they provide higher MOR and MOE values. The mean values for the MOR of LDPE-bonded plywood samples were: 82.96–93.57 MPa for beech veneer; 63.70–102.65 MPa for birch veneer; 113.55–129.45 MPa for hornbeam veneer; and 60.47–63.80 MPa for poplar veneer. The bending strength of beech, birch, hornbeam and poplar solid woods are 104, 110, 128 and 68 MPa, respectively [39]. Therefore, it was found that the plywood samples had MOR values similar to those of the solid wood. In another study [40], the statistically significant effects of veneer wood species on some properties of LVL were also noted.

It has been well established [39] that there is a linear relationship between MOR and density, that is, the MOR increases with increasing density. However, based on the results presented in Figure 6, we can conclude that the relationship between MOR and the density of plywood samples is more complex. That is, the MOR of the sample depends not only on its density, but also on the thickness of the veneer used, the amount of applied polymer, the thickness of the plywood samples, etc.

This explains why the samples of birch and poplar plywood with a minimum LDPE film thickness of 50 μm had almost the same bending strength. In this case, not only the densities of the veneer and plywood, but also the thicknesses of the veneer and plywood influenced the bending strength. The thickness of the film at 50 μm can be neglected. These parameters were smaller for poplar plywood compared to birch plywood. The same applies to beech and birch plywood. Beech veneer had a lower density (605 kg/m^3) and less thickness compared to birch veneer (655 kg/m^3). Beech plywood, which had a lower density and thickness, showed greater bending strength than birch plywood, which had a higher density and thickness.

This is confirmed by previous studies [41–43] that showed that plywood samples made using thin veneer had a higher bending strength compared to samples made using thick veneer. The decrease in strength may be due to the fact that the thicker the veneer, the deeper and the longer the lathe checks will be [1,43]. In addition, there is a linear relationship between MOR and the thickness of plywood samples [41]. MOR decreases with an increase in the thickness of the samples [41,44].

3.4. Shear Strength of Plywood Samples

Shear strength is one of the most important mechanical properties of plywood. Figure 8 shows the effects of the wood species and thickness of LDPE film on the shear strength of plywood samples. The obtained mean values of shear strength ranged from 1.54 to 1.77 MPa for beech veneer, from 1.13 to 1.36 MPa for birch veneer, from 0.94 to 1.31 MPa for hornbeam veneer and from 1.03 to 1.44 MPa for poplar veneer. All shear strength values met the requirements of EN 314-2 standards [34] for Class 1 (dry conditions) plywood. However, hornbeam samples bonded with 80 μm thickness film failed (0.94 MPa). Among the studied wood species, the highest shear strength mean values correspond to the plywood samples made with birch and beech veneers, that is, 1.22 and 1.64 MPa, respectively. The shear strength mean values of the samples from hornbeam and poplar veneer were the smallest at 1.09 and 1.19 MPa, respectively. The shear strength values for the samples made with birch and poplar veneers differ insignificantly based on Duncan's test ($p > 0.05$).

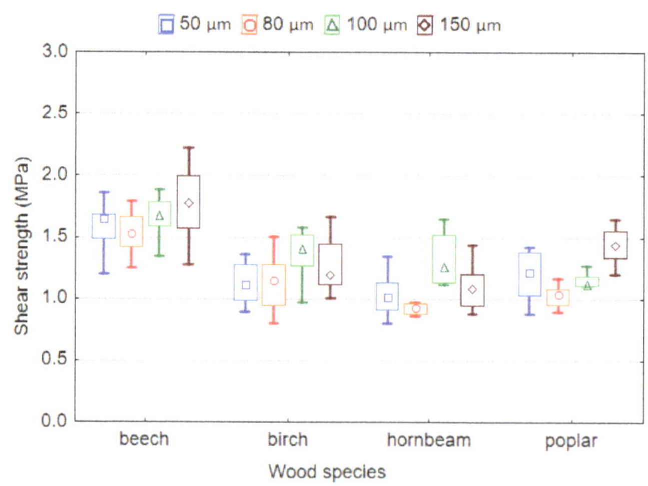

Figure 8. Shear strength of plywood bonded by LDPE film.

The results obtained in this study contradict the results of other researchers who have indicated that wood species with high density have higher shear strength values [30,45]. It is known that the formation of bonding strength largely depends on the amount of plastic film, its penetration of the wood cavities, and the formation of mechanical locks [46]. Apparently, the similar values of shear strength in the plywood samples made from poplar, birch and hornbeam veneers can be explained by the different thickness and porosity of the wood veneer used. Given the same film thickness, the amount of polymer that will penetrate per unit volume of poplar veneer (veneer density 390 kg/m^3 and thickness 0.75 mm) will be much greater than the amount of polymer that will penetrate per unit volume of birch (veneer density 655 kg/m^3 and thickness 1.55 mm) or hornbeam veneers (veneer density 730 kg/m^3 and thickness 1.50 mm). In addition, less polymer will penetrate the cavities of birch and hornbeam veneers because their porosity is lower than poplar veneer. It is well known that better and deeper penetration of polymers into wood creates stronger bonds in the samples [8,13,18,19,21,24,37].

In addition, previous studies have shown that not only the density but also the thickness of the veneer also significantly affects the bonding strength [41,43]. They found that increasing veneer thickness usually leads to a decrease in bonding strength. Therefore, as we can see from Figure 8, plywood made of poplar veneer of lower density and thickness shows the similar bonding strength as plywood made of birch and hornbeam veneer, which have higher density and thickness.

The lowest shear strength values of 1.19 MPa and 1.24 MPa correspond to the samples bonded by LDPE film with a thickness of 80 and 50 μm, respectively, while the highest shear strength values of 1.36 MPa and 1.37 MPa correspond to the samples bonded by LDPE film of 100 and 150 μm, respectively. However, based on Duncan's test the values of shear strength in the samples bonded by LDPE film with a thickness of 80 and 50 μm, as well as 100 and 150 μm, differ insignificantly ($p > 0.05$).

The changes in shear strength might be related to the penetration of the LDPE polymer in the wood. Wood is a porous material, and thus penetration of adhesive plays an important role in wood adhesion. The adhesive must penetrate the wood before it is cured to provide adequate mechanical interlocking [17]. The lowest shear strength in the plywood samples bonded with LDPE film of 50 and 80 μm is due to the low adhesive spread rate for these thicknesses, which was equivalent to 46.0 and 73.6 g/m^2, respectively. This is a lower amount of adhesive than the amount used in practice for liquid thermosetting adhesives. The molten LDPE film was pressed into the vessels and cracks of the veneer during the pressing process. Therefore, when less molten film remains between the sheets of the veneer, this causes a reduction in the bonding strength [11,21]. Increasing the thickness of the film, and hence the amount of polymer, leads to increased bonding strength. Therefore, it is recommended that LDPE films with a thickness of 100 μm and more be used. An obvious increase in the bonding strength with an increase in the number of HDPE layers was observed by Chang et al. [17].

3.5. Water Absorption and Thickness Swelling of Plywood Samples

Figure 9 shows the influence of the wood species and thickness of the LDPE film adhesive on the WA and TS of plywood samples. The WA and TS increases with an increase in the soaking time. The smallest values of WA and TS were observed during soaking for 2 h, and then the WA and TS increased. ANOVA analysis showed that both variables significantly ($p < 0.05$) affect WA and TS. According to the F values, the wood species had the greatest effect on the WA (2 h) and TS (2 h) ($F = 127.757$ and $F = 50.127$, respectively) and WA (24 h) and TS (24 h) ($F = 883.623$ and $F = 162.324$, respectively) of the plywood samples, while the thickness of the plastic film affected the WA (2 h) and TS (2 h) ($F = 54.642$ and $F = 3.953$, respectively) and WA (24 h) and TS (24 h) ($F = 53.434$ and $F = 2.941$, respectively) to a lesser extent.

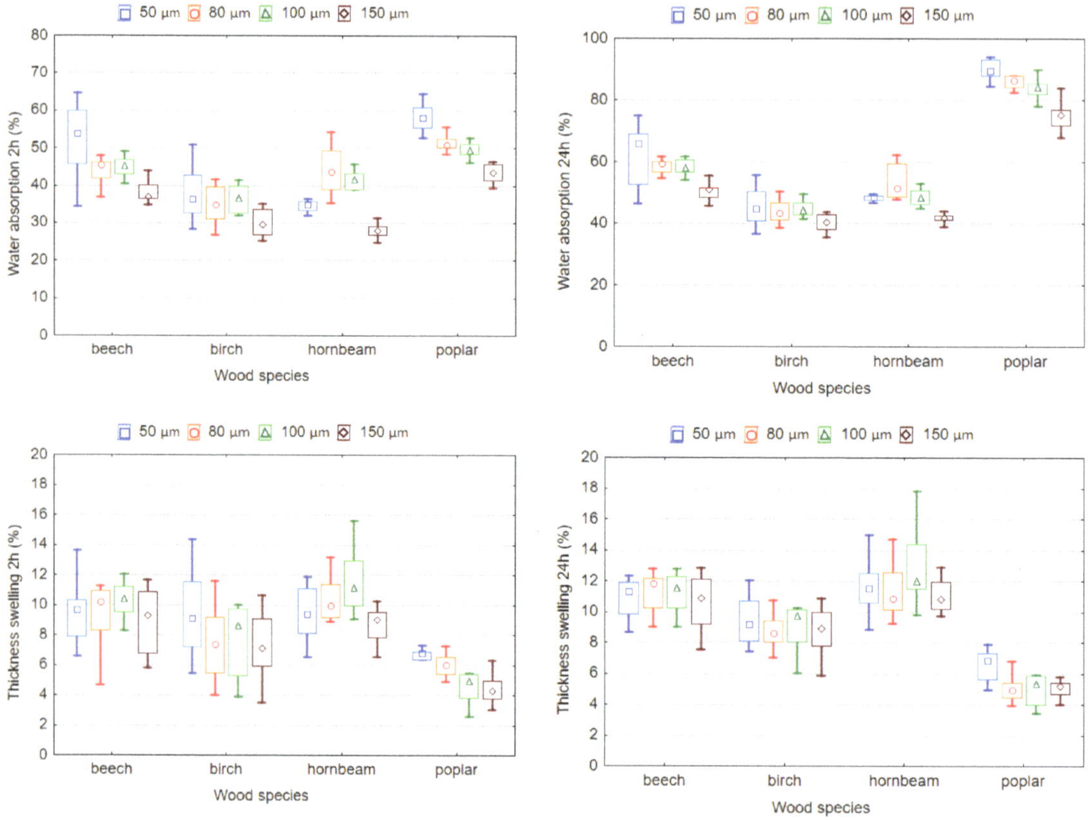

Figure 9. Water absorption and thickness swelling of plywood bonded by LDPE film.

The highest WA values after 2 and 24 h of soaking in water were found in the plywood samples from beech and poplar veneers, while the lowest corresponded to the hornbeam and birch veneers. The lowest value of WA (24 h) was recorded in samples made with birch veneer (43.7%), with an increase in samples made with hornbeam (48.0%) and beech (57.0%) veneers, and the highest was recorded in samples made with poplar veneer (84.0%). A different trend was observed for TS (24 h). The lowest TS (24 h) values corresponded to the samples made with poplar veneer (5.4%), while the highest TS (24 h) corresponded to the samples made with birch, beech or hornbeam veneers (9.0%, 10.9% and 11.7%, respectively). The values of TS (24 h) for birch and beech veneer do not differ significantly ($p > 0.05$).

The obtained results are explained by the differences in the anatomical structure of the wood that was used. It can be seen from Figure 9 that the WA of plywood samples decreases with an increase in the density of the wood veneers. This is because the number of pores (internal voids) in the samples with higher density is less than in the samples with lower density. Therefore, the WA of plywood samples from high-density wood veneers occurs mainly due to the penetration of moisture into the cell walls, and this process is much slower than filling the pores with moisture [39]. Plywood samples from poplar veneers had the lowest percentage of TS because more polymer penetrated per unit volume of this veneer than per unit volume of birch and hornbeam veneers. The obtained values of WA are in good agreement with the commonly accepted statement that WA is related to density [47]. The higher density results in a lower number of empty spaces, and consequently, lower WA.

Plywood samples produced with a higher LDPE film thickness showed lower percentages of WA and TS compared to the plywood samples produced with a lower film thickness. The lowest TS and WA values correspond to the samples bonded with LDPE film with a 150 μm thickness. The LDPE film with a 50 μm thickness showed the worst TS and WA due to the low consumption of such a film thickness. This causes insufficient filling of wood cavities with molten polymer. The amount of adhesive has been found to have a similar effect on the TS and WA of wood materials such as particleboard or medium-density fiberboard [48]. Increasing the amount of adhesive prevents the penetration of water into the panels and thus reduces TS and WA. The values of WA (24 h) in the samples bonded with a film thickness of 50 and 80 μm differ insignificantly ($p > 0.05$). The differences between the values of TS (24 h) in the samples bonded with a film thickness of 50, 80 and 100 μm as well as 80, 100 and 150 μm were insignificant ($p > 0.05$).

4. Conclusions

LDPE film of different thicknesses (50, 80, 100 and 150 μm) was successfully used for the bonding of plywood using veneer from various wood species—beech, birch, hornbeam and poplar. The wood species significantly affects the physical and mechanical properties of plywood. High-density beech, birch and hornbeam wood veneers provide the plywood with higher density, MOR, MOE and TS but lower WA than low-density poplar veneer. By contrast, plywood samples made using poplar veneer showed bonding strengths identical to samples made using birch and hornbeam veneers. This is due to not only the effect of wood density, but also the thickness of the veneer and the amount of polymer that penetrates into the wood. The plywood samples manufactured with thin veneers had better bonding strength than those manufactured with thick veneers. The beech plywood samples had the highest mechanical properties values. The properties of the plywood samples increased with an increase in the thickness of the LDPE film from 50 to 150 μm. A thin veneer package heats up faster than a thick veneer package. The thickness of the plastic film, due to its small value, did not affect the heating rate of the veneer package. Environmentally-friendly thermoplastic-bonded plywood panels can be used successfully in indoor applications.

Author Contributions: Conceptualization, P.B.; methodology, P.B., I.K. and O.C.; investigation, O.C., I.K., N.B. and O.B.; writing—original draft preparation, P.B.; writing—review and editing, P.B., I.K. and N.B. All authors have read and agreed to the published version of the manuscript.

Funding: This research received no external funding.

Institutional Review Board Statement: Not applicable.

Informed Consent Statement: Not applicable.

Data Availability Statement: The data that support the findings of this study are available upon reasonable request from the authors.

Acknowledgments: The authors express their sincere thanks to V. Biruk and B. Stupnytsky for their help in performing the experiments.

Conflicts of Interest: The authors declare no conflict of interest.

References

1. Shi, S.; Walker, J. Wood-based composites: Plywood and veneer-based products. Chapter 11. In *Primary Wood Processing: Principles and Practice*, 2nd ed.; John, C.F., Walker, Eds.; Springer: Dordrecht, The Netherlands, 2006; pp. 391–426.
2. Koprda, Š.; Balogh, Z.; Magdin, M.; Reichel, J.; Molnár, G. The Possibility of Creating a Low-Cost Laser Engraver CNC Machine Prototype with Platform Arduino. *Acta Polytech. Hung.* **2020**, *17*, 181–198. [CrossRef]
3. FAO Yearbook of Forest Products 2018. Available online: http://www.fao.org/3/cb0513m/CB0513M.pdf (accessed on 15 July 2021).
4. Dunky, M. Adhesives in the wood industry. In *Handbook of Adhesive Technology*, 2nd ed.; Revised and Expanded; Pizzi, A., Mittal, K.L., Eds.; Marcel Dekker Inc.: New York, NY, USA; Basel, Switzerland, 2003; 71p. [CrossRef]

5. Frihart, C.R.; Hunt, C.G. Adhesives with Wood Materials: Bond Formation and Performance. In *Wood Handbook—Wood as an Engineering Material*; General Technical Report FPL-GTR-190; Department of Agriculture, Forest Service, Forest Products Laboratory: Madison, WI, USA, 2010; Chapter 10.
6. World Health Organization—International Agency for Research on Cancer. *Monographs on the Evaluation of Carcinogenic Risk to Humans*; Formaldehyde, 2–Butoxyethanol and 1–tert–Butoxypropan–2–ol; World Health Organization—International Agency for Research on Cancer: Lyon, France, 2006; Volume 88.
7. Pizzi, A.; Papadopoulos, A.N.; Policardi, F. Wood Composites and Their Polymer Binders. *Polymers* **2020**, *12*, 1115. [CrossRef]
8. Kajaks, J.; Reihmane, S.; Grinbergs, U.; Kalnins, K. Use of innovative environmentally friendly adhesives for wood veneer bonding. *Proc. Est. Acad. Sci.* **2012**, *61*, 207–211. [CrossRef]
9. Oh, Y.S. Use of polyethylene as an additive in plywood adhesive. *J. Korean Wood Sci. Technol.* **1998**, *26*, 14–18.
10. Follrich, J.; Muller, U.; Gindl, W. Effects of thermal modification on the adhesion between spruce wood (*Picea abies Karst.*) and a thermoplastic polymer. *Holz Als Roh-Und Werkst.* **2006**, *64*, 373–376. [CrossRef]
11. Bekhta, P.; Müller, M.; Hunko, I. Properties of Thermoplastic-Bonded Plywood: Effects of the Wood Species and Types of the Thermoplastic Films. *Polymers* **2020**, *12*, 2582. [CrossRef] [PubMed]
12. Tang, L.; Zhang, Z.G.; Qi, J.; Zhao, J.R.; Feng, Y. The preparation and application of a new formaldehyde-free adhesive for plywood. *Int. J. Adhes. Adhes.* **2011**, *31*, 507–512. [CrossRef]
13. Fang, L.; Chang, L.; Guo, W.; Ren, Y.; Wang, Z. Preparation and characterization of wood-plastic plywood bonded with high density polyethylene film. *Eur. J. Wood Prod.* **2013**, *71*, 739–746. [CrossRef]
14. Fang, L.; Chang, L.; Guo, W.-J.; Chen, Y.; Wang, Z. Influence of silane surface modification of veneer on interfacial adhesion of wood–plastic plywood. *Appl. Surf. Sci.* **2014**, *288*, 682–689. [CrossRef]
15. Fang, L.; Xiong, X.; Wang, X.; Chen, H.; Mo, X. Effects of surface modification methods on mechanical and interfacial properties of high-density polyethylene-bonded wood veneer composites. *J. Wood Sci.* **2017**, *63*, 65–73. [CrossRef]
16. De Barros Lustosa, E.C.; Del Menezzi, C.H.S.; de Melo, R.R. Production and properties of a new wood laminated veneer/high-density polyethylene composite board. *Mater. Res.* **2015**, *18*, 994–999. [CrossRef]
17. Chang, L.; Guo, W.; Tang, Q. Assessing the tensile shear strength and interfacial bonding mechanism of poplar plywood with high-density polyethylene films as adhesive. *BioResources* **2017**, *12*, 571–585. [CrossRef]
18. Chang, L.; Tang, Q.; Gao, L.; Fang, L.; Wang, Z.; Guo, W. Fabrication and characterization of HDPE resins as adhesives in plywood. *Eur. J. Wood Prod.* **2018**, *76*, 325–335. [CrossRef]
19. Song, W.; Wei, W.; Ren, C.; Zhang, S. Developing and evaluating composites based on plantation eucalyptus rotary-cut veneer and high-density polyethylene film as novel building materials. *BioResources* **2016**, *11*, 3318–3331. [CrossRef]
20. Song, W.; Wei, W.; Ren, C.; Zhang, S. Effect of heat treatment or alkali treatment of veneers on the mechanical properties of eucalyptus veneer/polyethylene film plywood composites. *BioResources* **2017**, *12*, 8683–8703. [CrossRef]
21. Bekhta, P.; Sedliačik, J. Environmentally-Friendly High-Density Polyethylene-Bonded Plywood Panels. *Polymers* **2019**, *11*, 1166. [CrossRef] [PubMed]
22. Demirkir, C.; Öztürk, H.; Çolakoğlu, G. Effects of press parameters on some technological properties of polystyrene composite plywood. *Kast. Univ. J. For. Fac.* **2017**, *17*, 517–522. [CrossRef]
23. Borysiuk, P.; Mamiński, M.Ł.; Parzuchowski, P.; Zado, A. Application of polystyrene as binder for veneers bonding—The effect of pressing parameters. *Eur. J. Wood Prod.* **2010**, *68*, 487–489. [CrossRef]
24. Goto, T.; Saiki, H.; Onishi, H. Studies on wood gluing. XIII: Gluability and scanning electron microscopic study of wood-polypropylene bonding. *Wood Sci. Technol.* **1982**, *16*, 293–303. [CrossRef]
25. Song, W.; Wei, W.; Li, X.; Zhang, S. Utilization of polypropylene film as an adhesive to prepare formaldehyde-free, weather-resistant plywood-like composites: Process optimization, performance evaluation, and interface modification. *BioResources* **2017**, *12*, 228–254. [CrossRef]
26. Song, W.; Wei, W.; Wang, D.; Zhang, S. Preparation and properties of new plywood composites made from surface modified veneers and polyvinyl chloride films. *BioResources* **2017**, *12*, 8320–8339. [CrossRef]
27. Chen, Z.; Wang, C.; Cao, Y.; Zhang, S.; Song, W. Effect of Adhesive Content and Modification Method on Physical and Mechanical Properties of Eucalyptus Veneer–Poly-β-Hydroxybutyrate Film Composites. *For. Prod. J.* **2018**, *68*, 419–429. [CrossRef]
28. Cui, T.; Song, K.; Zhang, S. Research on utilizing recycled plastic to make environment-friendly plywood. *For. Stud. China* **2010**, *12*, 218–222. [CrossRef]
29. Niska, K.; Sain, M. *Wood-Polymer Composites*; Woodhead Publishing, Ltd.: Cambridge, UK, 2008; 384p.
30. Bal, B.C.; Bektas, I. Some mechanical properties of plywood produced from eucalyptus, beech, and poplar veneer. *Maderas-Cienc. Tecnol.* **2014**, *16*, 99–108. [CrossRef]
31. *EN 323*; Wood-Based Panels—Determination of Density, European Committee for Standardization: Brussels, Belgium, 1993.
32. *EN 310*; Wood-Based Panels—Determination of Modulus of Elasticity in Bending and of Bending Strength, European Committee for Standartiztion: Brussels, Belgium, 1993.
33. *EN 314-1*; Plywood—Bonding Quality—Part 1: Test Methods, European Committee for Standardization: Brussels, Belgium, 2004.
34. *EN 314-2*; Plywood—Bonding Quality—Part 2: Requirements, European Committee for Standardization: Brussels, Belgium, 1993.
35. *EN 317*; Particleboards and Fibreboards. Determination of Swelling in Thickness after Immersion in Water, European Committee for Standardization: Brussels, Belgium, 1993.

36. Bekhta, P.; Salca, E.-A. Influence of veneer densification on the shear strength and temper-ature behavior inside the plywood during hot press. *Constr. Build. Mater.* **2018**, *162*, 20–26. [CrossRef]
37. Smith, M.J.; Dai, H.; Ramani, K. Wood-thermoplastic adhesive interface—Method of characterization and results. *Int. J. Adhes. Adhes.* **2002**, *22*, 197–204. [CrossRef]
38. Shukla, S.R.; Kamdem, D.P. Properties of laboratory made yellow poplar (*Liriodendron tulipifera*) laminated veneer lumber: Effect of the adhesives. *Eur. J. Wood Prod.* **2009**, *67*, 397–405. [CrossRef]
39. Ugolev, B.N. *Drevesinovedenie i Lesnoe Tovarovedenie [Wood Science and Forest Commodity Science]*; GOU VPO MGUL: Moscow, Russia, 2007; 351p.
40. Aydın, İ.; Çolak, S.; Çolakoğlu, G.; Salih, E. A comparative study on some physical and mechanical properties of laminated veneer lumber (LVL) produced from beech (*Fagus orientalis* L.) and eucalyptus (*Eucalyptus camaldulensis* dehn.) veneers. *Holz Roh Werkst.* **2004**, *62*, 218–220. [CrossRef]
41. Bekhta, P.; Salca, E.-A.; Lunguleasa, A. Some properties of plywood panels manufactured from combinations of thermally densified and non-densified veneers of different thicknesses in one structure. *J. Build. Eng.* **2020**, *29*, 101116. [CrossRef]
42. Daoui, A.; Descamps, C.; Marchal, R.; Zerizer, A. Influence of veneer quality on beech LVL mechanical properties. *Maderas Cienc. Tecnol.* **2011**, *13*, 69–83. [CrossRef]
43. Darmawan, W.; Nandika, D.; Massijaya, Y.; Kabe, A.; Rahayu, I.; Denaud, L.; Ozarska, B. Lathe check characteristics of fast growing sengon veneers and theireffect on LVL glue-bond and bending strength. *J. Mater. Proces. Technol.* **2015**, *215*, 181–188. [CrossRef]
44. de Melo, R.R.; Del Menezzi, C.H.S. Influence of veneer thickness on the properties of LVL from Parica (Schizolobium amazonicum) plantation trees. *Eur. J. Wood Prod.* **2014**, *72*, 191–198. [CrossRef]
45. Örs, Y.; Çolakoğlu, G.; Aydın, İ.; Çolak, S. Comparison of some technical properties of plywood produced from beech, okoume and poplar rotary cut veneers in different combinations. *J. Polytech.* **2002**, *5*, 257–265.
46. Kajaks, J.A.; Bakradze, G.G.; Viksne, A.V.; Reihmane, S.A.; Kalnins, M.M.; Krutohvostov, R. The use of polyolefins-based hot melts for wood bonding. *Mech. Compos. Mater.* **2009**, *45*, 643–650. [CrossRef]
47. Del Menezzi, C.H.S.; Tomaselli, I. Contact thermal post-treatment of oriented strandboard to improve dimensional stability: A preliminary study. *Holz Als Roh-Und Werkst.* **2006**, *64*, 212–217. [CrossRef]
48. Halligan, A.F. A review of thickness swelling in particleboard. *Wood Sci. Technol.* **1970**, *4*, 301–312. [CrossRef]

polymers MDPI

Review

Research Progress of Wood-Based Panels Made of Thermoplastics as Wood Adhesives

Xianfeng Mo [1], Xinhao Zhang [2], Lu Fang [1,2,*] and Yu Zhang [2]

[1] Co-Innovation Center of Efficient Processing and Utilization of Forest Resources, Nanjing Forestry University, Nanjing 210037, China; moxianfeng47@hotmail.com

[2] College of Furnishings and Industrial Design, Nanjing Forestry University, Nanjing 210037, China; zhangxinhaonjfu@163.com (X.Z.); ZY19980427@163.com (Y.Z.)

* Correspondence: fanglu@njfu.edu.cn

Abstract: When thermoplastic resins such as polyethylene (PE) and polypropylene (PP) are selected as wood adhesives to bond wood particles (fibers, chips, veneers) by using the hot-pressing technique, the formaldehyde emission issue that has long existed in the wood-based panel industry can be effectively solved. In this study, in general, thermoplastic-bonded wood-based panels presented relatively higher mechanical properties and better water resistance and machinability than the conventional urea–formaldehyde resin-bonded wood-based panels. However, the bonding structure of the wood and thermoplastic materials was unstable at high temperatures. Compared with the wood–plastic composites manufactured by the extruding or injection molding methods, thermoplastic-bonded wood-based panels have the advantages of larger size, a wider raw material range and higher production efficiency. The processing technology, bonding mechanism and the performance of thermoplastic-bonded wood-based panels are comprehensively summarized and reviewed in this paper. Meanwhile, the existing problems of this new kind of panel and their future development trends are also highlighted, which can provide the wood industry with foundations and guidelines for using thermoplastics as environmentally friendly adhesives and effectively solving indoor pollution problems.

Keywords: thermoplastic resin; formaldehyde-free adhesive; hot-pressing technique; wood-based panels

Citation: Mo, X.; Zhang, X.; Fang, L.; Zhang, Y. Research Progress of Wood-Based Panels Made of Thermoplastics as Wood Adhesives. *Polymers* **2022**, *14*, 98. https://doi.org/10.3390/polym14010098

Academic Editor: Pavlo Bekhta

Received: 22 November 2021
Accepted: 25 December 2021
Published: 28 December 2021

1. Introduction

The development of the wood-based panel industry has played a prominent role in alleviating the contradiction between the supply and demand of wood products, protecting forest resources and the ecological environment [1–3]. Plywood, fiberboard and particleboard are the three predominant types of wood-based panels and have been widely developed as alternative materials for furniture manufacturing, decorative panel production, wood structure building boards and other fields because of their stable physical properties, wide sources of raw materials and good end-use properties [4]. However, wood adhesives, the essential components predominantly used in manufacturing wood composites, are mainly formaldehyde-based materials, such as urea–formaldehyde (UF) resin, phenol–formaldehyde (PF) resin and melamine urea–formaldehyde (MUF) resin [5,6]. These adhesives have the advantages of good adhesion and technological maturity, but they release formaldehyde when producing and using the wood composites [7]. In 2004, the International Agency for Research on Cancer reclassified formaldehyde from "probably carcinogenic to humans" to "carcinogenic to humans". The European Union, the USA, China and Japan now have legislation regulating the allowed levels of formaldehyde emission from wood-based products [8]. These events culminated with the implementation of the most stringent formaldehyde release standard in 2017. The new allowable emission levels vary by product, but the limits (not to be exceeded) are in the range of 50 to 100 ppb [9], which put forward much higher requirements regarding wood adhesives. Significant efforts have been dedicated to reducing synthetic (anthropogenic) CH_2O emission from the resins, principally by reducing the F/U mole ratio and modifying the formulation [10,11] and adding CH_2O scavengers [12]. However, these cannot solve the problem from the root cause. Alternative "no-added-formaldehyde" (NAF) resins, such as those based upon soybean, starch and lignin, have also captured great attention [13,14]. However, they only occupy minor positions in this vast market due to their poor water resistance, mobility and stability [15–17].

In recent years, the use of thermoplastic resins to bond wood materials has become a promising method to manufacture wood-based panels due to its advantages of environmental friendliness, good flexibility, water resistance and processability [18–22]. The study of thermoplastics as wood adhesives began in the 1990s and was first proposed by Dr. Han of Kyoto University [23]. He indicated that PP and modified PP have the merit of unique adhesion performance when combined with wood components. The study of the use of various thermoplastics as wood adhesives in China began in the mid-1990s and was proposed by Wang from the Chinese Academy of Forestry [24]. In 2005, the research results of thermoplastic-bonded plywood obtained the authorization of China's national invention patent, and it has completed independent intellectual property rights. In 2006, it was awarded as China's national key new product. The preparation methods and performance of the thermoplastic-bonded wood-based panels have been continuously studied and developed in recent years [25–35]. This paper aims to summarize the recent progress in thermoplastic-bonded wood-based composites, focusing on their processing technology, bonding mechanism and physical–mechanical performance. Meanwhile, the existing problems and the future development trends of thermoplastics as wood adhesives are highlighted.

2. Processing Technology of Thermoplastic-Bonded Wood-Based Panels

2.1. *Raw Materials*

Under certain pressure, melted thermoplastics have the merit of unique adhesion performance when combined with wood or other plant components. Both virgin and recycled thermoplastics can be used for compounding with wood, but the melting temperature of the selected thermoplastics must be lower than the degradation temperature of wood (200 °C). Thermoplastic resins that have been studied included polypropylene (PP), polyethylene (PE), polyvinyl chloride (PVC), polystyrene (PS) [35] and poly-β-hydroxybutyrate (PHB) [36]. Thermoplastics are solvent-free and have various forms, such as granules, powders, sheets, films, wires and blocks. Various forms of wood, such as wood veneer, wood particle and wood fiber, can be directly bonded with thermoplastics. In most of the existing studies, thermoplastic films were preferentially selected to bond wood veneers to prepare wood–plastic plywood [29,33–35]. When bonding wood particles, both thermoplastic films and thermoplastic powders have been popular choices [27,31,32].

2.2. *Processing Technology of Wood–Plastic Plywood*

Similar to the processing methods of commercially available medium-density fiberboard (MDF), particleboard or plywood, thermoplastic-bonded wood-based panels can be manufactured by the traditional hot-pressing technique [24]. There is no gluing stage in the whole production process, but the panel must be cold-pressed immediately after it is taken out of the hot press [35–38].

The manufacturing process of wood–plastic plywood is relatively simple, as shown in Figure 1. The setup requires only plastic films between every two wood veneers, and the material is then fabricated using combined hot pressing and cold pressing. During the preparation of wood–plastic plywood, the thermoplastic film is heated and melted into the porous structure of the wood veneer, and then cooled and solidified to form the "glue nail" microstructure interface phase (Figure 2). In the heating stage, thermoplastics experience a transition from liquid to solid, playing the role of wood adhesive. Both the hot-pressing conditions and wood surface properties affect the bonding strength. Due to the characteristics of the thermoplastic film, shrinkage inevitably occurs during the formation of the wood–plastic interface. The shrinkage residual stress must be compensated in the cold-pressing stage; otherwise, the interfacial debonding failure of composites can easily occur under an external load [39,40].

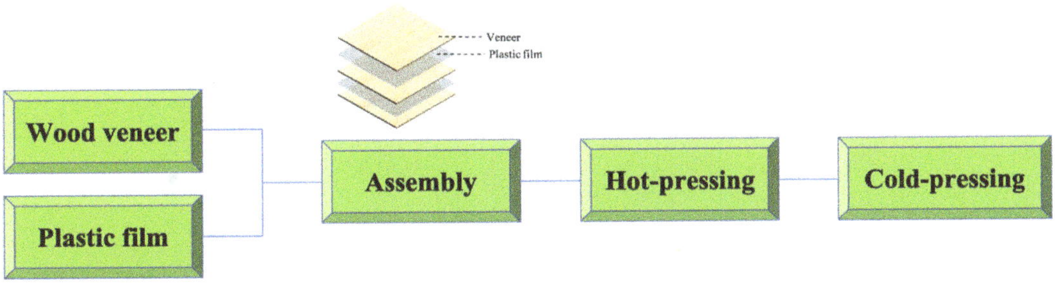

Figure 1. The scheme of wood–plastic plywood fabrication.

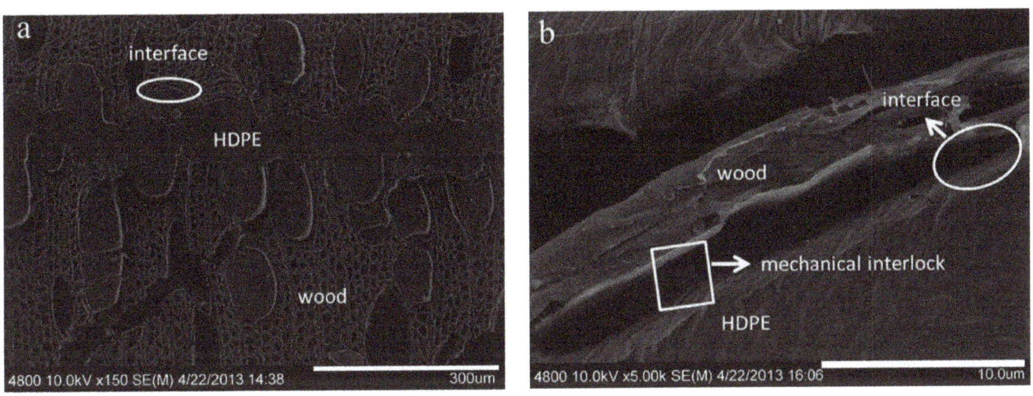

Figure 2. Derived from Ref [26]. Bonding interface between HDPE film and wood veneer: (**a**) magnification ×150; (**b**) magnification ×5000.

2.3. Processing Technology of Wood–Plastic Particleboard

The combination of wood flour and thermoplastics by extrusion or injection methods has become a standard practice. In wood–plastic composites (WPCs) manufactured by these two technologies, thermoplastics serve as a continuous phase, while wood flour is a dispersed phase to provide strength and stiffness [41]. WPCs can be made into solid products with various shapes, sizes and colors, and their products have been widely used in industry [42]. However, the use of the extrusion process is quite limited for large-sized particles due to its high machinery costs [43], and multi-dimensional products can only be prepared by injection molding [44,45]. To overcome these problems, the concept of bonding wood particles with the use of thermoplastics by the hot-pressing technique has been developed (as shown in Figure 3) [46–50]. Panels prepared by this method are called wood–plastic particleboard. On the one hand, it has the advantages of higher productivity and a lower pressing pressure requirement. The proportion of wood components can reach up to 80% by volume using this process, and the natural wood structure can be maintained. On the other hand, this method can obtain large-sized boards with a length of 2440 mm, a width of 1220 mm and a thickness of 3 to 40 mm [51].

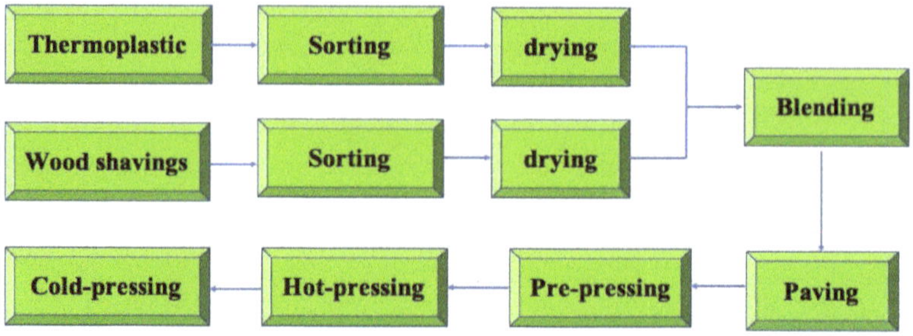

Figure 3. The scheme of wood–plastic particleboard fabrication.

For wood–plastic particleboard, the matrix can be ordinary wood shavings or other shavings prepared from agricultural residues such as bamboo, straw or cotton stalk [52]. Due to the wide range of raw materials, obtaining a very uniform blend of shavings and thermoplastic mixture has become one of the key points in the manufacturing technology of wood–plastic particleboard [53,54]. It is very important to find the proper moisture content, particle size and particle shape of the raw materials. A general rule for mixing is that large pieces of wood shavings should be matched with large pieces of thermoplastic, and small pieces of shavings matched with small pieces of thermoplastic. Increasing the fiber aspect ratio can produce a larger contact area between the fiber and plastic matrix, which is able to improve the composites' performance [55,56]. Qi [31,57] prepared cotton stalk–HDPE-oriented composites. It was found that the mechanical properties of composites prepared with long cotton stalk bundles were 2 to 3 times higher than those prepared with short or granular cotton stalk bundles. In the meantime, it is also necessary to focus on the processing cost when recycled plastic is utilized. For commonly used recycled plastics, such as PE and PP, directly sorting, cleaning and pulverizing before use is an ideal method.

Since no adhesive with initial tack is added in the manufacturing process of wood–plastic particleboard, and wood particles and thermoplastics are only physically mixed, their combinations are very scattered. Therefore, ensuring that the is mixture formed into a uniform mat and effectively transported during the assembly process has become another challenge for wood–plastic particleboard manufacturing [58]. Li et al. [59] prepared wood-fiber-reinforced PP composites by flat hot-pressing and compression molding. Results showed that when the wood fiber content reached 80%, the composites displayed a good flexural modulus and impact strength for both of the methods. Compared with the flat hot-pressing method, composites produced by the compression molding method had higher density and a better flexural modulus. Meanwhile, the composites manufactured by the flat hot-pressing method showed better surface wettability and impact strength. It is also worth noting that efficient bonding of the wood–thermoplastic system can be achieved only when the thermoplastic binder is sufficiently plasticized to be easily extruded into the pores of the wood shavings. Therefore, a low-pressure preheating step is necessary for the production of wood–plastic particleboard [60]. In order to utilize agricultural and forestry wastes, He et al. [52] prepared PP wood–plastic composites filled with rice straw powder, rice husk powder, wood powder and bamboo powder, as well as their mixture, using mixed compression molding and layer compression molding, respectively. For the mixed compression molding method, PP was first blended with the powders and then molded. For the layered compression molding, the powders were directly layered on the PP film. The results showed that the mechanical properties, water absorption and moisture absorption performance of the PP composites prepared by mixed compression molding were better than those of layered molding PP composites. The filler can be well-distributed in the melted PP matrix when using the mixed compression molding. By contrast, the wood-based panels prepared by the layered compression molding have a poor interface between the thermoplastic and the fillers.

3. Performance of Thermoplastic-Bonded Wood-Based Panels

Fabrication of thermoplastic-bonded wood-based panels with the hot-pressing methods in the laboratory resembles the standard procedure of composite production in industry. Its products are closely comparable to commercial plywood, MDF and particleboard. Compared with natural-based adhesive-bonded wood-based panels, this can offer improved water resistance and relatively simple processing technology.

3.1. Environmental Performance

Thermoplastics play the role of wood adhesives in the manufacturing process of thermoplastic-bonded wood-based composites. Since the traditional formaldehyde adhesive has been completely replaced, the most remarkable feature of thermoplastic-glued wood-based panels is their environmental friendliness. Table 1 presents the free formaldehyde content of wood–plastic particleboard as determined by two different test methods [58]. The values measured by these two methods are far lower than that in the "formaldehyde release limit of wood-based decorative materials and their products" standard. It is worth noting that the presence of a small amount of formaldehyde arises from the wood itself, and is called biogenic formaldehyde [61]. All organic wood components contribute to biological formaldehyde's generation and emission. A great deal of evidence has shown that this kind of formaldehyde mainly arises from lignin. At the same time, the total volatile organic compound (TVOC) release rate of thermoplastic resin wood-based panels was found to be only 0.01 mg/(m^2·h), which is also far below the value specified in the relevant standards [62,63].

Table 1. Environmental performance of wood–plastic particleboard.

Test Item/Test Method	Test Value	Standard Value
Formaldehyde/perforator test	0.2 mg/100 g	≤9 mg/100 g
Formaldehyde/desiccator test	0.2 mg/L	≤1.5 mg/L
TVOC/72 h emission rate	0.01 mg/(m^2·h)	≤0.5 mg/(m^2·h)

3.2. Physical and Mechanical Properties

3.2.1. Water Resistance

Thermoplastic resins, especially polyolefin plastics, are usually hydrophobic materials. They do not absorb water, which gives thermoplastic-bonded wood-based composites excellent water resistance. In the bonding process, thermoplastics can not only be filled in the porous structure of wood, but also cover part of the hygroscopic wood surface. The speed of water entering the wood is therefore slowed down, and the moisture penetrating into the wood is decreased accordingly [25,36,50,64]. Wang et al. [58] tested the water resistance of several particleboards. It was found that both the 2 h and 24 h thickness swelling (TS) of the thermoplastic-bonded particleboards were lower than 3%. After soaking in boiling water for 2 h, the internal bonding strength (IBS) of thermoplastic-bonded particleboards was higher than the traditional sliced veneer-decorated wood-based panels. Fang et al. [25] found that the water resistance of plywood bonded with PE film was much better than that of urea–formaldehyde (UF) resin-bonded panels. After soaking for 168 h, the water absorption and thickness expansion of PE film plywood were 85.8% and 7.7%, respectively, which are 18.8% and 4.9% lower than those of UF resin plywood. Due to their low expansion rate and good dimensional stability in a humid environment, thermoplastic-bonded wood composites will have extensive application prospects in cement formwork and outdoor packaging. As shown in Table 2, the larger the amount of thermoplastic resin used, the better the water resistance of the wood-based panel. However, it will have an adverse impact on the mechanical properties of the panel if the usage is excessive. For wood–plastic particleboards, most authors indicated that the panel achieves optimal resistance to bending forces when the content of wood particles is in the range of 40% to 60%. For wood–plastic plywood, when the film thickness is greater than 0.1 mm (adhesive dosage is around 100 g/m^2 in double glue line), it can meet the strength requirement for interior products (type II).

Table 2. Water resistance of some thermoplastic-bonded wood-based panels [25,33,34,65].

Particleboard Type	Adhesive Dosage	24 h TS/%	IBS/MPa
Recycled PE-bonded poplar particleboard	30~70%	0.8~8.0	1.0~2.5
Recycled PP-bonded poplar particleboard	30~70%	1.9~12.0	1.1~2.5
Recycled PS-bonded poplar particleboard	30~70%	1.3~17.4	4.0~4.4
PE film-bonded poplar plywood	184 g/m^2	5.9	Type II
PP film-bonded eucalyptus plywood	150 g/m^2	6.8	Type I
PVC film-bonded eucalyptus plywood	320 g/m^2	—	Type II
UF resin-bonded poplar plywood	320 g/m^2	6.5	Type II
UF resin-bonded eucalyptus plywood	320 g/m^2	7.5	Type II

3.2.2. Mechanical Properties

Thermoplastics have the merit of unique adhesion performance when composited with wood components [66–69]. Since there is no chemical reaction between thermoplastics and wood, mechanical interlock is the main bonding mechanism of thermoplastic-bonded wood composites. When a specific thermoplastic is selected as the wood adhesive, the appropriate hot-pressing temperature must be determined [22,25,35,70]. The hot-pressing temperature affects the thermoplastic viscosity. It is believed that the lower the viscosity, the deeper the penetration depth [71]. Generally, the hot-press temperature should be 15 to 35 °C higher than the melting temperature of the thermoplastic so that it can fully flow into and between the wood cell lumens and voids, helping to form more glue nails. Luedtke et al. [72] showed that poly(lactic acid) (PLA) bonded plywood prepared at >160 °C had higher tensile strength than those formed at 140 °C, which was associated with a thinner bondline and greater PLA migration away from the bondline, consistent with lower melt viscosity under high temperatures. The deeper PLA penetration depth contributed to the better physical interlocking of the PLA within the wood ultrastructure. However, it should be noted that the thickness of the adhesive bondline also has a great impact on the quality and performance of the adhesive. Over-penetration may occur if the hot-pressing temperature is too high, which will deteriorate the strength of thermoplastic-bonded wood composites [38,70]. Due to the high viscosity of thermoplastics, they usually take a long time to reach the fully molten state, and then gradually spread out on the veneer surface to complete the penetration step. Fang et al. [37] prepared formaldehyde-free plywood using PE film as a wood adhesive and found that the hot-pressing time and hot-pressing temperature have an interactive effect on the plywood tensile shear strength.

The properties of thermoplastic-bonded wood-based panels are closely related to the thermoplastic type and their dosage, as shown in Table 3 [25,36,64,66,73]. PE is one of the most widely used thermoplastic polymers and has a very simple structure. It has been widely studied in the wood industry recently because of its low price and good water resistance [18,25,29,30,57,70,74]. Results showed that the stiffness of thermoplastic PE is lower than that of thermosetting UF resin, but its plasticity can give PE film-bonded plywood stronger bending failure resistance and therefore a higher MOR value. The bonding strength and elastic modulus the PE film-bonded poplar plywood were found to be similar to those of UF resin-bonded poplar plywood (the difference was less than 1%). In addition, due to the high-temperature softening property of PE and its poor compatibility with wood, the prepared PE plywood can only meet the strength requirements of indoor materials (type II). When the outdoor material test standard was performed (three-cycle treatment: soaking in boiling water for 4 h, then drying at 63 °C for 20 h and soaking in boiling water for 4 h again), the PE film was completely separated from the wood. PP has better high-temperature resistance compared with PE, but higher energy consumption is required for preparing the panels. Li et al. [75] fabricated eucalyptus plywood using PP film as an adhesive. The tensile shear strength was 1.4 MPa after outdoor aging treatment (type I, the specified value of the standard is 0.7 MPa). Xia [32,76] prepared oriented cotton stalk–PP boards and oriented cotton stalk–HDPE boards under the same processing conditions (hot-press temperature 185 °C, hot-press time 15 min, board density 0.7g/cm^3). Results showed that the mechanical properties and water resistance of the two directional composite boards were better than those of conventional MDF and particleboard. The 24 h TS value of both boards was less than 3%. When the amount of thermoplastic resin was fixed at 15%, the static flexural strength, elastic modulus and internal bonding strength of the oriented cotton stalk–PP boards were 60.60, 5074.4 and 1.48 MPa, respectively, which are 33%, 13% and 13% higher than those of PE composites. When the mass fraction of thermoplastic resin exceeded 15%, the mechanical properties of the two composites decreased, and the water resistance was further improved. Polyvinyl chloride (PVC) film has the characteristics of lower price and wider availability compared with PP and PE [33,77,78]. It has also been successfully applied to the manufacturing of plywood.

Table 3. Mechanical properties of plywood bonded with different adhesives [22,30,35,36,69].

Adhesives	Hot-Pressing Temperature (°C)	Type II Bonding Strength (MPa)	Type I Bonding Strength (MPa)	MOR (MPa)	MOE (MPa)
PE film	152	1.5	0	82.8	7480
PP film	180	1.9	1.4	109.3	13,890
PVC film	183	1.1	0.5	65.1	8600
PS	140	1.25	—	79.7	5253
PHB	170	1.19	—	58.3	6001
UF resins	115~120	1.3~1.4	0	75.7~94.5	7520~13,019

4. Technical Problems of Wood-Based Panel Made of Thermoplastic Resin Adhesive

The physical–mechanical properties of this novel formaldehyde-free wood-based panel can meet the strength requirements for different applications by controlling the process conditions. Their technological and environmental benefits confirm their suitability to gradually replace the traditional plywood, particleboard or fiberboard. However, thermoplastics have the characteristics of melting and softening at high temperatures. When the temperature approaches or exceeds the melting point of the thermoplastic, the bonding structure between the thermoplastic and wood cannot be maintained, and the strength of the panel will be lost [25,79]. When thermoplastics are used as adhesives, their incompatibility with wood can present another major problem [18,26,29,36,47]. As shown in Figure 2b, there was a large gap at the interface. Therefore, enhancing the compatibility between thermoplastic and wood raw materials and improving the high-temperature resistance will be the most important work in the future. Multiple review studies related to the improvement of the wood fiber–plastic interfacial interaction have been published [41,80–82]. The most popular and most widely described methods are thermal treatment, plasma treatment, alkali treatment, silanization, acetylation, maleation, acrylation or isocyanate treatment. For wood–plastic particleboard, various modification methods that have been successfully applied in WPCs are essentially applicable. Meanwhile, for a wood veneer–plastic film bonding system, the interfacial methods are relatively limited due to the size limitation. Here, we simply focus on some effective methods for wood–plastic plywood.

Giving the wood veneer or plastic film new chemical properties has been proven to be the most effective solution [30,83,84]. Zhou et al. [29,85] grafted oxygen- and nitrogen-containing functional groups onto the surfaces of plastic bags using an industrial atmospheric dielectric-barrier discharge plasma system. Due to the enhanced surface polarity and wettability, formaldehyde-free plywood with a high bonding strength (0.82 MPa) was successfully fabricated. Fang et al. [26] introduced the bifunctional structure of vinyltrimethoxysilane hydrolysate into a wood veneer and successfully established a chemical bridge between the wood veneer and PE film. The bonding interface gap decreased to a single molecular scale because of the silane surface treatment, resulting in an increase in tensile shear strength by 293.2%. The temperature point at which the panel strength sharply decreased raised from 140 to 180 °C, indicating that the interfacial improvement was also of benefit to the thermal stability of the composites.

Physical modification has also attracted extensive attention because of its unique advantages. In order to promote the penetration of thermoplastics in wood, some scholars have used a self-made hole rolling machine to roll dense blind holes on the surface of the wood veneer [86]. This method helped to form a more stable "dendritic glue nail" microstructure and contributed to improving the bonding strength and dimensional stability of the PE film-bonded wood-based panel. The perforation of the thermoplastic film was also conducive to increasing the mechanical interlocking between the plastic and wood [87]. This is because the heat transfer rate was accelerated during the hot pressing. However, the efficiency of these methods is low, and the self-strength of wood veneer or plastic films can easily be reduced. Thermal treatment will not only degrade hemicellulose in wood, but also lead to the migration or volatilization of various extractives to the wood surface [88]. Therefore, an alternative modification method has been explored as a strategy to improve the compatibility of the wood veneer and hydrophobic thermoplastic without contaminating the environment. However, the high temperature usually reduced the elastic modulus of wood. The key technology of thermal treatment was to select appropriate parameters, such as treating temperature and treating medium, so as to effectively change the surface properties of wood without reducing its mechanical properties. To date, there have been several reports of improved dimensional stability, water resistance and biological resistance [89–91], but little improvement in flexural properties has been reported. Future work should focus on exploring environmentally friendly modification methods, with low energy consumption and low cost, that are conducive to industrial production. At the same time, the mechanism of various modification methods should be further clarified to add new content to the development of high-performance green wood adhesives.

5. Conclusions

Thermoplastic-bonded wood-based panels prepared using the hot-pressing technique have the characteristics of no formaldehyde emission and excellent water resistance. This has become an ideal alternative material for furniture, flooring decoration and so on. To date, a large number of systematic studies have been carried out on these novel panels, but there are still some problems that require further attention.

(1) As for the types of thermoplastics, virgin and single plastics are preferred, among which PP and PE are the most studied. The hot-pressing temperature for preparing the composites is relatively high (above 140 °C). In order to alleviate the defects caused by plastic shrinkage, cooling forming equipment needs to be added for the preparation of thermoplastic-bonded wood-based panels. In general, the manufacturing cost of thermoplastic-bonded wood-based panels is higher than that of traditional trialdehyde-bonded panels.

(2) The incompatibility between thermoplastic and wood has an adverse impact on the mechanical properties. Therefore, thermoplastic-bonded wood-based panels can only find use in some nonstructural applications. To date, numerous modification methods have been tested and proven to counter the incompatibility between hydrophilic wood and hydrophobic thermoplastics, but some of their enhancement mechanisms need to be further clarified.

(3) For thermoplastic-bonded particleboard or fiberboard, obtaining uniformly mixed materials at the lowest cost, ensuring the uniform paving and molding of mixtures and selecting the cold-pressing parameters according to different manufacturing processes still lack data and theoretical support.

(4) The performance evaluation of wood-based panels made with thermoplastic resin adhesives mainly focuses on indexes such as bonding strength, bending resistance and water resistance, while there is a lack of research on the flame retardance, mildew resistance and ultraviolet aging resistance properties.

Author Contributions: Conceptualization, X.M. and L.F.; Formal analysis, X.M. and L.F.; Methodology, X.M. and X.Z.; Resources, L.F. and X.Z.; Writing—original draft, L.F.; Writing—review and editing, L.F. and X.Z.; Investigation, Y.Z. All authors have read and agreed to the published version of the manuscript.

Funding: This research was funded by the Natural Science Foundation of Jiangsu Province (CN) (No. BK20150881), the Priority Academic Program Development of Jiangsu Higher Education Institutions (PAPD), the Innovation Fund for Young Scholars of Nanjing Forestry University, China (No. CX2015020), the Starting Foundation of Nanjing Forestry University, China (No. GXL024) and the College Students Innovation and Entrepreneurship Training Program Project of Nanjing Forestry University (2020NFUSPITP0159).

Institutional Review Board Statement: Not applicable.

Informed Consent Statement: Not applicable.

Data Availability Statement: Not applicable.

Acknowledgments: The authors are grateful to the Advanced Analysis and Testing Center of Nanjing Forestry University.

Conflicts of Interest: The authors declare no conflict of interest.

References

1. Wang, J.; Cao, X.; Liu, H. A Review of the Long-term Effects of Humidity on the Mechanical Properties of Wood and Wood-based Products. *Eur. J. Wood Prod.* **2021**, *79*, 245–259. [CrossRef]
2. Papadopoulos, A. Advances in Wood Composites. *Polymers* **2019**, *12*, 48. [CrossRef] [PubMed]
3. Švajlenka, J.; Kozlovská, M. Evaluation of the Efficiency and Sustainability of Timber-based Construction. *J. Clean. Prod.* **2020**, *259*, 120835. [CrossRef]
4. Liu, R.; Liu, M.; Qu, Y.; Huang, A.; Ma, E. Dynamic Moisture Sorption and Formaldehyde Emission Behavior of Three Kinds of Wood-based Panels. *Eur. J. Wood Prod.* **2018**, *76*, 1037–1044. [CrossRef]
5. Treu, A.; Bredesen, R.; Bongers, F. Enhanced Bonding of Acetylated Wood with an MUF-based Adhesive and a Resorcinol-formaldehyde-based Primer. *Holzforschung* **2020**, *74*, 382–390. [CrossRef]
6. Tang, Q.; Tan, H.; Wei, Q.; Cui, H.; Guo, W.; Zhou, G. Preparation of Low Formaldehyde Emission of High-density Fiberboard and Its Properties. *J. For. Eng.* **2019**, *4*, 26–30.
7. Pizzi, A.; Mittal, K.I. *Handbook of Adhesive Technology, Revised and Expanded*; Taylor & Francis: Oxfordshire, UK, 2003.
8. Salthammer, T.; Mentese, S.; Marutzky, R. Formaldehyde in the Indoor Environment. *Chem. Rev.* **2010**, *110*, 2536–2572. [CrossRef]
9. *Formaldehyde Emission Standards for Composite Wood Products*; RIN 2070-AJ44; Environmental Protection Agency (EPA): Washington, DC, USA, 2016.
10. Park, B.; Lee, S.; Roh, J. Effects of Formaldehyde/Urea Mole Ratio and Melamine Content on the Hydrolytic Stability of Cured Urea-melamine-formaldehyde Resin. *Eur. J. Wood Prod.* **2009**, *67*, 121–123. [CrossRef]
11. Que, Z.; Furuno, T.; Katoh, S.; Nishino, Y. Effects of Urea-formaldehyde Resin Mole Ratio on the Properties of Particleboard. *Build. Environ.* **2007**, *42*, 1257–1263. [CrossRef]
12. Hematabadi, H.; Behrooz, R.; Shakibi, A.; Arabi, M. The Reduction of Indoor Air. Formaldehyde from Wood Based Composites Using Urea Treatment for Building Materials. *Constr. Build. Mater.* **2012**, *28*, 743–746. [CrossRef]
13. Pang, H.; Wang, Y.; Chang, Z.; Xia, C.; Han, C.; Liu, H.; Li, J.; Zhang, S.; Cai, L.; Huang, Z. Soy Meal Adhesive with High Strength and Water Resistance Via Carboxymethylated Wood Fiber-induced Crosslinking. *Cellulose* **2021**, *28*, 3569–3584. [CrossRef]
14. Karthuser, J.; Biziks, V.; Mai, C.; Militz, H. Lignin and Lignin-Derived Compounds for Wood Applications—A Review. *Molecules* **2021**, *26*, 2533. [CrossRef]
15. Pizzi, A. Recent Developments in Eco-efficient Bio-based Adhesives for Wood Bonding: Opportunities and Issues. *J. Adhes. Sci. Technol.* **2006**, *20*, 829–846. [CrossRef]
16. Yang, I.; Kuo, M.; Myers, D.; Pu, A. Comparison of Protein-based Adhesive Resins for Wood Composites. *J. Wood Sci.* **2006**, *52*, 503–508. [CrossRef]
17. Sellers, T. Wood Adhesive Innovations and Applications in North America. *Forest Prod. J.* **2001**, *51*, 12–22.
18. Fang, L.; Chang, L.; Guo, W.; Wang, Z. Overview on Adhesives and Relevant Modification for Environmentally-Friendly Plywood. *China Wood Ind.* **2012**, *26*, 22–27.
19. Wilkowski, J.; Borysiuk, P.; Górski, J.; Czarniak, P. Analysis of relative machinability indexes of wood particle boards bonded with waste thermoplastics. *Drewno* **2013**, *56*, 139–144.
20. Grinbergs, U.; Kajaks, J.; Reihmane, S. Usage of Ecologically Perspective Adhesives for Wood Bonding. *Sci. J. Riga Techol. Univ. Mater. Sci. Appl. Chem.* **2010**, *22*, 114–117.
21. Kajaks, J.; Reihmane, S.; Grinbergs, U.; Kalnins, K. Use of innovative environmentally friendly adhesives for wood veneer bonding. *Proc. Est. Acad. Sci.* **2012**, *61*, 207–211. [CrossRef]

22. Kajaks, J.; Bakradze, G.; Viksne, A.; Reihmane, S.; Kalnins, M.; Krutohvostov, R. The use of polyolefins-based hot melts for wood bonding. *Mech. Compos. Mater.* **2009**, *45*, 643–650. [CrossRef]
23. Han, G.S. Preparation and Physical Properties of Moldable Wood-Plastic Composites. Ph.D. Thesis, Kyoto University, Kyoto, Japan, 1990.
24. Wang, Z.; Guo, W. Environment-Friendly Plywood Production Process. CN Patent 1103284C, 5 December 2000.
25. Fang, L.; Chang, L.; Guo, W.; Ren, Y.; Wang, Z. Preparation and Characterization of Wood-plastic Plywood Bonded with High Density Polyethylene Film. *Eur. J. Wood Prod.* **2013**, *71*, 739–746. [CrossRef]
26. Fang, L.; Xiong, X.; Wang, X.; Chen, H.; Mo, X. Effects of Surface Modification Methods on Mechanical and Interfacial Properties of High-density Polyethylene-bonded Wood Veneer Composites. *J. Wood. Sci.* **2017**, *63*, 65–73. [CrossRef]
27. Lin, T.; Guo, W.; Gao, L.; Chang, L.; Wang, Z. Effect of Cotton Stalk Pattern and Screening Size on Properties of Cotton Stalk/Recycled Plastic Composite Panels. *J. For. Environ.* **2011**, *31*, 5.
28. Lin, T.; Guo, W.; Gao, L.; Wang, Z. Effect of Particle Size on Properties of Cotton Stalk-Recycled Plastic Composites. *J. Southwest For. Univ.* **2011**, *31*, 78–82.
29. Yu, P.; Zhang, W.; Chen, M.; Zhou, X. Plasma-treated Thermoplastic Resin Film as Adhesive for Preparing Environmentally-friendly Plywood. *J. For. Eng.* **2020**, *5*, 41–47.
30. Tang, L.; Zhang, Z.; Qi, J.; Zhao, J.; Feng, Y. The Preparation and Application of a New Formaldehyde-free Adhesive for Plywood. *Int. J. Adhes. Adhes.* **2011**, *31*, 507–512. [CrossRef]
31. Qi, C.S. Fabrication of Oriented Biomass-High Density Polyethylene Composites using Hot Pressing Process and Its Molding Mechanism. Ph.D. Thesis, Northwest A&F University, Xianyang, China, 2013.
32. Xia, N. Fabrication and Manufacturing Mechanism of Oriented Cotton Stalk-Polypropylene Film Boards. Ph.D. Thesis, Northwest A&F University, Xianyang, China, 2013.
33. Wang, D.D. A Study of the Manufacturing and Strengthening Mechanism of Composite Materials Made from Eucalyptus Veneer and Polyvinyl Chloride (PVC) Film. Ph.D. Thesis, Beijing Forestry University, Beijing, China, 2016.
34. Li, X.F. Research on the Pressing Crafts and Strengthening Mechanism of Composite Materials from Eucalyptus Veneer and Polypropylene (PP) Film. Ph.D. Thesis, Beijing Forestry University, Beijing, China, 2015.
35. Demirkir, C.; Öztürk, H.; Çolakoğlu, G. Effects of Press Parameters on Some Technological Properties of Polystyrene Composite Plywood. *Kast. Univ. J. For. Fac.* **2017**, *17*, 517–522.
36. Chen, Z.; Wang, C.; Cao, Y.; Zhang, S.; Song, W. Effect of Adhesive Content and Modification Method on Physical and Mechanical Properties of Eucalyptus Veneer–Poly-β-Hydroxybutyrate Film Composites. *For. Prod. J.* **2018**, *68*, 419–429. [CrossRef]
37. Fang, L.; Chang, L.; Guo, W.Z. Analysis of Optimization Manufacturing Technology of Poplar Plywood Glued with High Density Polyethylene Film. *China Wood Ind.* **2013**, *27*, 17–20.
38. Fang, L.; Zeng, J.; Liao, X.; Zou, Y.; Shen, J. Tensile Shear Strength and Microscopic Characterization of Veneer Bonding Interface with Polyethylene Film as Wood Adhesive. *Sci. Adv. Mater.* **2019**, *11*, 1223–1231. [CrossRef]
39. Kucher, M.; Yakovleva, O.; Chyzhyk, J. Thermal Expansion and Shrinkage of Unidirectional Composites at Elevated Temperatures. *Strength Mater.* **2020**, *52*, 790–797. [CrossRef]
40. Peng, X.; Zhang, Z. Hot-pressing Composite Curling Deformation Characteristics of Plastic Film-reinforced Pliable Decorative Sliced Veneer. *Compos. Sci. Technol.* **2018**, *157*, 40–47. [CrossRef]
41. Wang, H.; Zhang, X.; Guo, S.; Liu, T. A Review of Coextruded Wood–plastic Composite. *Polym. Compos.* **2021**, *42*, 4174–4186. [CrossRef]
42. Zhang, L.; Chen, Z.; Dong, H.; Fu, S.; Ma, L.; Yang, X. Wood Plastic Composites Based Wood Wall's Structure and Thermal Insulation Performance. *J. Bioresour. Bioprod.* **2021**, *6*, 65–74. [CrossRef]
43. Chen, H.; Chen, T.; Hsu, C. Effects of wood particle size and mixing ratios of HDPE on the properties of the composites. *Eur. J. Wood Prod.* **2006**, *64*, 172–177. [CrossRef]
44. Ayrilmis, N.; Benthien, J.; Thoemen, H.; White, R. Effects of fire retardants on physical, mechanical, and fire properties of flat-pressed WPCs. *Eur. J. Wood Prod.* **2012**, *70*, 215–224. [CrossRef]
45. Benthien, J.; Thoemen, H. Effects of Raw Materials and Process Parameters on the Physical and Mechanical Properties of Flat Pressed WPC Panels. *Compos. Part A Appl. Sci.* **2012**, *43*, 570–576. [CrossRef]
46. Ayrilmis, N.; Jarusombuti, S. Flat-pressed Wood Plastic Composite as an Alternative to Conventional Wood-based Panels. *J. Compos. Mater.* **2011**, *45*, 103–112. [CrossRef]
47. Friedrich, D. Thermoplastic Moulding of Wood-Polymer Composites (WPC): A Review on Physical and Mechanical Behaviour under Hot-pressing Technique. *Compos. Struct.* **2021**, *262*, 113649. [CrossRef]
48. Borysiuk, P.; Zbieć, M.; Jenczyk-Tołłoczko, I.; Jabłoński, M. Thermoplastic bonded composite chipboard Part 1—Mechanical properties. In Proceedings of the 8th International Science Conference: "Chip and Chipless Woodworking Processes", Zvolen, Slovakia, 6–8 September 2012; pp. 393–398.
49. Zbieć, M.; Borysiuk, P.; Mazurek, A. Thermoplastic bonded composite chipboard Part 2—Machining tests. In Proceedings of the 8th International Science Conference: "Chip and Chipless Woodworking Processes", Zvolen, Slovakia, 6–8 September 2012; pp. 399–405.
50. Borysiuk, P.; Wilkowski, J.; Krajewski, K.; Auriga, R.; Skomorucha, A.; Auriga, A. Selected Properties of Flat-pressed Wood-polymer Composites for High Humidity Conditions. *Bioresources* **2020**, *15*, 5156–5178. [CrossRef]

51. Satyanarayana, K.; Arizaga, G.; Wypych, F. Biodegradable Composites Based on Lignocellulosic Fibers—An Overview. *Prog. Polym. Sci.* **2009**, *34*, 982–1021. [CrossRef]
52. He, C.; Hou, R.; Xue, J.; Zhu, D. Performances of PP Wood-plastic Composites with Different Processing Methods. *Trans. CSAE* **2012**, *28*, 145–150.
53. Schmidt, H.; Benthien, J.; Thoemen, H. Processing and Flexural Properties of Surface Reinforced Flat Pressed WPC Panels. *Eur. J. Wood Prod.* **2013**, *71*, 591–597. [CrossRef]
54. Guo, W.; Wang, Z. Application and Production Technology of Wood Plastic Particleboard. *China Wood-Based Panels* **2014**, *7*, 1–4.
55. Rahman, K.; Islam, M.; Ratul, S.; Dana, N.; Musa, S.; Hannan, M. Properties of Flat-pressed Wood Plastic Composites as A Function of Particle Size and Mixing Ratio. *J. Wood Sci.* **2018**, *64*, 279–286. [CrossRef]
56. Migneault, S.; Koubaa, A.; Erchiqui, F.; Chaala, A.; Englund, K.; Wolcott, M. Effects of Processing Method and Fiber Size on the Structure and Properties of Wood-plastic Composites. *Compos. Part A Appl. S.* **2009**, *40*, 80–85. [CrossRef]
57. Qi, C.; Guo, K.; Gu, R.; Liu, Y. Preparation Technology of Oriented Cotton Stalk Bunches/High Density Polyethylene Composite Panels. *Trans. CSAE* **2010**, *26*, 265–300.
58. Wang, Z.; Guo, W.; Gao, L. Properties, Applications and Development Trends of Wood-plastic Composite Particleboard. *China Wood-Based Panels* **2005**, *5*, 12–16.
59. Li, Z.; Wang, W. Preparation and Properties of Polypropylene Based Composites with High Wood Fibers Content. *J. For. Eng.* **2017**, *2*, 9–15.
60. Borysiuk, P.; Mamiński, M.; Parzuchowski, P.; Zado, A. Application of Polystyrene as Binder for Veneers Bonding—The Effect of Pressing Parameters. *Eur. J. Wood Prod.* **2010**, *68*, 487–489. [CrossRef]
61. Tasooji, M.; Wan, G.; Lewis, G.; Wise, H.; Frazier, C. Biogenic Formaldehyde: Content and Heat Generation in the Wood of Three Tree Species. *ACS Sustain. Chem. Eng.* **2017**, *5*, 4243–4248. [CrossRef]
62. *Indoor Decorating and Refurbishing Materials—Limit of Formaldehyde Emission of Wood-based Panels and Finishing Products*; GB 18580-2001; Standardization Administration: Beijing, China, 2001.
63. *Technical Requirement for Environmental Labeling Products Wood Based Panels and Finishing Products*; HJ 571-2010; Standardization Administration: Beijing, China, 2010.
64. Xia, N.; Chen, X.; Guo, K. Optimal Film Content Improving Mechanical and Water Absorption Properties of Oriented Cotton Stalk-plastic Boards. *Trans. CSAE* **2015**, *31*, 303–309.
65. Wang, Z.; Zhao, X.; Guo, W. Process Factors and Performances of Recycled Plastic-Wood Fiber Composites. *J. Beijing For. Univ.* **2005**, *27*, 1–5.
66. Bekhta, P.; Müller, M.; Hunko, I. Properties of Thermoplastic-Bonded Plywood: Effects of the Wood Species and Types of the Thermoplastic Films. *Polymers* **2020**, *12*, 2582. [CrossRef]
67. Sun, Y.; Guo, L.; Liu, Y.; Wang, W.; Song, Y. Glue Wood Veneer to Wood-fiber-high-density-polyethylene Composite. *Int. J. Adhes. Adhes.* **2019**, *95*, 102444. [CrossRef]
68. Liu, Y.; Li, X.; Wang, W.; Sun, Y.; Wang, H. Decorated Wood Fiber/high Density Polyethylene Composites with Thermoplastic Film as Adhesives. *Int. J. Adhes. Adhes.* **2019**, *95*, 102391. [CrossRef]
69. Shen, T.; Sun, Y.; Liu, Y.; Wang, W.; Shan, W. Study on Preparation and Adhesion Property of Veneer Overlaid Wood Flour/high Density Polyethylene Composites. *J. For. Eng.* **2020**, *5*, 101–107.
70. Chang, L.; Tang, Q.; Gao, L.; Fang, L.; Wang, Z.; Guo, W. Fabrication and Characterization of HDPE Resins as Adhesives in Plywood. *Eur. J. Wood Prod.* **2016**, *76*, 325–335. [CrossRef]
71. Kamke, F.; Lee, J. Adhesive Penetration in Wood—A review. *Wood Fiber Sci.* **2007**, *39*, 205–220.
72. Luedtke, J.; Gaugler, M.; Grigsby, W.; Krause, A. Understanding the Development of Interfacial Bonding within PLA/wood-based Thermoplastic Sandwich Composites. *Ind. Crop. Prod.* **2019**, *127*, 129–134. [CrossRef]
73. Fang, L.; Yin, Y.; Han, Y.; Chang, L.; Wu, Z. Effects of Number of Film Layers on Properties of Thermoplastic Bonded Plywood. *J. For. Eng.* **2016**, *1*, 45–50.
74. Bekhta, P.; Sedliačik, J. Environmentally-Friendly High-Density Polyethylene-Bonded Plywood Panels. *Polymers* **2019**, *11*, 1166. [CrossRef]
75. Li, X.; Ren, C.; Wei, W.; Zhang, S. Pressing Crafts and Mechanical Properties of Eucalyptus Veneer/Polypropylene (PP) Film Composites. *J. Northeast For. Univ.* **2015**, *43*, 87–90.
76. Xia, N.; Chen, Q.; Guo, K. Fabrication Technology Optimization of Oriented Cotton Stalk-polypropylene Film Boards. *Trans. CSAE* **2015**, *31*, 308–314.
77. Gao, Y.L. Process Optimization of Plywood Made of PVC Film and Eucalyptus Veneer. Master's Thesis, Fujian Agriculture and Forestry University, Fuzhou, China, 2018.
78. Wu, Y. Process Optimization of PVC Film and Eucalyptus Veneer to Prepare Composite Plywood. *Int. Wood Ind.* **2020**, *50*, 62–67.
79. Tian, F.; Chen, L.; Xu, X. Dynamical Mechanical Properties of Wood-High Density Polyethylene Composites Filled with Recycled Rubber. *J. Bioresour. Bioprod.* **2021**, *6*, 152–159. [CrossRef]
80. Peng, X.; Zhang, Z.; Zhao, L. Analysis of Raman Spectroscopy and XPS of Plasma Modified Polypropylene Decorative Film. *J. For. Eng.* **2020**, *5*, 45–51.
81. George, J.; Sreekala, M.; Thomas, S. A Review on Interface Modification and Characterization of Natural Fiber Reinforced Plastic Composites. *Polym. Eng. Sci.* **2001**, *41*, 1471–1485. [CrossRef]

82. Mohit, H.; Arul Mozhi Selvan, V. A comprehensive review on surface modifcation, structure interface and bonding mechanism of plant cellulose fiber reinforced polymer based composites. *Compos. Interfaces* **2018**, *25*, 629–667. [CrossRef]
83. Xia, N.; Li, F.; Liu, K.; Shao, Y.; Xing, S.; Guo, K.Q. Effects of Coupling Agents on the Properties of Cotton Stalk-polypropylene Film Boards. *Bioresources* **2020**, *15*, 8617–8630. [CrossRef]
84. Wei, S.; Zhang, S.; Fei, B.; Zhao, R. Effect of Monomer Type on Polydopamine Modification of Bamboo Flour and the Resulting Interfacial Properties of Bamboo Plastic Composites. *Ind. Crop. Prod.* **2021**, *171*, 113874.
85. Zhou, X.; Cao, Y.; Yang, K.; Yu, P.; Chen, W.; Wang, S.; Chen, M. Clean Plasma Modification for Recycling waste Plastic Bags: From Improving Interfacial Adhesion with Wood Towards Fabricating Formaldehyde-free Plywood. *J. Clean. Prod.* **2020**, *269*, 122196. [CrossRef]
86. Ye, C.; Yang, W.; Xu, J.; Chen, Z.; Liao, R.; Zhong, Z. Hot-pressing Technology of Formaldehyde-free Plywood Made by Veneers with Rolling Holes and High-density Polyethylene Film. *J. Northwest A&F Univ. (Nat. Sci. Ed.)* **2015**, *43*, 76–82.
87. Li, Z.; Qi, X.; Gao, Y.; Zhou, Y.; Chen, N.; Zeng, Q.; Fan, M.; Rao, J. Effect of PVC Film Pretreatment on Performance and Lamination of Wood-plastic Composite Plywood. *RSC Adv.* **2019**, *9*, 21530–21538. [CrossRef]
88. Gu, L.; Ding, T.; Jiang, N. Development of Wood Heat Treatment Research and Industrialization. *J. For. Eng.* **2019**, *4*, 1–11.
89. Ayrilmis, N.; Jarusombuti, S.; Fueangvivat, V.; Bauchongko, P. Effect of Thermal-Treatment of Wood Fibres on Properties of Flat-pressed Wood Plastic Composites. *Polym. Degrad. Stabil.* **2011**, *96*, 818–822. [CrossRef]
90. Follrich, J.; Muller, U.; Gindl, W. Effects of Thermal Modification on the Adhesion between Spruce Wood and a Thermoplastic Polymer. *Eur. J. Wood Prod.* **2006**, *64*, 373–376. [CrossRef]
91. Fang, L.; Chang, L.; Guo, W.; Chen, Y.; Wang, Z. Effect of High Temperature on Properties of Wood/Plastic Plywood. *China Wood Ind.* **2014**, *28*, 5–8.

polymers

MDPI

Review

A Critical Review on Wood-Based Polymer Composites: Processing, Properties, and Prospects

Manickam Ramesh [1,*], Lakshminarasimhan Rajeshkumar [2], Ganesan Sasikala [3], Devarajan Balaji [2], Arunachalam Saravanakumar [4], Venkateswaran Bhuvaneswari [2] and Ramasamy Bhoopathi [5]

1 Department of Mechanical Engineering, KIT-Kalaignarkarunanidhi Institute of Technology, Coimbatore 641402, Tamil Nadu, India
2 Department of Mechanical Engineering, KPR Institute of Engineering and Technology, Coimbatore 641407, Tamil Nadu, India; lrkln27@gmail.com (L.R.); balaji.ntu@gmail.com (D.B.); bhuvanashankar82@gmail.com (V.B.)
3 Department of Mathematics, SRM Valliammai Engineering College, Kattankulathur, Kanchipuram 603203, Tamil Nadu, India; sasigmath83@gmail.com
4 Department of Mechanical Engineering, K.S.R.M College of Engineering, Kadapa 516003, Andhra Pradesh, India; skumar3011@gmail.com
5 Department of Mechanical Engineering, Sri Sairam Engineering College, Chennai 600044, Tamil Nadu, India; bhoopathir.mech@gmail.com
* Correspondence: mramesh97@gmail.com

Abstract: Waste recycling is one of the key aspects in current day studies to boost the country's circular economy. Recycling wood from construction and demolished structures and combining it with plastics forms wood-polymer composites (WPC) which have a very wide scope of usage. Such recycled composites have very low environmental impact in terms of abiotic potential, global warming potential, and greenhouse potential. Processing of WPCs can be easily done with predetermined strength values that correspond to its end application. Yet, the usage of conventional polymer composite manufacturing techniques such as injection molding and extrusion has very limited scope. Many rheological characterization techniques are being followed to evaluate the influence of formulation and process parameters over the quality of final WPCs. It will be very much interesting to carry out a review on the material formulation of WPCs and additives used. Manufacturing of wood composites can also be made by using bio-based adhesives such as lignin, tannin, and so on. Nuances in complete replacement of synthetic adhesives as bio-based adhesives are also discussed by various researchers which can be done only by complete understanding of formulating factors of bio-based adhesives. Wood composites play a significant role in many non-structural and structural applications such as construction, floorings, windows, and door panels. The current review focuses on the processing of WPCs along with additives such as wood flour and various properties of WPCs such as mechanical, structural, and morphological properties. Applications of wood-based composites in various sectors such as automotive, marine, defense, and structural applications are also highlighted in this review.

Keywords: wood composites; biodegradable polymers; wood flour; wood adhesives; characterization; processing; applications

Citation: Ramesh, M.; Rajeshkumar, L.; Sasikala, G.; Balaji, D.; Saravanakumar, A.; Bhuvaneswari, V.; Bhoopathi, R. A Critical Review on Wood-Based Polymer Composites: Processing, Properties, and Prospects. *Polymers* **2022**, *14*, 589. https://doi.org/10.3390/polym14030589

Academic Editor: Pavlo Bekhta

Received: 31 December 2021
Accepted: 24 January 2022
Published: 31 January 2022

1. Introduction

In contrast to other plants, wood is produced by a perceiving and adapting organism, that is, the wooden body in a living tree is able to carry liquid from the foundations up to the top and to maintain structural performance of the slender branch by absorbing bending stresses caused primarily by breeze [1,2]. This results in the creation of the complicated vertical wood structure, which has an assigned role that is individual to each body. Cell membranes that are compact and the reduction of the bridge cell interior hole area are the elements for attaining structural performance [3]. Figure 1 shows the different zoom elements of the hierarchical wood structure depicting the cell wall arrangement at the nanoscale.

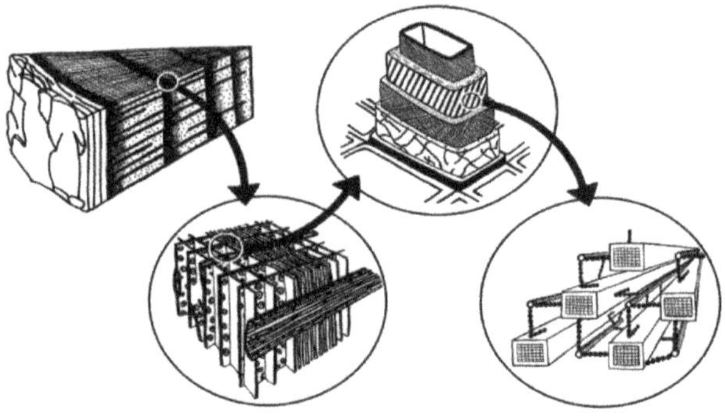

Figure 1. Hierarchical wood structure, Reproduced from [3], Wiley Online Library, 2018.

Hardwood composite materials are thermosets that incorporate a small timber in the form of powder or microfibers. They are used in a variety of applications of the dispersed phase [4]. Wood-polymer composites (WPC) are substances or items made up of one or more natural materials or flours but one or a mixture of polymers, such as polyamide, rayon, or latex. Their cheap and superior efficiency, as well as their elevated sustainable development, low moisture absorption, sturdiness against ecological impact such as insects and fungi when compared to timber, high-dimensional data stability over their entire life, and high relative stiffness, have attracted the attention of manufacturers and researchers in recent decades [5]. Natural materials and flours are derived from a variety of sources and contain a variety of polymers. As a replacement for natural wood (for example, in fencing, flooring, and decking), they are especially helpful in wet workplaces or anywhere where wood fibers interacted with liquid: the water-insoluble polypropylene (PP) separates as well as defending the soluble fibers, increasing durability and lowering the need for servicing initiatives, at least to a certain degree [6,7]. Moreover, their applicability in the fields of sound and the motor industry are remarkable. The key benefits of using hardwood as filler are the cost savings and the increase in the environmental properties of the resultant composite that it provides. The first point is self-explanatory, as hardwood particles are often low-cost materials that are commonly derived from farming or commercial leftovers. Furthermore, a significant fraction of quasi substance of coal origin is being replaced by a more environmentally friendly ingredient, often as high as 60 wt. or 70 wt.% of a more eco sustainable element, whereas if monomer is indeed a naturally derived recyclable plastic, the ecological advantage can be enhanced yet further [8–11].

Fibers as supplements, on the other hand, have some disadvantages. Firstly, there is a restricted selection of matrixes to choose from. Since the natural fiber constituents, lignin and its derivatives (hemicellulose and cellulose), deteriorate in the appearance of oxygen at temperatures of 100, 195, and 250 °C, respectively, it is broadly acknowledged that only plastics with melting points below 195 °C are suitable for use in this implementation [12,13]. In reality, the only applicants are the so-called necessities, which include polyethylene (PE) in both its dense population as well as medium carbon different versions, polycaprolactone (PCL), polyvinyl chloride (plastic), styrofoam (PS), as well as its PCL, including such injection molded plastic as well as acrylonitrile-styrene acrylate (ASA) and PE. As an extra option, some biodegradable plastics, such as polylactic acid (PLA), can be used as matrix materials, and a thermoplastic polyurethane polymer has also been successfully employed as a matrix material. Of course, if cotton or cellulose nanocrystals were utilized in place of natural fibers, a greater variety of polymeric matrix options would be available [14–16].

Stiffness and strength are only satisfactory in terms of composite when expensive coupling agents are incorporated into the material formulation; these additives are required for enhancing interoperability between both the wood chips as well as the polyurethane that also might otherwise not communicate whatever pertinent synthetic resemblances [17,18]. This is done in order to create a load-transferring contact among the natural fibers and that is effective. Because WPCs are often brittle, toughening agents including such styrene and butadiene rubber (SBR), ethylene-propylene monomers leather (EPDM), or plastic elastomers are frequently included in their composition [19–22]. Great affection among the wood fiber and the polymer matrix is crucial to enhance the physical characteristics of wood plastic composites. However, due to the intrinsic characteristics of wood products and the wettability of polymers, the two factors are entirely inconsistent with one another. It is possible to enhance the compatibility of a composite by adjusting the polymer or wood fiber in either a physical or chemical manner, or by applying compatibilizers/coupling substances [23]. Behavior and bonding chemicals have the potential to dramatically increase the efficacy of WPC. Using correctly recruited pairing entities, few authors claim that WPC tensile and flexural strengths can be increased by 2 to 4 times, depending on the test method, stiffness can be increased by up to 40%, impact resistance can be at least doubled, and density and water absorption can be increased by 2 to 4 points of time, relying on the duration of absorption [24].

Because WPCs are naturally composed of a variety of materials, it is possible to speculate that repurposing mixed waste plastics within composites will be easier than reusing them as true raw resources, and that this will result in environmental benefits compared to using traditional woman polymers [25,26]. Use of recyclable plastic in WPCs in manufacturing better recycled goods has also piqued the interest of numerous researchers, who believe that biodegradable polymers have great potential for being used in new products. However, while it has been demonstrated that behavior modification can improve the characteristics of wood plastic composites [27,28], little attention has been paid to analyzing the potential and consequences of several binder. Inside the presence of additive waste plastics, the immiscibility of diverse polymers, along with the inherent compatibility of natural wood and the polymer matrix, creates a challenge in the utilization of mix bioplastics in WPCs. Compatibilizers utilized in WPC, on the other hand, are mainly maleated/maleic grafting backsplash [29,30], which have shown better characteristics in WPCs manufactured from mixes of PE and PP. As previously mentioned, styles can vary; bonded polyolefins can also be employed again for crosslinking of copolymers, which suggests that using terephthalate compounds in WPCs derived from a combination of plastics waste might result in elevated behavior [31–33].

Another disadvantage of WPCs is their strong level of combustibility, which must be handled in order to prevent fires. Flame retardants (FRs) are incorporated into WPCs during the blending procedure, and this is the most efficient technique for modifying the flame-resistant qualities of WPCs [34,35]. In the field of flame retardants, ammonium polyphosphate (APP) is a standard, highly effective, and extensively utilized atmosphere flame retardant that has been utilized to improve the fire-retardant qualities of wood plastic composites (WPCs) [36,37]. In terms of improving the fire-resistant effectiveness of APP on WPCs, substances such as elastic graphene sheets, SiO_2, or $CaCO_3$ have been merged with APP to obtain effective fire retardancy for WPCs. Expandable graphite, SiO_2, or $CaCO_3$ were also merged with APP to obtain optimal burn resistance for WPCs [38]. Other inorganic additives such as aluminum powder and magnesium sulphate have been reported to be effective fire retardants for WPCs in addition to APP. Additional organic flame retardants have been developed to enhance the barrier characteristics of WPCs by enhancing the carbonaceous production. A large dosage of mechanical or biological flame retardants described above, on the other hand, is typically required to achieve an acceptable fire-resistance level, as stated previously [39–41]. The incompatibility of these flame retardants with the polymer matrices resulted in a decrease for both impact strength hardness, as well as an increase in fracture toughness, in the resulting WPCs. As a result, the technical advances to balance the flame retardancy of WPCs with their other qualities are critical. In prior studies, it has been claimed that the introduction of nano-fillers can enhance the flame retardancy of polymer composites while also improving the mechanical characteristics and heat resistance of those composites at the same time [34,42,43].

There is no doubt that wood-based plastic goods have significant environmental benefits that can be quantified [44]. Furthermore, materials derived from wood waste have the potential to produce low carbon-based products while putting less strain on forests [45]. One option for lowering the environmental impact of plastics and timber debris is to use these wastes to make WPC goods, such as pallets, which are made of wood fiber and polymer. However, in order to evaluate the sustainability impact of WPC pallets, a comprehensive life cycle analysis must be performed. Aside from that, it is crucial to keep in mind that various commodities have varying lifespans, recycling abilities, and recycling potential. It is defined by the International Organization for Standardization (ISO) as being one of the environmental management techniques that addresses the social and environmental risks, ecological consequences throughout a product's life cycle from raw material obtaining through manufacturing, utilizing edge treatment, recycling, and disposal practices [46–48].

In construction components, including flooring planks and screens as well as automotive components, WPCs are extensively employed [49–51]. Aside from these, some more particular applications for WPCs have indeed been identified and tried in recent years, decking, cladding, paneling, fencing, and furniture, to name a few examples. For example, WPCs have been used in the manufacturing of pallets [52]. Due to their wide range of applications, WPCs can be considered value-added materials when contrasted with other synthetic structures such as fiber cement composites. Cement-bonded composites, for example, have been employed primarily as building materials in recent decades [53,54]. Construction and demolition wood wastes (CDWs) have demonstrated their potential as a food grain feedstock for the development of WPCs. Mineral wool, in contrast to the hardwood and plastics components of CDW, has been discovered to be a viable primary resource for the production of WPCs [55]. As a result, the manufacturing of WPCs could contribute to the achievement of CDW's material recovery aim. When CDW is used in the manufacture of WPC, traditional proper disposal operations and techniques such as landfills and incinerators are eliminated. Moreover, the use of CDW as a raw resource for WPC manufacture represents a tangible step towards resource utilization because it eliminates the need for virgin recycled plastic and timber [56,57]

Some other applications of WPC can be found in the construction industry. According to the author's estimate, terrace boards account for 85% of this in Europe, the rest is mainly processed to façade panels. As these are linear products, they are extruded. WPC decking is mostly made of PP or PE, and in the case of façades the plastic-matrix is mainly PVC-based [58]. The chlorine content gives the compound a fire classification one level better, namely B1 according to German DIN 4102 [59], which is advantageous for façade applications. The modulus of elasticity is also higher, which counteracts the deformation of the material under heat. In recent years, WPC has also become interesting for other industries, namely the packaging and automotive sectors [60]. These are under increasing pressure to minimize the consumption of fossil-plastics. WPC saves petrochemical polymers, and this in the amount of the wood-fibers added. This leads to an almost balanced cradle-to-gate CO_2-balance. The reason for this is that during their growth, the wood-fibers already absorbed exactly the amount of harmful greenhouse gases from the atmosphere, which is later emitted for production in the form of energy and fuel consumption. WPC is therefore almost climate neutral. The balance is all the more positive the longer the WPC remains in an application [61–63].

Building products have a comparatively much longer service life, which in the case of façades is 30 to 50 years [64]. If the material is even recycled at the end of its life, which is already theoretically possible with WPC today but has not yet been implemented industrially, then this has an even more positive effect on the conservation of resources. In addition to technical criteria, sustainability-related aspects could be another reason for building planners to select the innovative WPC material for outdoor purposes, such as the manufacturing of façades [65]. The current review emphasizes the various processing methods of WPCs, properties of WPCs, and wood-based adhesives, their characterization methods, and applications of WPCs. Life cycle assessment and cost analysis of recycled wood composites have also been discussed in this article.

2. Processing of WPCs

A thermoplastic or thermoset polymeric matrix does seem to have been reinforced with wood in WPCs which are also two-phase materials. Based on availability as well as desired characteristics of the finished products, the reinforced wood can take any form (fibers, particles, sawdust, or flakes of wood) [66]. In addition to palm leaves and agricultural waste, WPCs can be made from a variety of materials [67]. WPCs might not have been ecologically responsible based on the polymer utilized to combine wood. Due to the lack of widespread use of thermoset polymers in WPCs, thermoplastics and their recyclability seem to be more favorable environmentally [66,68]. The addition of heat was necessary for the wetting of wood of thermoplastic in order to augment the adhesion of the two materials. The melting but rather softening temperature of hard to use plastic should not surpass the decomposition temperature of wood in the thermal blending process. Polyolefins, including polystyrene, higher and lower PE, PP, and polyvinyl chloride, have been the only thermoplastics that meet the above criteria. The compounding process is also challenged by ensuring that polymer melt disperses all through the wood particles in a uniform manner. A product's appearance and finishing quality are influenced by how well it is mixed. In case of short fiber-reinforced composites, the extent of blending is a critical parameter that can have a negative effect on the mechanical characteristics of the finished products [69,70]. As a result of ineffective or excessive mixing, the wood and plastic fibers are damaged severely. To produce WPCs, thermoplastic methods including single screw extrusion, twin-screw extrusion, injection molding, and so on are the primary methods. In addition to solid or hollow body sections and coverings, films, and foams, these items are also available [71].

First, the components (wood/thermoplastic) are combined, then they are communicated to the second stage, where various methodologies are used to form WPCs in a range of forms other than consistent profiles. Regarding the finished products' privacy properties, each technique has its own set of privacy parameters. During the production of a one-piece pallet, extrusion has a benefit over injection molding because injection molding burns wood fibers due to excessive heat generated during the high-speed injection. The extrusion process, on the other hand, generates less heat because of its lower shear. Extrusion is well-versed in a diverse variety of die designs when it comes to finished goods shapes [72,73].

The characteristics of the composite materials have been studied in relation to a number of variables, including wood concentrations, coupling agents, as well as impact modifiers [74]. The first step was to parch the WF at a temperature of 50 °C for 12 h, resulting in a water content of 20–30% of the oven-dry particle mass. The WFs were indeed ground in a rotational grinder even without the addition of any extra water after parching. Since being ground to something like a fine powder, the WF was retained in a 60-mesh screen. WF was therefore desiccated in such a furnace at 100 °C with one day to remove the moisture (1–2%) before being used in composite fabrication. After that, a 60-mesh screen was used to collect the fine powdered WFs. Prior to fabrication, the WF was allowed to dry in a furnace at 100 °C for one day to eliminate any remaining water content of 1–2% [75]. Processed in a 30 mm co-rotating twin-screw extruder, the WF, matrix, and coupling agents had a length-to-diameter proportion of 30:1. For each zone, the barrels were kept at 170, 180, 185, or 190 °C. The extruder die was kept at a temperature of 200 °C. After being extruded, the samples were placed in a water bath as well as pelletized. Prior to actual processing, the pellets were deposited inside a closed bag and thereafter left to dry to a water content of 1–2% in an oven. First, from the feed region here to the die zone, samples were heated to 180–200 °C. Process time was about 20 s and pressure was 4–5 MPa with the specimens. As per ASTM D 618 standards, the specimens then seem to have been influenced at 23 °C as well as a humidity level of 50% [76,77].

WFs were blended with coupling agent and matrix in a high velocity mixer after being oven dried at 80 °C for 24 h. Various quantities of WFs, the required quantity of PP, and the coupling agent were mixed together to prepare the WPCs. Then, the granules of WPCs were again oven-dried before molding at 80 °C for 24 h [78]. For the preparation of WFCs, a twin-screw extruder, as shown in Figure 2, with aspect ratio of 40, screw diameter of 2 mm, a screw speed of 150 rpm, and a material input of 1 kg/h was set and used. A temperature of 180 °C was set between hopper and die as process temperature. Before processing the PP, WF, and fire retardant in an extruder, they were homogenously mixed. As per JIS K7113 standards, an end-gated mold was used in a screw injection-molding machine to prepare the tensile bars which were dumbbell-shaped at a temperature of 210 °C [79].

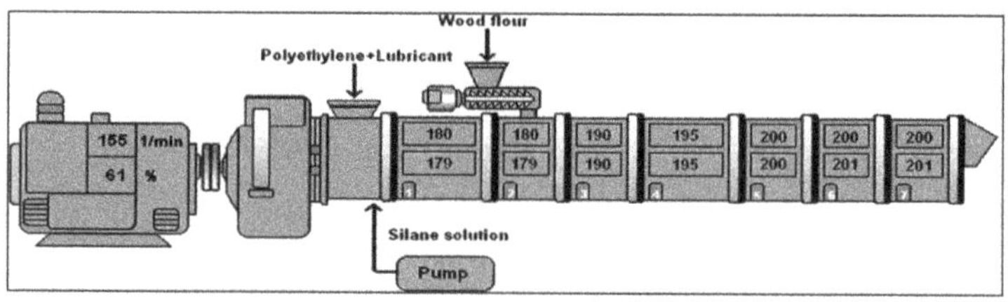

Figure 2. Extruder machine for WPC manufacturing.

Five minutes after achieving a constant torque, the inclusion of WF began. Three equal portions of WF were added over the course of two minutes. In intervals of 10 to 25 min, the kneading was kept up until a continuous torque was achieved. After a few minutes, the temperature rose to 150 °C and the combination was left to respond quickly for 20 min. After trying to cool, the WPC was cut into granules with a radius of about 3 mm [80]. The WFC was synthesized in an extruder using a temperature spectrum of 190–230 °C for the compounding process. Injection molding was used to create the compounded samples [81]. Gravimetric-type material feeder-fitted extruders were used to prepare the composites. A twin-screw slide feeder was used to enforce the WF into polymer melt while a peristaltic pump was used to feed the silane solution. In order to prevent the evaporation of water and other reactants in the melt, all atmospheric-pressure ventilations and vacuum ventilations were made impassable. Peroxides disintegrate as radicals when the silane solution passed in through the extruder thus splicing the silane into the composites. Apart from the rate of decomposition of dicumyl-peroxide, the extruder temperature also depended upon the ability to attain sufficient compounding of the material. For a residence time of 55–60 s, the speed of the screw was maintained at 155 rpm and the range of temperature was 180–200 °C. Almost 97% of dicumyl-peroxide was decomposed, which can be approximated theoretically to five half-life times if an actual process temperature for melting was kept around 195 °C. The extruded profile of the WPC was hot-pressed to a thickness of 2.5 mm at 135 °C for 15 s so as to improve the surface texture by straightening it out and was stored either in a thermal bath at 90 °C or at room temperature at 20 °C. The simulated thermal bath, whose relative humidity was almost 100% while that of the ambient atmosphere was 30%, was in the form of a plastic box that contained wires and gratings placed in an oven. The bottom of the thermal bath was filled continuously with water as it gets evaporated often. The composites, immediately after processing, were tested for their ability to cross-link. Cross-linking was prevented by the low temperature in the freezer as hydrolysis was slowed down [82].

2.1. Hot-Pressing Technique

The literature on hot-pressing shows that this process is mostly used to produce laboratory samples. Maximum temperatures should not exceed 210 °C and the pressure can be up to 50 tons on specimens with average dimensions of up to 300 mm × 300 mm.

With increasing temperature, pressure, and duration, the contact angle increases and water absorption declines. With higher material density, the mechanical strength also increases and the WPC surface appears more uniform. This method is not only suitable for the production of WPC parts from compound granulates but could also be used for the subsequent conditioning of prefabricated WPC sheets. Obviously, the material properties can be further optimized by post-operative hot-pressing under higher pressure, which would have a positive effect on an outdoor application. Some authors reported a separate process of heating up the material in a hot-press and then compressing it in a cold press. In order to use this process for the purpose of industrial "thermoforming" of WPC semi-finishings, several individual processes can obviously be linked together in a similar way. In this respect, this process appears to be superior to the hot-water- and UV-conditioning described above [83].

The hot-pressing method is the heating of WPC-pellets in a metal mold of a certain thickness, which approximately corresponds to the thickness of the sample. Charlet et al. [84] used this method to produce resin-bonded wood-fiber boards of 100 mm × 100 mm × 2 mm, which were produced under 60 kN pressure and 80 °C temperature by 2 h until the resin binder solidified. Ghani and Ahmad [85] used a hot–cold pressing process. They first pulverized WPC compounds made of rice-husks and PE and pressed them into a 14 mm × 14 mm × 3 mm mold at 140 °C and for 14 min to form small test plates, which were then compacted and cooled as "cold press" at room temperature under 1.8 tons of pressure [86]. For hot-pressing, it is important to insert a wax paper between the granules and the hot metal plates. Whether this has a distorting effect on the subsequent surface quality is not known from the literature. Benthien and Thoemen [87] used temperatures of 190 °C to 210 °C for PP-WPC in the pellet press, but temperatures above this level decompose the wood-fiber, which would have a negative effect on the mechanical strength.

2.2. Use of Additives

In addition to wood and thermoplastic, WPCs also contain small amounts of coupling agents and lubricants. These additives help the composites work better. By serving a variety of functions, additives seek to improve the final product's properties. The interfacial bonding among both hydrophobic polymers and hydrophilic woods is critical in the manufacturing of WPCs. Coupling agents because stress is converted here between matrix and fibers at the interface, the composites' intact mechanical properties are dependent on interfacial bonding. Chemical treatment is one way to improve the bonding; coupling agents such as alkali, acetyl, and silane are commonly used. Fiber treatment with silanes is possible due to the fact that one end of the silane chain reacts with hydrophilic fibers, while the other prefers to respond with hydrophobic polymers, resulting in a chemical bridge between the two ends [88,89]. Hydrolysis of alkoxy clusters on silane to water first produces silanol (Si–OH) groups, which could then respond with hydroxyl groups here on the fiber surface to form hydrogen or covalent bonds. These silanes have been studied the most extensively because of the wide range of applications that they have. Covalent bonds between silane and the matrix have been shown to enhance the hydrophobicity of natural fibers and WPCs' strength [69]. WPCs are primarily made using lubricants as additives. In order to ensure a uniform movement of the melt through to the equipment as well as a mold during the production of WPCs, the lubricant must be added to the WPC mixture during melting. This is due to the plastic's high viscosity. Because excess lubrication can lead to a lack of wood/plastic interface linkage, the characteristic is that the amount of lubrication should be kept to a minimum in order to avoid this problem and to select the proper type of handling for wood fiber composites. The oxidized PE lubricant meets the lubricant requirements in many polymer systems [90,91]. Table 1 enlists various lubricants, additives, and coupling agents used during the processing of WPCs.

Table 1. Use of coupling agents, additives, and lubricants for manufacturing WPCs.

Sample No.	Wood Reinforcement	Matrix Material	Content of Wood (wt.%)	Coupling Agent	Lubricant and Additives	Ref.
1.	Poplar wood	High density polyethylene (HDPE)	38–58	Modified PE (3%)	Cellulose nanoparticle additives (5%)	[92]
2.	Beech wood	HDPE	20–50	Modified PE (5%)	-	[93]
3.	Pine wood	HDPE	35 & 40	Modified PE (3%)	Commercial lubricant (2–10%)	[94]
4.	Eucalyptus wood	HDPE	55	Modified PE (7%)	Commercial lubricant (2–10%)	[95]
5.	Pine wood	HDPE	40 & 50	Ester (5%)	-	[96]
6.	Maple wood	HDPE	30–60	-	Thermoplastic silicone additive	[97]
7.	Maple-pine wood hybrid	HDPE	40 & 60	-	-	[98]
8.	Maple wood	HDPE	55–60	Modified PE (2%)	Ethylene bis-stearamide additive (1%)	[99]
9.	Maple wood	HDPE	30–40	-	Zinc stearate additive (2%)	[100]
10.	Lignocel wood	Low density PE (LDPE)	10, 75	-	-	[101]
11.	Birch wood	PP	40	-	Struktol additive (1.9–3.12%)	[102]
12.	Lignocel wood	PP	15, 80	-	-	[101]
13.	White fir wood	PP	35	-	Thermoplastic elastomer additives (50%)	[103]
14.	Pine-firewood hybrid	PP	50, 80	-	-	[104]
15.	Wood flour	PP	30	Modified PP (1%)	-	[105]
16.	Wood flour	PP	50	Modified PP (3%)	Thermoplastic silicone additive (1%)	[106]
17.	Bagasse of wood	Acrylonitrile styrene acrylate	50	-	Styrene butadiene rubber additives (15%)	[107]
18.	Poplar wood	Polystyrene	10–50	-	-	[108]
19.	Whitewood flour	Polyvinyl alcohol	30	-	-	[109]
20.	Wood flour	Vegetable oil derived biopolymer	20	-	-	[110]
21.	Poplar wood	Thermoplastic polyurethane	10–40	Polyolefin grafted maleic anhydride, chitosan and diphenyl methyl propane diisocyanate (5%)	-	[111]
22.	Spruce wood	PLA	10–30	-	-	[112]
23.	Pine wood	PP	20–60	Modified PP (4–17%)	-	[113]
24.	Pine wood	PP	30	Modified PP (7–9%)	Struktol additive (1–2%)	[114]
25.	Wood flour	PP	45	Modified PP (5%)	Stearic acid lubricant (3–5%)	[115]
26.	Wood flour	PP	10–45	Modified PP (2%)	-	[116]
27.	Bamboo wood	PP	30–50	Modified PP (3–5%)	-	[117]
28.	Russian fir wood	PP	50–70	-	-	[118]
29.	Poplar wood	PP and HDPE hybrid	70	Maelic anhydride (0.5–2%)	-	[119]
30.	Poplar wood	LDPE	40	Modified PE (4%)	-	[120]
31.	Birch wood	LDPE	-	-	Ethylene propylene diene monomer rubber lubricant (2%)	[121]
32.	Aspen wood	HDPE	-	Grafter wax and PE (4–8%)	-	[122]
33.	Maple wood	HDPE	30 and 40	Silane (3.5%)	Mineral additives (5%)	[123]
34.	Bamboo wood flour	HDPE	10–60	Modified PE (1%)	Paraffin lubricant (7%) and organic halides (1%)	[124]
35.	Maple wood	HDPE	70	-	-	[125]
36.	Spruce wood	HDPE	20–65	Modified PE (3%)	-	[126]
37.	Mangrove wood	HDPE	10–30	-	-	[127]
38.	Eucalyptus wood	HDPE	50	Modified PE (7%)	-	[128]

2.3. *Wood Composites from Bioadhesives*

In recent times, an excellent and updated review of wood composites as well as the polymer binders used during their production, and an updated review of wood composites and the polymer binders used in their production was published. Figure 3 shows the review's focus on the most critical aspects to keep an eye out for when creating high-quality wood composites and the leading binders currently used in the industry. Even though traditional oil-derived adhesives still dominate the wood composite industry for reasons of supply, the review found that progress has been almost unbelievable, and developments dictated by intellectual ferment are caused by a number of external constraints. Formaldehyde and other toxins are being reduced or eliminated by stricter government regulations, consumer awareness, and the resulting drive of the industry to favor more environmentally friendly materials, and eventually, the drive of the industry to decrease or even minimize their own dependence on petrochemicals, owing to the real or envisioned future reduction in oil reserves and the rise in the value of raw materials for purely transgenic products [129,130].

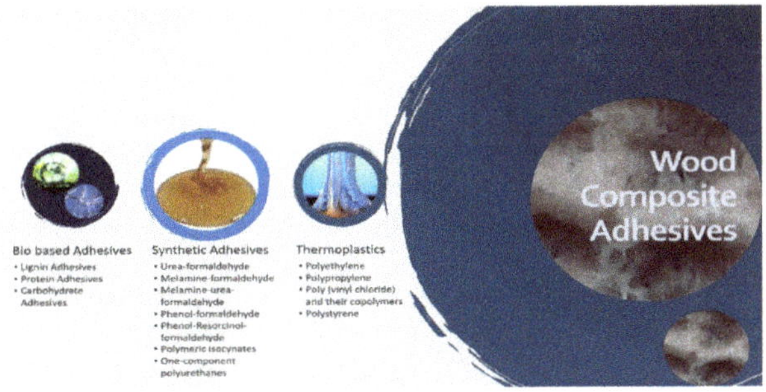

Figure 3. Adhesives for wood composites.

As a wood-based composite adhesive, UF resin has been reactively blended with various starch concentrations, esterified starch, and oxidized starch. UF-starch blends have been found to improve water resistance, formaldehyde emission, and low-brittleness properties [131–133]. UF adhesive strength has been found to be comparable to synthetic resin adhesive systems in esterified-starch combined UF and free formaldehyde content was less than 0.3%. The starch adhesive is safe for the human body and could be implemented to wood adhesion. When the UF resin, as well as modified starch, interact, they construct a net structure, which improves the water resistance of starch glue and reduces drying time [134,135]. The formaldehyde emissions can be reduced and the cost was maintained even though starch was used to replace some of the UF in the above systems [136].

Different cross-linkers have been used to enhance the performance characteristics of UF-starch combined glues. The cross-linker isocyanate was used to modify starch adhesives. Carboxymethylcellulose (CMC) and starch adhesive is polyvinyl alcohol, borax, and system for wood composites, isocyanate could be used as a cross-linker, as well as different solid components, isocyanate additions, as well as additives such as PVA and acrylic emulsion have all been researched for their effects on starch adhesive bonding strength as well as water resistance [137–140]. To increase the starch's water resistance and bonding strength, additives as well as isocyanates have been added. Cross-linking the cornstarch-UF blend framework with hexamethoxymethylmelamine (HMMM) resulted in an ecologically responsible wood adhesive. Several of the commercially available urea-formaldehyde plywood adhesives for the inside use have mechanical characteristics similar to this one [141,142].

2.3.1. Lignin

Traditionally, the majority of the world's lignin supply has come from pulping waste. Pulp, as well as paper mills, uses these lignin-derived fragments as boiler fuel because their value is so low. They have a wide range of structural units, from near-native to severely degraded, in their composition. In order to improve the adhesive characteristics of the derived adhesive, modifications as well as cross-linking must be made to the lignin structure [143]. Lignin is made up of C_6C_3 phenolic units that are linked together by cross-linking. Chemical organic molecules in lignin include hydroxyl, methoxyl, and carbonyl groups. Lignin chemical groups can be identified using Fourier transform infrared (FTIR) as well as liquid chromatography, elemental composition, pyrolysis-GC/MS, UV spectroscopy, and wet chemistry methodologies such as methoxyl content analysis as well as nitrobenzene oxidations [144–146]. Correlation between the 1H and 13C atoms remaining residues of lignin are very often found to contain cellulose, carbohydrate, and protein contaminants that can be detected using 2D NMR instead of 1D NMR [147,148].

Several studies have examined the use of lignin-based adhesives. Wood particles' self-bonding properties as well as their improved performance through enzyme treatment have both been extensively studied in the literature [149–151]. Kai et al. [152] examined how lignin functionalization could be used to develop sustainable materials such as biopolymers as reinforcement fillers, antioxidants, and UV adsorbents, antimicrobial agents, and carbon precursors as well as materials for tissue engineering and gene therapy. It is reported recently the fundamental understanding of lignin solubilization using SANS and NMR, which will be very useful for analyzing functional lignin for adhesive synthesizing. Zhao et al. [153] recently disclosed similar findings. PF (phenol-formaldehyde) resins, which are used to make plywood, have traditionally included lignin as a partial replacement for the phenol in the resin. Resins with lignin in their chemical structure have lower reactivity, which is an issue when shorter cure times are required. Even though 50% of kraft lignin could provide excellent outcomes in terms of resin viscosity, storage stability, as well as bonding capacity, the pressing time would have to be increased by 30%.

Lignin derived from those to include more species of wood (bamboo, eucalyptus) has no effect on the bonding properties of plywood when used in commercial mixes, but only 15% of the phenol in commercial mixes can be replaced by lignin. Some researchers have claimed that lignin, glyoxal, as well as pMDI or tannin can completely replace PF resin in fiberboard production [154,155]. MDI isocyanates respond with lignin's phenolic as well as aliphatic hydroxyl groups to form urethane clusters. Lignosulphonate was combined with glyoxal and pMDI by Mansouri et al. [156]. It is possible to make particleboards with strong internal interfacial adhesion using lignosulphonate-glyoxylate lignin. While glyoxal is not toxic, it is less responsive than formaldehyde and therefore less useful. Lignin cross-linking mechanisms shown in Figure 4 resulted in self-bonding within the molecules.

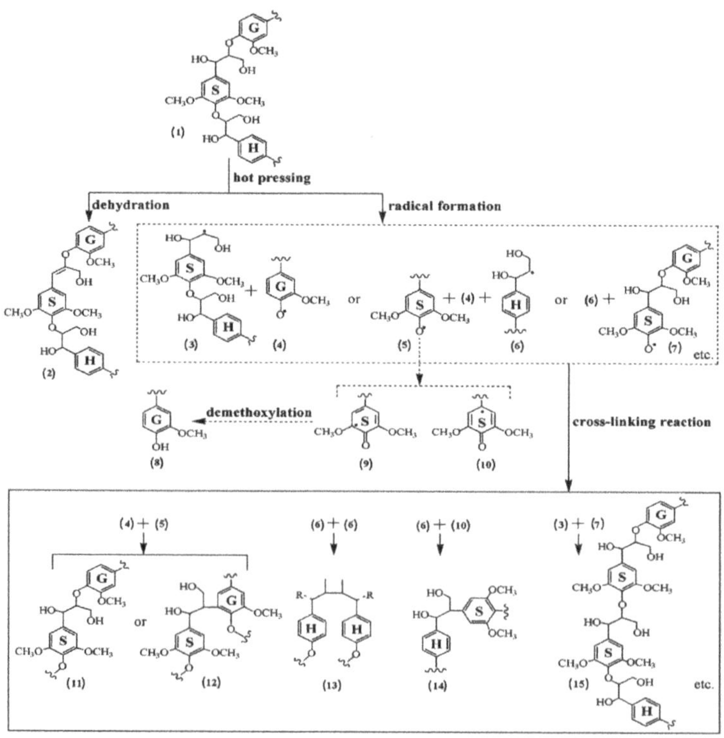

Figure 4. Cross-linking mechanisms for lignin, Reproduced from [157], ACS Publications, 2017.

2.3.2. Tannin

Hydrolysable tannins as well as condensed tannins are two types of natural polyphenols called tannins that are hydrolysable and condensed, respectively [158]. Many plant species contain tannins, and only very few have that much accumulation to make extraction worthwhile. There are a wide variety of plants that contain tannins, including pine, oak, and chestnut, as well as wattle, eucalyptus, myrtle, maple, birch, and willow. The adhesive characteristics of tannin extracts can be affected by different extraction methods. Powdered tannins are the end result of extracting plant material, purifying the isolates, and spray drying it [159]. Pectins, sugars, amino acids, other polymeric carbohydrates, and some other substances are also extracted [160].

In the production of phenol-formaldehyde resin, hydrolysable tannins have really been effectively used more as a substitute material for phenol (up to 50%) [161]. Nevertheless, their low macromolecular structure, low phenol substitution level, limited global production, and relatively high price end up making them less interesting in comparison to compressed tannins [154,162]. Tannin concentrates with a yearly output of more than 200,000 tons account for as much as 90% of the world's advertising tannin manufacturing [160]. Because of the polyphenol content as well as formaldehyde reaction of these reactive polyphenols, tannins can be used as wood-based panel glue. With formaldehyde as a cross-linker, tannins could be used alone or in a mixture with amino-plastic and phenolic resins to form adhesives. For interior grade MDF, tannins can be used to replace some or all of the phenol in phenol-urea-formaldehyde resins as well as generate 100% tannin resins [163,164].

Diverse tannin sources and timing of tannic inclusion in the production chain of MDF manufacture (for example, during deliberation of wood shavings) have been proven to have an impact on the features of resultant panels [131]. Gonultas et al. [165] and Ucar et al. [166] created and evaluated glue compositions for wooden purposes that were made with a variety of binder and tannin granules from Turkish red pine bark. By analyzing the formaldehyde interaction including both concentrated and hydrolyzed tannin, Ozacar et al. [167] discovered a thermoplastic tannin-based hardwood glue solution that may be used to bond wood. Some research has concentrated on generating resins that are completely clear of formaldehyde by mixing tannins with the other bio-based materials, such as protein, for instance [168]. With the development of particleboard adhesive that is based on tannins, which are extracted from industrial lignocellulosic wastes such as chestnut shell and chestnut bur, as well as eucalyptus bark and leaves, Santos and colleagues [169] explored the feasibility of simply eliminating formaldehyde from adhesive formulations. Nath et al. [170] investigated the qualities and prospects for the manufacturing of particleboard using a tannin-based adhesive derived from mangrove species, and their findings were published in Nature Communications. Cui et al. [171] used cellulose nanofibers in tannin-based glue for plywood manufacture, and they found a significant effect on the mechanical qualities of the panels generated as a result of their experiment.

Formaldehyde emissions from tannin adhesives are quite low due to the phenolic composition of the adhesive. The uses of non-emitting hardeners, as well as tannins which are healed by auto-condensation in the lack of aldehydes, have been shown to further reduce emissions [172]. The autocatalytic hardening of reactive tannins, such as procyanidins, can occur without the use of an external catalyst in the presence of these tannins. It is possible for softer tannins, such as condensed tannins, to undergo auto condensation if a little quantity of alkaline SiO_2 is present at a high pH, but this is rare. During the process of auto-condensation, the O_1–C_2 connection of the phenolic recurrent cycle is broken, which causes auto-condensation among the reactive C_2 of the free link and free spots in the phenolic unit of some other polymer [154,173]. Auto-condensation may therefore happen at ambient temperature when the pH of the tannin adhesive is elevated, resulting in an increase in the stiffness of the tannin adhesive according to the pH requirements of different tannins; for example, mimosa tannins' auto-condensation happens at alkaline pH, and *Acacia Nilotica* spp. tannins' auto-condensation occurs at their initial acidic pH [174].

2.3.3. Starch

Starch is one among the naturally occurring polymeric materials and has gained research focus in various applications such as additives, food packaging, adhesives, and papermaking industries owing to its advantageous features such as cheap cost, better adhesive nature, abundant availability, and renewability [175–178]. Starch is made of two different polysaccharides such as amylopectin and amylose while these two polysaccharides are constructed with glucose of various sizes and shapes. Properties of the wood adhesives are governed by the ratio of amylopectin and amylose present in the starch which in turn depends on the biological origin of the starch [176]. Properties of starch-based wood adhesives were influenced by hydrogen bonding and these bonds were relatively weaker when compared with the chemical bonds. Poor resistance towards water absorption is also due to the easy bond formation of hydrogen bonds with water molecules. Hence, modification of starch for improving its resistance towards water is important so that it can be widely used in adhesive-based applications [179]. It was stated in many research works that the wear resistance and bond strength of starch-based wood adhesives improved when they were combined with compounds like formaldehyde, tannin, polyvinyl alcohol and isocyanates.

Few studies focused on preparing and grafting starch-based wood adhesives with vinyl acetate while ammonium persulfate was used as an initiator [141,180]. It was stated in the study that graft efficiency played a significant role and influenced the bond performance of starch-based wood adhesive [181]. Esterification of starch is the simple and conventional chemical modification which converts the hydroxyl group of the starch into an ester group so that the hydrophobic nature of the starch was enhanced. Some authors prepared an esterified corn starch by initializing a reaction using maleic anhydride and cross-linking the same using polyisocyanate pre-polymer. It was noticed from the results that the optimal pre-polymer proportion was 10% by weight which resulted in 12 and 5 MPa of dry and wet shear strengths, respectively [182]. Other researchers prepared the starch-based wood adhesive using an auxiliary agent along with blocked isocyanate. A few researchers synthesized eco-friendly starch-based bioadhesive for wood panel manufacturing through the grafting of vinyl acetate in corn starch and making the compound to undergo cross-linking polymerization reaction with methylol acrylamide. Hydrophobicity of the above obtained adhesive was measured to be more than 1 MPa which was attributed to the complex network structure formation and enhanced density of the density due to cross-linking [177,183].

In some other studies, the reaction of urea-formaldehyde resin with various proportions of bio-based starch, oxidized and esterified starch to formulate wood and wood-composite adhesives. It was noticed from the tests that the blending of starch with urea-formaldehyde systems rendered lesser formaldehyde emissions and low adhesive brittleness [184,185]. Meanwhile, blending urea-formaldehyde with oxidized starch rendered an adhesive with resistance towards high temperature, oil, chemical, and aging, and better insulation properties [186]. A few researchers tried to enhance the properties of the starch-based adhesives by adding suitable additives or fillers into them. In some of the experiments, silica nanoparticles were used as fillers in starch-based adhesive to enhance the bonding strength, hydrophobicity, and thermal stability. It was noted from the results that addition of 10% by weight of silica nanoparticles resulted in shear strength values of 2.98 and 5.12 MPa in wet and dry conditions, respectively. From all the above discussions, it could be seen that starch-based adhesives can be a potential bio-based adhesive for the processing of WPCs. Table 2 lists various types of bio-based adhesives used for the manufacturing of WPCs along with their processing parameters and strength values [187,188].

2.3.4. Soy Protein-Based Adhesives

Soy protein has been used as an adhesive since the early period but its utilization as a bio-based adhesive began only during late 1930s. After being commercialized, soy protein adhesives were used as adhesives for paper and wood and in paints and coatings, it was used as a binder. Caustic treatment was used for the denaturation of soy-protein adhesive when it was used as an adhesive for manufacturing plywood. However, the main disadvantages when they were caustic-treated include less solid content, higher hydrophilicity, low biological stability, and smaller pot lives which restricted their usage only in interior applications [189]. In the early 1970s, all the soy-protein based adhesives were replaced by a mix of soy-protein adhesives with synthetic adhesives such as urea-formaldehyde and phenol-formaldehyde. These adhesives exhibited better resistance towards moisture and higher strength but they were less biodegradable.

Usually, it was seen that soy-protein adhesives were relatively cheap with lower pressing temperature and ease of handling. They developed a bond among the WPCs with relatively higher moisture content. It was also noticed that soy protein adhesives had lesser pot lives and high viscosities, thus resulting in the formation of WPCs with lesser resistance towards water, low strength, and lower biological degradation [190]. Soy protein adhesives were also sensitive towards changes in pressing conditions, pH, temperature, and ionic strength while the properties of the adhesives were greatly governed by the content of protein present in it [191]. Viscosity of the soy protein adhesives were reduced by protein hydrolysis or by the usage of lower content of solids during its processing. Hydrolysis results in disintegration of macromolecules of protein into small fragments, thus ending up in lower bond strengths and this treatment can be carried out either by enzymatic reaction or hot caustic treatments [192]. Usage of soy protein adhesives to fabricate particle boards from straws instead of fabricating them from woods was the least studied in much of the research. Since straw surface contains wax and silica in considerable quantities, they are more hydrophobic and hence soy protein adhesives suit better to them when compared with formaldehyde-based adhesives [193].

Table 2. Processing parameters of bioadhesives for WPC manufacturing.

Sample No.	Wood Panel Material	Main Adhesive Element	Time of Pressing (min)	Pressing Temperature (°C)	Density (kg/m³)	Bending Strength (MPa)	Shear Strength (MPa)	Ref.
1.	Particle board	Lignin	8	200	710	0.38	-	[194]
2.	Particle wood	Corn starch	-	-	-	-	7.5	[195]
3.	Particle wood	Corn starch	1640	30	-	0.35	-	[196]
4.	Particle wood	Kraft lignin/Soy protein	10	180	-	-	6.5	[197]
5.	High density fiberboard	Corn residue and cationic starch	8	235	1130	0.45	-	[198]
6.	Particle board	Pine-based tannin	8	200	-	0.99	-	[199]
7.	Particle board	Maritime pine-based tannin	8	230	670–690	0.46–0.52	-	[200]
8.	Particle wood	Modified soy protein and Sorghum lignin	11	212	-	-	6.22	[201]
9.	Particle wood	Modified soy protein and Extruded sorghum lignin	11	212	-	-	5.78	[201]

Table 2. *Cont.*

Sample No.	Wood Panel Material	Main Adhesive Element	Time of Pressing (min)	Pressing Temperature (°C)	Density (kg/m^3)	Bending Strength (MPa)	Shear Strength (MPa)	Ref.
10.	High density fiberboard	Lignin from softwood kraft	17	180	1345	0.67	-	[202]
11.	Particle board	Mimosa tannin and pine-based tannin	8	200	717	0.36	-	[203]
12.	Particle board	Tannic acid powder	21	210	1060	0.51	-	[204]
13.	Particle wood	Commercially condensed tannin	7	150	471	-	0.99	[205]
14.	Particle board	Citric acid sucrose	11	210	910	0.38	-	[206]
15.	Medium density fiber board	Modified mimosa tannin	6.5–8.5	165–190	750–850	0.7	-	[207]
16.	Particle board	Organosolved lignin	5	228	720	0.81	-	[208]
17.	Particle wood	Water washed cottonseed meal	22	110	-	-	4.56	[209]

Polyamides are the most frequently used agent for inducing cross-linking reaction in soy protein adhesives. Yet, due to the availability of low solid content and high viscosity in polyamides, alternative agents for curing were also addressed in various studies [191,193]. Some studies reported the usage of a new curing agent obtained from ammonium hydroxide in water and epichlorohydrin while manufacturing plywood panels from soy flour. The hydrophobic nature of the adhesive was enhanced by the incorporation of NaOH during the adhesive formation. It was noted from the results that ionic bonding was formed from the modified soy protein adhesive when the adhesion process happened in the WPCs at higher pH values [197,210,211]. Figure 5 shows the schematic of cross-linking mechanisms of soy protein adhesives adopted from natural mussel and gecko-based adhesives.

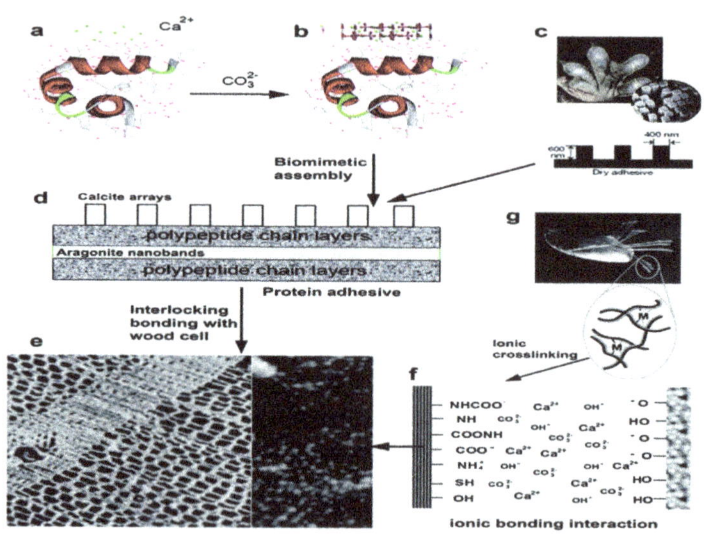

Figure 5. (**a**,**b**,**d**) Synthesis of soy protein adhesives, (**c**,**g**) gecko and mussel adhesive proteins, (**e**,**f**) mechanical and ionic crosslinking mechanisms at the surface, Reproduced from [211], Royal Society of Chemistry, 2017.

2.4. Post-Treatment of Wood Composites

In order to understand the application-based development of WPCs, it is very much essential to know about the various post-treatment processes involved in treating WPCs. Two predominant methods prevail: post-treatment using ultraviolet (UV) radiation and using boiling water. UV-lamps are also used to prove the aging resistance of WPC in climatic chambers. The research literature offers comparatively more studies on this subject, and they report a significant color change as fading and brightening of the material with a correspondingly long exposure time [212,213]. Chaochanchaikul et al. [214] derived a decreasing hydrophobicity of the WPC-surface from reduced contact angles after 720 h of artificial weathering under UV-lamps with 313 nm wavelength. This is at first in contradiction to Khan et al. [215] which, however, shows that an initial hydrophobicity of the plastic-surface is reversed with prolonged exposure, which can be explained by increasing photo-oxidation and roughness. Peng et al. [216] reported obvious micro-cracks on the surface in this context.

The research on UV-irradiated WPC shows that the material not only heats up as a result of the irradiation energy, but also has positive effects on essential mechanical material-properties, which comes from the cross-linking of polymer chains. This would even increase the performance of WPC in outdoor applications, at least during their early life-phase. More importantly, however, it is possible to reach temperatures at which plastic deformation is feasible under UV-exposure with correspondingly high dosage and exposure-time. In the research literature on UV-conditioned WPC to date, however, this has never been a central question, which is why more precise information on surface and material temperature as a function of intensity and duration is lacking. Investigations on this subject with the aim of a "post-heating" process would thus complement the existing literature. However, material heating by means of UV-lamps takes time and can have a damaging effect if carried out too intensively [217,218].

The current research literature on the boiling test in connection with WPC samples consistently reports maximum heating temperatures of 100 °C under atmospheric pressure and maximum duration of 2 h. This results in significantly higher water absorption and swelling of the material due to the hygric properties of the wood-fibers. There is also a destruction of the outer polymer layer, which favors the absorption of water by the fibers behind. How other mechanical and physical material properties react to hot water immersion is hardly reported [219]. At least in the case of roughness, a significant increase can be assumed. Due to the limited maximum temperature of 100 °C, or 120 °C under overpressure, sculptural forming of WPC appears to be more difficult. The small amount of information from the literature would justify further studies on the physical and time-dependent material behavior due to boiling water storage, but the long conditioning time does not seem very practical for the purpose of industrial thermoforming [220].

Conditioning WPC in boiling water is explicitly mentioned in the material testing standard DIN EN 15534-1 [221] and serves to prove the surface quality, especially for coating with synthetic resin-paints or bonding with veneers. The maximum temperature boiling water can reach is 100 °C, which initially limits the heating of WPC. Cooking under higher pressure, as usually used in vapor pressure-boilers when preparing food, generates temperatures of up to 120 °C. This would already allow plastic deformation of WPC. However, the research literature hardly reports about this conditioning process. Alnajjar et al. [222] carried out the boiling test on WPC-60Wood/40PE. The authors demonstrated 30% to 50% more water absorption of the WPC-samples by boiling for 2 h than by water immersion below room-temperature. Similar results are reported by Mohamed et al. [223] on WPC-70Spruce/30PE. The reasons for the increased water absorption were investigated in more detail by Li et al. [224] on WPC-50 Bamboo/50PE. The authors found that boiling damages the fiber-matrix bond, which promotes water absorption in deeper fibers and makes the surface rougher.

3. Properties of WPCs

An injection molding process and extrusion were used to prepare the WFCs. Investigation of mechanical and thermos-dynamic properties was carried out with WF loading as a function. Characterization techniques such as scanning electron microscopy (SEM) and differential scanning calorimetry (DSC) were used to interrelate the morphology of the fiber/matrix interface with the effects of fiber-silane thermo-chemical vapor deposition treatment for WF and a maleic anhydride (MA) graft co-polymerization technique. Theoretical models for the stiffness of the composites such as the Halpin–Tsai/Tsai–Pagano micro-mechanical model were used to compare the results in order to investigate the MA enforcement into fiber/matrix interface and the induced phenomenological effects of silane coupling agent [225].

3.1. Mechanical Properties

A possible solution to enhance the mechanical properties of natural fiber-reinforced composites, suggested by several researchers, is the hybridization of fibers with inorganic fillers. The mechanical properties of WPCs may not be significantly affected by fiber type, but it has also been reported that the type of fiber and the lignin, cellulose, and hemicelluloses content have a strong influence on mechanical properties. Addition of WF enhanced the mechanical properties of composites, but simultaneously it increased the burning speed of the materials [226–228].

3.1.1. Tensile and Flexural Properties

The mechanical properties such as tensile strength, elongation at break, toughness, fracture energy, and Young's modulus of the WPCs was measured with an Instron tester (Model: 4201), Zwick/Rowell model Z010, universal testing machine (UTM) following ASTM D638, ISO 527-2 standards with the speed of cross-head as 10 mm/min [229–231]. The standard dimension is 20 mm length, 12.5 mm width, and 3 mm thickness [230]. The impact strength of the specimen was tested by using an Impact tester (Model: TMI 43-01) following ASTM D 256 standards. According to ASTM D 790 and ASTM D 7264 standards, the flexural tests were carried out at room temperature with a load capacity of 10 kN in Zwick/Rowell model Z010 UTM [229,231]. According to ASTM D 790-86 standard, a specimen of dimension 70 × 12 × 3 mm^3 were prepared for a three-point bending test and compression test conducted at room temperature in UTM with a cross head speed of 1 mm/min and 0.5 mm/min respectively [229]. A cross-head speed of 2 mm/min, support span of 140 mm, and a square plate specimen of dimensions 200 mm × 30 mm × 10 mm were the requirements and parameters for a four-point flexural test. For evaluating Young's modulus through rate of strain, strain gauges were used. A constant temperature of 23 ± 1 °C and a relative humidity of 50 ± 5% were considered while taking all mechanical measurements [230,232].

The tensile strength, tensile modulus, and strain at break of the composites were analyzed. The product of cross-head speed and time gave displacement which was used to calculate the rate of strain [233]. All the prepared specimens had an approximate aspect ratio of 2 and were square in shape. Molybdenum sulfide wax was used to lubricate the precisely machined parallel faces of the specimen. The graph depicted a huge fall of strength and stiffness with respect to time for all compositions of WF. The interfacial adhesion between the fiber and matrix was majorly reduced due to the hydrophilic nature of the composites. This indirectly affects the stress transfer between the fiber and the matrix due to poor interfacial bonding. When a newer bond forms between cellulose and water molecules, the intra-molecular hydrogen bonding was diluted [82,234].

3.1.2. Impact Strength

How well a façade material absorbs impact loads from hail and storms can also be expressed by its impact strength according to DIN EN ISO 179-1 [235]. It describes the amount of impact energy in Joule, which the material can absorb before it breaks. The impact strength test is performed in the literature on WPC as the Charpy impact test on unnotched specimens. In this test, an oscillating hammer pendulum hits the sample with a certain kinetic energy and breaks it around its weak axis, i.e., its thickness. The fracture energy can be determined from Δh, which is the difference in pendulum height before and after the pendulum passage, and from the hammer weight m. Nevertheless, from the research literature, it is evident that the impact strength decreases with increasing fiber-content, but this reduction is assumed to be of a higher level in the case of hardwood [236,237]. With regard to thermoforming of WPC under hot-pressing, the following hypotheses appear interesting for an experimental clarification: H-10: Type-B specimens exhibit higher impact strength due to hardwood-fibers at comparatively lower share. H-11: Hot-pressing increases the impact strength in A- and B-specimens due to higher surface hardness. Confirming the hypotheses would mean that hot-pressing can compensate for the disadvantage in impact strength resulting from the use of cheaper softwood-fiber or higher fiber-content [238].

3.1.3. Surface Hardness

The surface hardness of a façade material is also a criterion of usability. This is because permanent defects can become visible on the building skin as cavities, especially due to mechanical influences, e.g., branches falling in storms. Similarly, shock-like impact of hard objects can cause cracks or even fractures in the façade. The Brinell hardness is described by DIN EN 1534 [239] using a Lloyd's test machine. Mostly, a 10 mm carbide ball and 300 kg force over 10 s on 50 mm × 50 mm samples is applied. After pressing the carbide ball into the surface of the material, the Brinell hardness is calculated from the measured force and the surface of penetration. The current research literature on hot-pressed WPC-specimens reports surface changes only in the context of microscopic examinations and describes the effect of fiber content on density and surface hardness [240]. After intensive literature research, there is no study which directly links the relevant parameters under hot-pressing with those of surface quality. Nevertheless, there are some sources which report WPC-investigations without aspects of temperature-related preconditioning. Hot-pressing could bring about an optimization here. Therefore, the following hypotheses should be tested in connection with surface quality: H-8: Type-A samples have a lower surface-hardness due to higher fiber content and softwood. H-9: Hot-pressing increases the surface-hardness of both types due to higher material density. If the surface strength is increased by hot-pressing, this increase can be used for a higher fiber content, which makes the WPC-compound in the façade more sustainable [238].

3.2. Physical Properties

When determining the decisive material properties of WPC for exterior applications, mechanical parameters are initially obvious. Previous research mainly reveals that for WPC, as a wood-based material in façades, mechanical strength is decisive, and in particular the bending and fastening mechanisms. However, these properties are mostly correlated with the hygric properties of WPC [241]. The high bio-part, which in the case of WPC is fixed with the wood fiber content, has a negative effect on the physical properties, and this in turn feeds back into the mechanical resistance-values. Finally, there are also optical criteria, which can be expressed in terms of the material's tendency to change color and shape, and which in turn are a result of hygric material-behavior. Obviously, physical parameters have a high informative value in material characterization towards "post-heating" WPC for thermoplastic forming. Few prominent physical properties such as water absorption capacity and material density play the main role when it comes to assessing the physical behavior of WPCs [238].

3.2.1. Water Absorption Behavior

Increased water absorption of WPC material can significantly reduce the durability, which makes the façade application inefficient. Whether and how much water a WPC-element absorbs depends on the surface quality. As already shown with contact angle method, the wettability gives an indication of the degree of permeability for moisture. However, investigations into water absorption must also show how quickly water diffuses through the material and when saturation occurs. The method for measuring water absorption is defined by test standard ASTM D 570, whereby water absorption is expressed as the %-actual change in sample-weight over time. Water absorption tests were conducted on the specimen which was oven-dried at 105 °C for 24 h. At a normal temperature of 23 ± 2 °C, the specimen was submerged in water for 2 h, after which they were wiped off with a cloth to remove the superficial water and then weighed. This process was repeated after 2 h again and weighed after 24 h [231,242]. Results of water immersion tests of MAPP and PP WF composites for 2 h and 24 h followed same pattern of increase in water absorption as the content of WF increases due to their hydrophilic nature. However, the results were just the opposite for the wood samples because of its hydrophobic nature whereas the water absorption could be more due to the presence of more spots as wood content increases. Attributes such as the presence of lumens, hydrogen bonding sites, and the interfacial gap between reinforcement and matrix were the main reason for water absorption in WF composites [243].

Immersion tests were conducted on rectangular specimens of dimension 35 mm × 12 mm × 3 mm to determine the moisture content as per DIN 52375, ASTM D 570-81 [230] standards. At least three specimens for each sample were taken for the test and their volume (from the specimen dimensions) and moisture content were determined [185]. Within a time of 24 h, different readings of weight gain were taken from the specimen soaked in distilled water at room temperature before which they were conditioned to reach a constant weight. When the readings reach saturation by giving the same value in three consecutive measurements, the final weight gain was calculated [230]. Different pressures were deliberately generated to vary the density of the pores and to demonstrate effects on hygric material properties. Correlations between water absorption and temperature, pressing force, and duration have not been investigated so far. Nevertheless, it seems plausible that, at least with increasing pressing force, the density will be higher, and the surface will seal itself at higher temperatures by closing cavities. Both should be reflected in reduced water absorption-coefficients. Therefore, the following hypothesis is formulated: H-4: Type-A zero samples show higher water absorption due to the larger fibre-content. H-5: Hot-pressing leads to reduced water absorption for both types due to closed surface pores [238,244].

3.2.2. Density

Density already played a major role in water absorption and seems to correlate negatively with this property. It is also regularly determined during physical tests on WPC-formulations. It was applied in a study by Bekhta et al. [245] as a dependent variable during hot-pressing of solid wood-veneers, and the authors demonstrated an increasing density with rising pressing temperature and force, which had a positive effect on the hydrophobicity of the veneers. Hoong and Paridah [246] achieved with both parameters higher bond-strengths between wood-chips and the resin-matrix in their explanatory model. When the pressing time was increased from 18 min to 20 min, the fracture modulus already grew by 25%. Research on the density of WPC was also conducted by Stark et al. [247] on WPC-50Pine/50PE, and in a similar way as the cited solid wood-veneers of Bekhta et al. [245]. Likewise, to Hoong and Paridah [246], a higher density led to higher strengths with respect to modulus of rupture and the stiffness. There is a significant difference in fiber-content between the sample types. Here, too, it is to be expected that Type-A, with 20% more fibers, will result in a lower density also because softwood is used. Regarding density, the following hypothesis can be formulated for a study: H-6: Type-A samples have a lower density due to the higher fiber content and softwood. H-7: Hot-pressing leads to a higher density for both types [238].

3.2.3. Contact Angle and Color Change Measurements

In the contact angle method, a liquid drop, usually 5 μL to 10 μL of distilled water, is applied to the surface of a solid sample, e.g., WPC. The angle between the liquid and solid surface is the contact angle. The hydrophobic property of the surface can be measured from its size. In the research literature on WPC, the contact angle is used as a measure of the weather resistance of WPC [248]. For angles $\theta \geq 90°$, the surface is considered non-wettable and water-repellent, which in the case of WPC indicates lower water absorption. Since this may be a temporary phenomenon, angle measurements must be carried out as a function of time. The wettability of the composites increased with increasing WF content. The incorporation of the coupling agent decreased the wettability of the composite specimens compared with untreated ones. The test result showed that WF could be utilized in the production of the filled composites because of the surface properties of the composites [84,249].

The research literature shows that the wettability of WPC increases with higher surface roughness and exposed wood-fibers, i.e., contact angles decrease. This is due to the hydrophilic properties of the wood-fibers and the increasing degradation of the polymer chains when exposed to UV-light under weathering. If hot-pressing now makes the WPC-surface smoother and denser and additionally causes thermal decomposition of wood-fibers close to the surface, as stated by Ayrilmis et al. [86], then the wettability should decrease with increasing temperature, pressing time, and pressure, and contact angles should become larger. This has a positive effect on a façade application with thermoplastic WPCs [250]. The following hypothesis can be formulated: H-2: Type-A samples have a higher wettability due to the larger fiber content. H-3: Hot-pressing leads to higher contact angles and lower wettability for both types [238].

According to relevant research literature, color changes in test specimens after conditioning are measured using the color space system (CIELAB). For this purpose, appropriate spectrophotometers are used, e.g., Minolta CM-2500d [25], StellarNet EPP2000 [212], or Macbeth Colour-Eye 7000 [213]. The CIELAB-method calculates color differences in brightness, which ranges from 0 (black) to 100 (white). The values range between −300 and +300 in each case, meaning the further away the value is from 0, the more intense the color tone. The total effect is expressed by index values. Most reported change-effects describe a fading of WPC test specimens, mostly due to UV-light [251]. In this regard, the characteristic value ΔE stands for the total alteration according to ISO 7724 [238,252].

3.2.4. Interaction Behavior

The properties of WPCs are highly dependent on the interaction between wood and polymer [95]. The interaction behavior of WPCs reinforced with linear LDPE was evaluated by Okasman [253]. To improve the interfacial bonding between the matrix and the wood flour, various compatibilizers were added. He conducted tensile testing, impact testing, and SEM analysis. From the results, he found that the mechanical strengths of the WPCs were improved significantly due to better interfacial bonding between wood flour with the LDPE matrix. The influence of the coupling agent and lubricants on the crystallization at interphase regions of wood/PP composites was studied and it was found that there is no difference in the kinetics of the crystal formation and also lubricants that tended to interfere with wood–PP coupling dispersed throughout the transcrystalline region around the fiber [254]. The enhanced dispersion of wood flour particles in the HDPE matrix, ensures efficient interfacial adhesion between filler and matrix which improve the properties significantly [255].

3.3. Thermal Properties

Evaluation of thermal properties of WPCs such as thermo-gravimetric analysis (TGA), differential scanning calorimetry (DSC), and flammability or fire retardancy behavior is currently considered to be important since most WPCs are subjected to outdoor applications and some temperature-influenced environments. The following sections deal with the various studies conducted on WPCs to assess their thermal behavior.

3.3.1. Thermo-Gravimetric Analysis

TGA was used to investigate the thermal behavior of the WF, matrix, and their resulting composites. The tests were conducted with 15 mg mass of sample under a flowing nitrogen atmosphere with a heating rate of 10 °C/min between 35 °C and 700 °C according to ASTME1131-08standards [256]. TGA/DTA measurements were performed on a Netzsch STA 409 TG analyzer. A sample was cut into small pieces and conditioned in vacuum at room temperature for 24 h. Degradation of WPCs usually occurs at a large temperature spectrum at both nitrogen and oxygen atmospheres. However, in contrast to the polymer matrix, some mass loss of WPCs in nitrogen atmosphere occurs in the temperature ranges 30–160 °C, 170–230 °C, and 410–590 °C while the mass loss of the polymer matrix occurs mainly from 310 °C–490 °C. Wood degradation in the temperature range from 200 to 350 °C is attributed to the disintegration of hemicellulose and cellulose while from 250 to 500 °C, it can be attributed to the disintegration of lignin [4,257]. Wood and polymer matrix degradation overlap between 200 °C and 350 °C which means that step separation for wood and matrix could not be achieved. However, when WF was degraded between 200 °C and 390 °C, then the matrix was degraded between 390 °C and 500 °C, and finally, by changing the inert atmosphere to oxidative atmosphere, the char of WPCs burned residue-free [256,258].

3.3.2. Differential Scanning Calorimetry

The melting and crystallization behaviors of WPCs were assessed through DSC using a Perkin Elmer, Mettler-Toledo DSC 821 calorimeter apparatus equipped with a cooling attachment, under a nitrogen atmosphere. The composite samples were heated from −50 °C to 185 °C at the rate of 3 °C/min before which 10–15 mg of the samples was preheated to 50 °C for 5 min. Afterwards, the samples were glazed 20 °C/min to room temperature. Two heating steps interspersed with a cooling step from 20 °C to 210 °C at a constant rate of 10 °C/min were carried out. The samples were analyzed in standard aluminum DSC pans [257,259]. Thermal properties of the composites such as fusion/crystallization enthalpies Hc and Hf and melty (Tf) and cristally (Tc) temperatures were examined using a non-isothermal DSC analyzer. Alternative heating and cooling of the samples was done between 210 °C and −60 °C in two stages; heating to 210 °C and cooling them to −60 °C as first stage and heating the samples again from −60 °C to 210 °C as second stage, at a rate of 10 °C per minute [260]. The heating process was performed twice; glass transition temperatures were obtained from the second run. DSC thermo-grams depicted the development of double melting peaks, transference of the melting temperature, and abridged heat of melting depicts the imperfect polymer crystal structure when WF is impregnated into the matrix. During the transition from molten to crystalline state, WF enables the fast formation of transcrystalline structures along the fiber surfaces which portrays WF as good nucleating agents. Due to WF impingement on bulk formed crystals and disordered crystal growth at the interface of matrix/fiber, some discontinuities originate in the matrix crystal structure which led to the reduction in whole polymer crystallinity [225].

3.3.3. Fire Retardancy Behavior

In engineering applications, fire retardancy is an important behavior if WPCs are considered as a material for construction. WPCs result as combustible and fire-prone materials since both wood and polymer materials are inherently combustible in nature. In order to mitigate this issue, enhancement of flame retardancy of WPCs deems prior importance. Usually, resistance towards fire can be achieved by ignition through fire or developing the resistance to the propagation of fire. Fire retardancy in a material can be developed basically by the formation of a barrier within the material [261]. It has been stated in much of the literature that such barriers act as an insulation within the material, thus mitigating the propagation of fire by reducing the rate of heat transfer. The flammability characteristics of the WFCs specify the fire performance of the material that is evaluated and classified during compounding process by flammability test [262]. In order to enable the fire retardancy, formation of phosphate of the material in presence of oxygen is important as most of the constituents in WPCs are considered to be the good sources for oxygen. Effective flame retardants such as chlorine and bromine halogenated compounds may be used but as they give out toxic gases during the reaction, they should be avoided. Studies by Sain et al. [263] showed that due to the addition of fire retardants, only a marginal drop in mechanical properties was found and to bring down the burning rate of WPCs to 50% without usage of flame retardant, 25% magnesium hydroxide was effective. According to Gracia et al. [264], self-extinguish materials were developed from PE-based composites due to the addition of fire retardants when reagents of hydroxide or phosphate were used. Experiments by cone calorimetry conducted by Stark et al. [265] arrived at a result that phosphate or hydroxide and various other fire retardants could be used in WFCs.

The influence of fire retardants on the composite's flammability was studied by two types of burning tests, namely horizontal and vertical tests. Before the test, the specimen was dried at 80 °C for 24 h. When the sample was positioned horizontally and when one of its ends was exposed to a natural gas fueled flame for 20 s, it is known as the horizontal test. Similarly, when the sample was positioned vertically and when one end of the sample was exposed to the flame for 10 s, then it is known as the vertical test. The height of flame and the angle of flame for horizontal and vertical tests were 10 mm, 30° and 20 mm, 30°, respectively. Burning time, which was measured, indicates the time of the flame to reach the second reference point from the first, which are 80 mm apart. In order to study the effect of dripping of the specimen from the fire source, a wire sheet was held at a distance of 10 mm under the specimen. If the specimen is self-extinguished or incombustible in the horizontal burning test, then it is subjected to the vertical burning test. Class of the sample was determined by the sample dripping, its burning state, and combustion time [266,267]. From all the above discussions, it could be stated that the fire retardancy of WPCs was improved by the addition of materials promoting fire resistance.

4. Life Cycle Assessment of WPCs

The ecological implications evaluation of various sources, which include WPCs, has been using the life cycle assessment (LCA) technique to analyze the possible prospective implications of mechanism [49,268]. A review of previous research on WPCs' LCA can be broken down into two categories: (i) studies trying to compare WPCs to other components (e.g., wood) as well as (ii) studies evaluating the ecological consequences of WPCs produced from various raw resources (original materials versus reprocessed materials). The environmental impact of WPCs produced from original as well as recycled (waste) materials has been studied by Sommerhuber et al. [269]. The primary unprocessed substances were virgin wood as well as waste wood, both of which contained HDPE as well as recycled HDPE. They discovered that WPCs produced from reprocessed as well as waste materials had lower environmental impacts than WPCs produced from virgin resources. WPCs made from virgin wood as well as pure glass fiber or reprocessed mineral wool, as well as WPCs made from virgin wood as well as virgin or recycled PP, were studied by Vantsi and Karki [270]. There were significant reductions in ecological implications of WPCs when recycled mineral wool was used rather than virgin glass fiber. These environmental impacts included global warming, acidification, and eutrophication, as well as abiotic depletion potential. Recycling PP has been found to reduce the risk of global warming and the depletion of abiotic resources. Previous studies on the evaluation of the ecological implications of WPCs, such as the ones cited above, focused on the ecological implications of WPC products rather than the ecological implications of WPC manufacturing as part of a CDW management solution [271].

Consequence LCA, which would be assumed to be critical in determining the impact of utilizing a specific pallet on the overall system has been omitted from previous studies, including those cited above. It is critical to look into the distinctions between attributional as well as consequential LCA findings and conclusions, as well as their applicability in situations involving waste recycling. The assumption that all pallets accomplish equally well throughout their life cycle was made in all previous studies. Pallets created of distinct substances have various life expectancies, repair times, and recycling rates, which were not regarded in any of the studies; from the initial to final stage, LCA incorporates the concept of end-of-life (EoL) care. LCA's final outcome could be significantly impacted by the EoL allocation's methodological differences [272,273].

Environmental implications of WPCs can be assessed using one of two LCA approaches. An attributional LCA (ALCA) examines the environmental implications of the physical flows that enter and exit the product's life cycle as well as its subsystems. Consequential LCA (CLCA) examines the production system as well as the systems connected to it that are expected to change during the production, consumption, and recycling of the item. Using the cradle-to-grave approach, the ALCA considers all aspects of a product's life cycle, from its initial use of raw resources from the ecosystem in the version of elementary flows, such as those generated by nature, to its final disposal, which includes emissions into the atmosphere and water, as well as the creation of waste [274,275]. Figures 6 and 7 shows the scope of ALCA and CLCA studies carried out on WPCs from their cradle to grave.

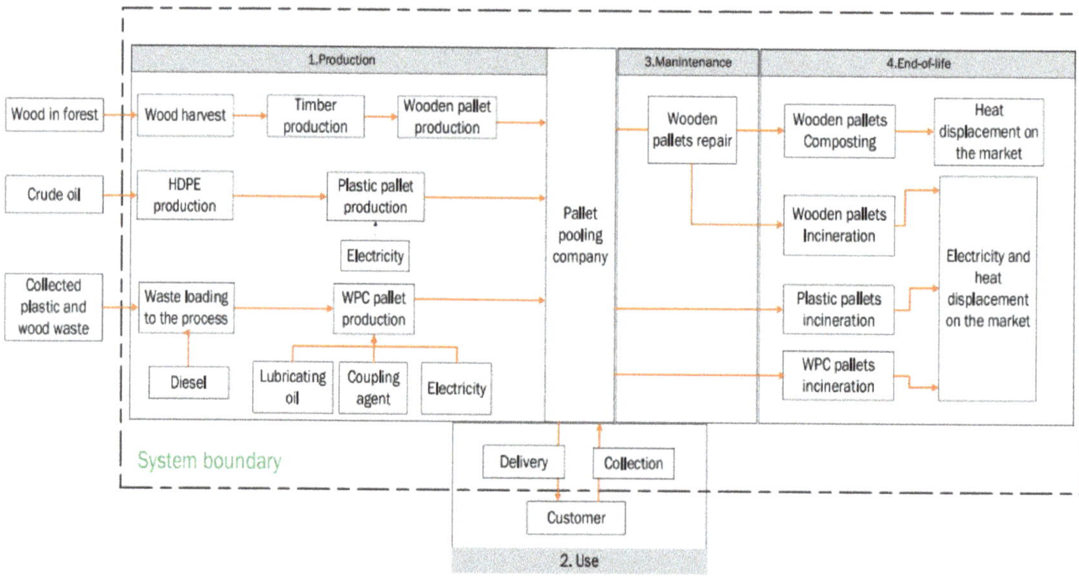

Figure 6. Scope of ALCA studies, Reproduced from [271], Springer, 2021.

Certain factors, such as raw material reliability (e.g., possible contaminants), accessibility, the requirement for the generated WPC, utilizing end-of-life stages for WPC-derived goods (e.g., wood and plastic), physical and mechanical characteristics of various WPC types, and improvement of the production methodology, were not taken into consideration in a few LCA analyses carried out. This opens the door to more investigation. It is fascinating to see how WPCs wind down at the end of their useful lives. Only in the manufacturing process can CDW WPCs be recycled [276]. EoL WPC material restoration is restricted because the crops are not yet prevalent. This raises the question as to whether this technique would then only extend the life among these substances with one cycle or if, with an advanced take-back framework, it might offer a technique for progressing toward that circular economy [277].

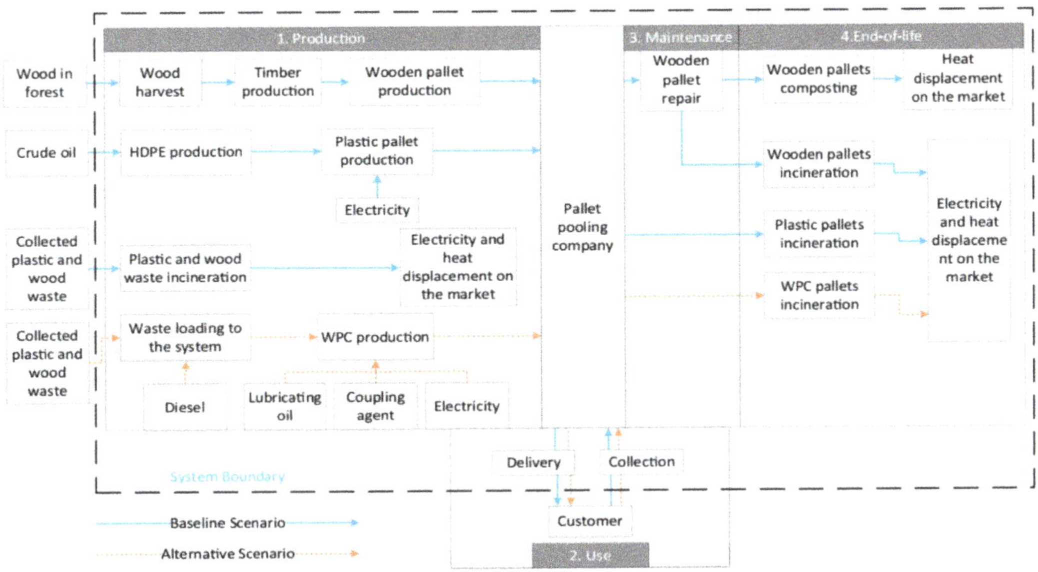

Figure 7. Scope of CLCA studies, Reproduced from [271], Springer, 2021.

According to a few LCA study results, WPCs outperformed individual wood-based as well as plastic materials. Nevertheless, the amount of plastic waste recirculated through WPC ingredients is dependent on the classification of plastic waste, the accessibility of recycled plastic, as well as the manufacturing facilities for WPC pallets. The recycling of mono-materials can be completed, however, by using non-pure plastics as well as fibers, of that kind as those found in mixed plastics as well as other WPC products. Material recovery as well as recyclability there at EoL affects the WPC pallet's circularity. Nearly half of WPC ingredients were created from wood waste and the other half from plastic waste. The environmental impact of this pallet's production is limited to the manufacturing and delivery of additives, electricity, and diesel because it is made from waste [278]. Circular economies, on the other hand, have been criticized for their zero-burden approach because waste is no longer seen as waste but as raw resources for those and more reasons [48]. As a result, the environmental impacts of WPC pallets might be exacerbated if plastic and wood have been assigned some of the burdens from their own previous life cycles.

WPC products' ecological implications are primarily caused by their alternate heat sources, energy source, production energy consumption, and weight. A significant part of the WPC product's life cycle is also played by carbon-neutral wood incineration as well as waste disposal with zero burdens. Several other studies have found that production WPCs can significantly reduce ecological consequences while also efficaciously managing CDW [279–281]. It was also concluded from other studies that the reduction of environmental impacts is highly feasible through the manufacturing of WPCs in order to effectively manage CDW.

5. Applications of WPCs

Practical applications of WPCs depend on the hydrophilic nature of cellulose, which results in moisture absorption and dispersion problems. Predominant applications of WPCs include construction, aerospace, and automobile industries but they are also applied for packaging, for the preparation of various household articles, furniture, office appliances, and other related items [282,283]. In most of these applications, they are used as structural materials, where the load-carrying capacity of the wood-based reinforcements play a significant role. This is determined by the particle characteristics of the reinforcing wood-based materials and by interfacial adhesion between the wood-based reinforcement and matrix [284].

WPCs are predominantly used in traffic materials, furniture building, construction, and military applications. WPCs are also becoming more significant materials in many other engineering applications. Moreover, WPCs are also used in civil, marine, and automotive engineering applications. Owing to the advancements in wood-based raw materials, their durability and mechanical characteristics during the production of WPCs, a few other outdoor applications were added to the list including utility poles, fences, decks, and building exterior wood works [285,286]. WPCs were also considered to be the most sustainable and quality-rich when compared with other traditional composites and individual wood-based materials. Wood-based flooring is considered to be the highest application volume of WPCs and it encompasses top veneers of laminated flooring, parquet flooring, and solid plank flooring. Majorly, walnut, maple, red oak, and ash are the types of woods used for composite flooring [287]. Patented sports applications of WPCs include laminated skis, club heads in golf courts, hockey sticks, baseball bats, bows, and gun stocks. With regards to musical instruments, WPCs are used for the manufacturing of bagpipe chanters, flute mouthpieces, wind instruments, and stringed instrument finger boards [288]. In order to manufacture scratch-resistant furniture, WPCs along with wood veneers were used. Products include tabletops, desks for writing, clock faces, plaques, and knife handles. Handrails are another class of small volume applications of WPCs. Handrails are commercial applications which are primarily installed in various locations such as malls, department stores, and airports, for their strength improvements, aesthetics, and ease of maintenance [289]. From all the above discussions, it could be stated that WPCs are predominantly used in low and medium density applications at a wider spectrum.

6. Conclusions

Wood polymer composites are considered to be a potential and promising material from environmental and economic viewpoints since they involve the use of recycled wood or wood flour. WPCs exhibited better strength and high marketability at a relatively low cost. Properties of WPCs were directly influenced by the processing and post-processing methods. WPCs fabricated with appropriate processing methods had lower contact angles, less color change, better physical and mechanical characteristics, and relatively higher decomposition temperatures. Addition of fillers and lubricants reduced the solid content in WPCs and reduced the viscosity, thus enhancing the bond strength of WPCs. As per most of the studies, WPCs were also manufactured using bio-based adhesives such as starch, tannin, lignin, and soy protein-based adhesives which were considered to be an environmentally friendly alternative for the conventional synthetic adhesives despite their demerits such as high viscosity, low bonding strength, and high hydrophilicity. Analyzing these demerits and overcoming them in a laboratory scale is still a challenging task to many researchers. Yet, the aforesaid bio-based adhesives could reduce the environmental mitigations caused in the form of formaldehyde emissions and formation of volatile organic compounds during the fabrication of WPCs. Life-cycle assessment studies were also carried out on WPCs to assess their environmental effects and from the results, it was found that the WPCs manufactured from bio-adhesives exhibited lesser effects when compared with synthetic resins such as urea-formaldehyde and phenol-formaldehyde. Further research to develop an assessment method for enhancing the durability and long-life of WPCs can be developed based on the conclusions arrived. Overall, it could be stated that the utilization of recycled wood-based materials to manufacture wood-polymer composites is highly viable with regards to eco-friendliness and contributes to the improvement circular economy. Meanwhile, it also saves the usage of virgin materials, thus enhancing the sustainability in production of composite materials.

Author Contributions: Conceptualization: M.R.; methodology: L.R. and D.B.; formal analysis and resources: A.S. and V.B.; writing—original draft preparation: M.R. and G.S.; writing—review and editing: L.R. and R.B. All authors have read and agreed to the published version of the manuscript.

Funding: This research received no external funding.

Institutional Review Board Statement: Not applicable.

Informed Consent Statement: Not applicable.

Data Availability Statement: Not applicable.

Conflicts of Interest: The authors declare no conflict of interest.

References

1. Taylor, A.M.; Gartner, B.L.; Morrell, J.J.; Tsunoda, K. Effects of heartwood extractive fractions of Thuja plicata and Chamaecyparis nootkatensis on wood degradation by termites or fungi. *J. Wood Sci.* **2006**, *52*, 147–153. [CrossRef]
2. Emons, A.M.C.; Höfte, H.; Mulder, B.M. Microtubules and cellulose microfibrils: How intimate is their relationship? *Trends Plant Sci.* **2007**, *12*, 279–281. [CrossRef] [PubMed]
3. Berglund, L.A.; Burgert, I. Bioinspired Wood Nanotechnology for Functional Materials. *Adv. Mater.* **2018**, *30*, e1704285. [CrossRef] [PubMed]
4. Clemons, C. Wood-plastic composites in the United States: The interfacing of two industries. *For. Prod. J.* **2002**, *52*, 10–18.
5. Rowell, R.M. *Handbook of Wood Chemistry and Wood Composites*; CRC Press: Boca Raton, FL, USA, 2005.
6. El-Haggar, S.M.; Kamel, M.A. Wood plastic composites. In *Advances in Composite Materials: Analysis of Natural and Man-Made Material*; In Tech: London, UK, 2011; pp. 325–344. [CrossRef]
7. Ghasem, J.M. Economic model assessment of wood–polymer composites production from agricultural wastes. *Ann. Biol. Res.* **2013**, *4*, 169–174.
8. Roig, I. Biocomposites for interior facades and partitions to improve air quality in new buildings and restorations. *Reinf. Plast.* **2018**, *62*, 270–274. [CrossRef]
9. Rajeshkumar, L. Biodegradable polymer blends and composites from renewable Resources. In *Biodegradable Polymer Blends and Composites*; Sanjay, M.R., Parameswaranpillai, J., Suchart, S., Ramesh, M., Eds.; Woodhead Publishing, Elsevier: Cambridge, UK, 2021; pp. 527–549. [CrossRef]
10. Huang, R.; Xiong, W.; Xu, X.; Wu, Q. Thermal expansion behavior of co-extruded wood-plastic composites with glass-fiber reinforced shells. *Bioresources* **2012**, *7*, 5514–5526. [CrossRef]
11. Ramesh, M. Wood flour filled thermoset composites. *Mater. Res. Found.* **2018**, *38*, 33–65.
12. Hietala, M.; Samuelsson, E.; Niinimäki, J.; Oksman, K. The effect of pre-softened wood chips on wood fibre aspect ratio and mechanical properties of wood–polymer composites. *Compos. Part A Appl. Sci. Manuf.* **2011**, *42*, 2110–2116. [CrossRef]
13. Aref, I.; Nasser, R.; Ali, I.; Al-Mefarrej, H.; Al-Zahrani, S. Effects of aqueous extraction on the performance and properties of polypropylene/wood composites from Phoenix dactylifera and Acacia tortilis wood. *J. Reinf. Plast. Compos.* **2013**, *32*, 476–489. [CrossRef]
14. Batra, R.; Gopinath, G.; Zheng, J. Damage and failure in low energy impact of fiber-reinforced polymeric composite laminates. *Compos. Struct.* **2012**, *94*, 540–547. [CrossRef]
15. Saravana Kumar, A.; Maivizhi Selvi, P.; Rajeshkumar, L. Delamination in drilling of sisal/banana reinforced composites produced by hand lay-up process. *Appl. Mech. Mater.* **2017**, *867*, 29–33. [CrossRef]
16. Hani, A.R.A.; Seang, C.T.; Ahmad, R.; Mariatti, J.M. Impact and flexural properties of imbalance plain woven coir and kenaf composite. *Appl. Mech. Mater.* **2012**, *271*, 81–85. [CrossRef]
17. Perisic, S.D.; Radovic, I.; Petrovic, M.; Marinković, A.; Stojanović, D.; Uskokovic, P.; Radojevic, V. Processing of hybrid wood plastic composite reinforced with short PET fibers. *Mater. Manuf. Process.* **2018**, *33*, 572–579. [CrossRef]
18. Huysman, S.; Debaveye, S.; Schaubroeck, T.; De Meester, S.; Ardente, F.; Mathieux, F.; Dewulf, J. The recyclability benefit rate of closed-loop and open-loop systems: A case study on plastic recycling in Flanders. *Resour. Conserv. Recycl.* **2015**, *101*, 53–60. [CrossRef]
19. Craciun, G.; Manaila, E.; Ighigeanu, D.; Stelescu, M.D. A method to improve the characteristics of EPDM rubber based eco-composites with electron beam. *Polymers* **2020**, *12*, 215. [CrossRef]
20. Bhuvaneswari, V.; Priyadharshini, M.; Deepa, C.; Balaji, D.; Rajeshkumar, L.; Ramesh, M. Deep learning for material synthesis and manufacturing systems: A review. *Mater. Today Proc.* **2021**, *46*, 3263–3269. [CrossRef]
21. Ramesh, M.; Deepa, C.; Selvan, M.T.; Rajeshkumar, L.; Balaji, D.; Bhuvaneswari, V. Mechanical and water absorption properties of Calotropis gigantea plant fibers reinforced polymer composites. *Mater. Today Proc.* **2021**, *46*, 3367–3372. [CrossRef]
22. Mansour, S.H.; El-Nashar, D.E.; Abd-El-Messieh, S.L. Effect of chemical treatment of wood flour on the properties of styrene butadiene rubber/polystyrene composites. *J. Appl. Polym. Sci.* **2006**, *102*, 5861–5870. [CrossRef]
23. Fabiyi, J.S.; McDonald, A.G. Effect of wood species on property and weathering performance of wood plastic composites. *Compos. Part A Appl. Sci. Manuf.* **2010**, *41*, 1434–1440. [CrossRef]

24. Guo, J.; Tang, Y.; Xu, Z. Wood plastic composite produced by nonmetals from pulverized waste printed circuit boards. *Environ. Sci. Technol.* **2010**, *44*, 463–468. [CrossRef] [PubMed]
25. Oladejo, K.O.; Omoniyi, T.E. Dimensional stability and mechanical properties of wood plastic composites produced from sawdust of Anogeissus leiocarpus (Ayin) with recycled polyethylene teraphthalate (PET) chips. *Eur. J. Appl. Eng. Sci. Res.* **2017**, *5*, 28–33.
26. Balaji, D.; Ramesh, M.; Kannan, T.; Deepan, S.; Bhuvaneswari, V.; Rajeshkumar, L. Experimental investigation on mechanical properties of banana/snake grass fiber reinforced hybrid composites. *Mater. Today Proc.* **2021**, *42*, 350–355. [CrossRef]
27. Feifel, S.; Stübs, O.; Seibert, K.; Hartl, J. Comparing wood–polymer composites with solid wood: The case of sustainability of terrace flooring. *Eur. J. Wood Wood Prod.* **2015**, *73*, 829–836. [CrossRef]
28. Toghyani, A.E.; Matthews, S.; Varis, J. Feasibility assessment of a wood-plastic composite post-production process: Cuttability. *Proc. Manuf.* **2018**, *25*, 271–278. [CrossRef]
29. Puttasukkha, J.; Khongtong, S.; Chaowana, P. Curing behavior and bonding performance of urea formaldehyde resin admixed with formaldehyde scavenger. *Wood Res.* **2015**, *60*, 645–654.
30. Horito, M.; Kurushima, N.; Ono, K.; Yazaki, Y. Plywood adhesives using PF resin with fibrillated bark slurry from radiata pine (Pinus radiata D. Don): Utilization of flavonoid compounds from bark and wood. *J. Wood Sci.* **2020**, *66*, 1–8. [CrossRef]
31. Chung, T.C. Synthesis of functional polyolefin copolymers with graft and block structures. *Prog. Polym. Sci.* **2002**, *27*, 39–85. [CrossRef]
32. Ramesh, M.; Deepa, C.; Rajeshkumar, L.; Tamil Selvan, M.; Balaji, D. Influence of fiber surface treatment on the tribological properties of Calotropis gigantea plant fiber reinforced polymer composites. *Polym. Compos.* **2021**, *42*, 4308–4317. [CrossRef]
33. Kim, S.; Lee, Y.K.; Kim, H.J.; Lee, H.H. Physico-mechanical properties of particleboards bonded with pine and wattle tannin-based adhesives. *J. Adhes. Sci. Technol.* **2003**, *17*, 1863–1875. [CrossRef]
34. Wang, W.; Zhang, W.; Zhang, S.; Li, J. Preparation and characterization of microencapsulated ammonium polyphosphate with UMF and its application in WPCs. *Constr. Build. Mater.* **2014**, *65*, 151–158. [CrossRef]
35. Bakar, M.A.; Ishak, Z.M.; Taib, R.M.; Rozman, H.D.; Jani, S.M. Flammability and mechanical properties of wood flour-filled polypropylene composites. *J. Appl. Polym. Sci.* **2010**, *116*, 2714–2722. [CrossRef]
36. Ayrilmis, N.; Akbulut, T.; Dundar, T.; White, R.H.; Mengeloglu, F.; Buyuksari, U.; Avci, E. Effect of boron and phosphate compounds on physical, mechanical, and fire properties of wood–polypropylene composites. *Constr. Build. Mater.* **2012**, *33*, 63–69. [CrossRef]
37. Seefeldt, H.; Braun, U.; Wagner, M.H. Residue stabilization in the fire retardancy of wood–plastic composites: Combination of ammonium polyphosphate, expandable graphite, and red phosphorus. *Macromol. Chem. Phys.* **2012**, *213*, 2370–2377. [CrossRef]
38. Kratofil Krehula, L.; Katančić, Z.; Marić, G.; Hrnjak-Murgić, Z. Study of fire retardancy and thermal and mechanical properties of HDPE-wood composites. *J. Wood Chem. Technol.* **2015**, *35*, 412–423. [CrossRef]
39. Madyaratri, E.W.; Ridho, M.R.; Aristri, M.A.; Lubis, M.A.R.; Iswanto, A.H.; Nawawi, D.S.; Antov, P.; Kristak, L.; Majlingová, A.; Fatriasari, W. Recent advances in the development of fire-resistant biocomposites—A review. *Polymers* **2022**, *14*, 362. [CrossRef]
40. Ramesh, M.; Rajeshkumar, L.; Deepa, C.; Tamil Selvan, M.; Kushvaha, V.; Asrofi, M. Impact of silane treatment on characterization of ipomoea staphylina plant fiber reinforced epoxy composites. *J. Nat. Fibers* **2021**, 1–12. [CrossRef]
41. Pan, M.; Mei, C.; Du, J.; Li, G. Synergistic effect of nano silicon dioxide and ammonium polyphosphate on flame retardancy of wood fiber–polyethylene composites. *Compos. Part A Appl. Sci. Manuf.* **2014**, *66*, 128–134. [CrossRef]
42. Fu, S.-Y.; Sun, Z.; Huang, P.; Li, Y.-Q.; Hu, N. Some basic aspects of polymer nanocomposites: A critical review. *Nano Mater. Sci.* **2019**, *1*, 2–30. [CrossRef]
43. Devarajan, B.; Saravanakumar, R.; Sivalingam, S.; Bhuvaneswari, V.; Karimi, F.; Rajeshkumar, L. Catalyst derived from wastes for biofuel production: A critical review and patent landscape analysis. *Appl. Nanosci.* **2021**, 1–25. [CrossRef]
44. WRAP. Plastics Market Situation Report. 2019. Available online: https://wrap.org.uk/resources/market-situation-reports/plastics-2019 (accessed on 24 November 2021).
45. WWF. Reuse and Recycling. 2016. Available online: http://forestsolutions.panda.org/solutions/reuse-and-recycling (accessed on 17 November 2021).
46. Cheng, H.; Chen, C.; Wu, S.; Mirza, Z.A.; Liu, Z. Emergy evaluation of cropping, poultry rearing, and fish raising systems in the drawdown zone of three gorges reservoir of China. *J. Clean. Prod.* **2017**, *144*, 559–571. [CrossRef]
47. Ramesh, M.; Deepa, C.; Niranjana, K.; Rajeshkumar, L.; Bhoopathi, R.; Balaji, D. Influence of Haritaki (*Terminalia chebula*) nano-powder on thermo-mechanical, water absorption and morphological properties of Tindora (*Coccinia grandis*) tendrils fiber reinforced epoxy composites. *J. Nat. Fibers* **2021**, *1–17*. [CrossRef]
48. Ilic, D.D.; Eriksson, O.; Ödlund, L.; Åberg, M. No zero-burden assumption in a circular economy. *J. Clean. Prod.* **2018**, *182*, 352–362. [CrossRef]
49. Bolin, C.A.; Smith, S. Life cycle assessment of ACQ-treated lumber with comparison to wood plastic composite decking. *J. Clean. Prod.* **2011**, *19*, 620–629. [CrossRef]
50. Sun, G.; Ibach, R.E.; Faillace, M.; Gnatowski, M.; Glaeser, J.A.; Haight, J. Laboratory and exterior decay of wood–plastic composite boards: Voids analysis and computed tomography. *Wood Mater. Sci. Eng.* **2017**, *12*, 263–278. [CrossRef]
51. Ramesh, M.; Rajeshkumar, L.; Balaji, D. Influence of process parameters on the properties of additively manufactured fiber-reinforced polymer composite materials: A review. *J. Mater. Eng. Perform.* **2021**, *30*, 4792–4807. [CrossRef]

52. Korol, J.; Burchart-Korol, D.; Pichlak, M. Expansion of environmental impact assessment for eco-efficiency evaluation of biocomposites for industrial application. *J. Clean. Prod.* **2016**, *113*, 144–152. [CrossRef]
53. Ashori, A.; Tabarsa, T.; Amosi, F. Evaluation of using waste timber railway sleepers in wood–cement composite materials. *Constr. Build. Mater.* **2012**, *27*, 126–129. [CrossRef]
54. Li, M.; Khelifa, M.; Khennane, A.; El Ganaoui, M. Structural response of cement-bonded wood composite panels as permanent formwork. *Compos. Struct.* **2019**, *209*, 13–22. [CrossRef]
55. Liikanen, M.; Grönman, K.; Deviatkin, I.; Havukainen, J.; Hyvärinen, M.; Kärki, T.; Varis, J.; Soukka, R.; Horttanainen, M. Construction and demolition waste as a raw material for wood polymer composites—Assessment of environmental impacts. *J. Clean. Prod.* **2019**, *225*, 716–727. [CrossRef]
56. Osburg, V.S.; Strack, M.; Toporowski, W. Consumer acceptance of Wood-Polymer Composites: A conjoint analytical approach with a focus on innovative and environmentally concerned consumers. *J. Clean. Prod.* **2016**, *110*, 180–190. [CrossRef]
57. Ramesh, M.; Rajeshkumar, L.; Bhoopathi, R. Carbon substrates: A review on fabrication, properties and applications. *Carbon Lett.* **2021**, *31*, 557–580. [CrossRef]
58. Partanen, A.; Carus, M. Wood and natural fiber composites current trend in consumer goods and automotive parts. *Reinf. Plast.* **2016**, *60*, 170–173. [CrossRef]
59. Friedrich, D. Welfare effects from eco-labeled crude oil preserving wood-polymer composites: A comprehensive literature review and case study. *J. Clean. Prod.* **2018**, *188*, 625–637. [CrossRef]
60. Ashori, A. Wood–plastic composites as promising green-composites for automotive industries. *Bioresour. Technol.* **2008**, *99*, 4661–4667. [CrossRef] [PubMed]
61. Friedrich, D.; Luible, A. Standard-compliant development of a design value for wood–plastic composite cladding: An application-oriented perspective. *Case Stud. Struct. Eng.* **2016**, *5*, 13–17. [CrossRef]
62. Ramesh, M.; Rajeshkumar, L.; Bhuvaneswari, V. Leaf fibres as reinforcements in green composites: A review on processing, properties and applications. *Emergent Mater.* **2021**, 1–25. [CrossRef]
63. Sahayaraj, A.F.; Muthukrishnan, M.; Ramesh, M.; Rajeshkumar, L. Effect of hybridization on properties of tamarind (Tamarindus indica L.) seed nano-powder incorporated jute-hemp fibers reinforced epoxy composites. *Polym. Compos.* **2021**, *42*, 6611–6620. [CrossRef]
64. Fortini, A.; Mazzanti, V. Combined effect of water uptake and temperature on wood polymer composites. *J. Appl. Polym. Sci.* **2018**, *135*, 46674. [CrossRef]
65. Zavadskas, E.K.; Antucheviciene, J.; Šaparauskas, J.; Turskis, Z. Multi-criteria assessment of facades' alternatives: Peculiarities of ranking methodology. *Proc. Eng.* **2013**, *57*, 107–112. [CrossRef]
66. Javier, C.S.; Sergio, A.R.; Roberto, Z.G.; Jorge, D.D. Optimization of the tensile and flexural strength of a wood-pet composite. *Ing. Investig. Tecnol.* **2015**, *16*, 105–112. [CrossRef]
67. Binhussain, M.A.; El-Tonsy, M.M. Palm leave and plastic waste wood compositefor out-door structures. *Constr. Build. Mater.* **2013**, *47*, 1431–1435. [CrossRef]
68. Aizat, A.G.; Paiman, B.; Lee, S.H.; Zaidon, A. Physico-mechanical properties and formaldehyde emission of rubberwood particleboard made with uf resin admixed with ammonium and aluminium-based hardeners. *Pertanika J. Sci. Technol.* **2019**, *27*, 473–488.
69. Pickering, K.; Efendy, M.A.; Le, T. A review of recent developments in natural fibre composites and their mechanical performance. *Compos. Part A Appl. Sci. Manuf.* **2016**, *83*, 98–112. [CrossRef]
70. Mohankumar, D.; Amarnath, V.; Bhuvaneswari, V.; Saran, S.P.; Saravanaraj, K.; Gogul, M.S.; Rajeshkumar, L. Extraction of plant based natural fibers–A mini review. In *IOP Conference Series: Materials Science and Engineering*; IOP Publishing: Bristol, UK, 2021; Volume 1145, p. 012023.
71. Wolcott, M.P.; Englund, K. A technology review of wood-plastic composites. In Proceedings of the 33rd International Particleboard/Composite Materials Symposium, Washington State University, Washington, DC, USA, 13–15 April 1999; pp. 103–111.
72. Rahman, K.S.; Islam, N.; Rahman, M.; Hannan, O.; Dungani, R.; Khalil, H.A. Flat-pressed wood plastic composites from sawdust and recycled polyethylene terephthalate (PET): Physical and mechanical properties. *SpringerPlus* **2013**, *2*, 629. [CrossRef] [PubMed]
73. Wilczyński, K.; Buziak, K.; Lewandowski, A.; Nastaj, A.; Wilczyński, K. Rheological basics for modeling of extrusion process of wood polymer composites. *Polymers* **2021**, *13*, 622. [CrossRef]
74. Maldas, D.; Kokta, B.V. Composite molded products based on recycled polypropylene and woodflour. *J. Thermoplast. Compos. Mater.* **1995**, *8*, 420–434. [CrossRef]
75. Kaymakci, A.; Ayrilmis, N. Surface roughness and wettability of polypropylene composites filled with fast-growing biomass: Paulownia elongata wood. *J. Compos. Mater.* **2014**, *48*, 951–957. [CrossRef]
76. Cavdar, A.D.; Mengeloğlu, F.; Karakus, K. Effect of boric acid and borax on mechanical, fire and thermal properties of wood flour filled high density polyethylene composites. *Measurement* **2015**, *60*, 6–12. [CrossRef]
77. Spinace, M.A.; Lambert, C.S.; Fermoselli, K.K.; De Paoli, M.A. Characterization of lignocellulosic curaua fibres. *Carbohydr. Polym.* **2009**, *77*, 47–53. [CrossRef]
78. Bledzki, A.K.; Faruk, O. Creep and impact properties of wood fibre–polypropylene composites: Influence of temperature and moisture content. *Compos. Sci. Technol.* **2004**, *64*, 693–700. [CrossRef]

79. Umemura, T.; Arao, Y.; Nakamura, S.; Tomita, Y.; Tanaka, T. Synergy effects of wood flour and fire retardants in flammability of wood-plastic composites. *Energy Proc.* **2014**, *56*, 48–56. [CrossRef]
80. Nörnberg, B.; Borchardt, E.; Luinstra, G.A.; Fromm, J. Wood plastic composites from poly(propylene carbonate) and poplar wood flour – Mechanical, thermal and morphological properties. *Eur. Polym. J.* **2014**, *51*, 167–176. [CrossRef]
81. Sain, M.M.; Balatinecz, J.; Law, S. Creep fatigue in engineered wood fiber and plastic compositions. *J. Appl. Polym. Sci.* **2000**, *77*, 260–268. [CrossRef]
82. Grubbström, G.; Oksman, K. Influence of wood flour moisture content on the degree of silane-crosslinking and its relationship to structure–property relations of wood–thermoplastic composites. *Compos. Sci. Technol.* **2009**, *69*, 1045–1050. [CrossRef]
83. Oyj, O. Environmental product declaration according to ISO 14025. In *Cold Rolled Stainless Steel*; Eco Europe Organization: Berlin, Germany, 2011.
84. Charlet, K.; Saulnier, F.; Dubois, M.; Béakou, A. Improvement of wood polymer composite mechanical properties by direct fluorination. *Mater. Des.* **2015**, *74*, 61–66. [CrossRef]
85. Ab Ghani, M.H.; Ahmad, S. The comparison of water absorption analysis between counterrotating and corotating twin-screw extruders with different antioxidants content in wood plastic composites. *Adv. Mater. Sci. Eng.* **2011**, *2011*, 1–4. [CrossRef]
86. Ayrilmis, N.; Benthien, J.T.; Thoemen, H. Effects of formulation variables on surface properties of wood plastic composites. *Compos. Part B Eng.* **2012**, *43*, 325–331. [CrossRef]
87. Benthien, J.T.; Thoemen, H. Effects of raw materials and process parameters on the physical and mechanical properties of flat pressed WPC panels. *Compos. Part A: Appl. Sci. Manuf.* **2012**, *43*, 570–576. [CrossRef]
88. Koohestani, B.; Darban, A.K.; Mokhtari, P.; Yilmaz, E.; Darezereshki, E. Comparison of different natural fiber treatments: A literature review. *Int. J. Environ. Sci. Technol.* **2019**, *16*, 629–642. [CrossRef]
89. Fathi, B.; Harirforoush, M.; Foruzanmehr, M.; Elkoun, S.; Robert, M. Effect of TEMPO oxidation of flax fibers on the grafting efficiency of silane coupling agents. *J. Mater. Sci.* **2017**, *52*, 10624–10636. [CrossRef]
90. Chauhan, V.; Kärki, T.; Varis, J. Review of natural fiber-reinforced engineering plastic composites, their applications in the transportation sector and processing techniques. *J. Thermoplast. Compos. Mater.* **2019**. [CrossRef]
91. Akdogan, A.; Vanli, A.S. *Wood-Reinforced Polymer Composites, Wood in Civil Engineering*; Concu, G., Ed.; IntechOpen: London, UK, 2017. [CrossRef]
92. Li, J.; Li, D.; Song, Z.; Shang, S.; Guo, Y. Preparation and properties of wood plastic composite reinforced by ultralong cellulose nanofibers. *Polym. Compos.* **2016**, *37*, 1206–1215. [CrossRef]
93. Carrino, L.; Ciliberto, S.; Giorleo, G.; Prisco, U. Effect of filler content and temperature on steady-state shear flow of wood/high density polyethylene composites. *Polym. Compos.* **2011**, *32*, 796–809. [CrossRef]
94. Adhikary, K.B.; Park, C.B.; Islam, M.R.; Rizvi, G.M. Effects of lubricant content on extrusion processing and mechanical properties of wood flour–high density polyethylene composites. *J. Thermoplast. Compos. Mater.* **2011**, *24*, 155–171. [CrossRef]
95. Mazzanti, V.; Mollica, F. A review of wood polymer composites rheology and its implications for processing. *Polymers* **2020**, *12*, 2304. [CrossRef] [PubMed]
96. Santi, C.R.; Hage, E.; Vlachopoulos, J.; Correa, C.A. Rheology and processing of HDPE/wood flour composites. *Int. Polym. Process.* **2009**, *24*, 346–353. [CrossRef]
97. Hristov, V. Melt flow instabilities of wood polymer composites. *Compos. Interfaces* **2009**, *16*, 731–750. [CrossRef]
98. Li, T.Q.; Wolcott, M.P. Rheology of HDPE–wood composites. I. Steady state shear and extensional flow. *Compos. Part A* **2004**, *35*, 303–311. [CrossRef]
99. Li, T.Q.; Wolcott, M.P. Rheology of wood plastics melt. Part 1. Capillary rheometry of HDPE filled with maple. *Polym. Eng. Sci.* **2005**, *45*, 549–559. [CrossRef]
100. Li, T.Q.; Wolcott, M.P. Rheology of wood plastics melt, part 2: Effects of lubricating systems in HDPE/maple composites. *Polym. Eng. Sci.* **2006**, *46*, 464–473. [CrossRef]
101. Kajaks, J.; Kalnins, K.; Matvejs, J. Rheological properties of wood-plastic composites based on polypropylene and birch wood plywood production residues. *Key Eng. Mater.* **2018**, *762*, 226–230. [CrossRef]
102. Laufer, N.; Hansmann, H.; Koch, M. Rheological characterization of the flow behavior of wood plastic composites in consideration of different volume fractions of wood. *J. Phys. Conf. Ser.* **2017**, *790*, 012017. [CrossRef]
103. Mazzanti, V.; Malagutti, L.; Blanchard, M.; Yi, S.; Mollica, F. In-line rheological properties of rubber toughened wood polymer composites. *IOP Conf. Ser. Mater. Sci. Eng.* **2019**, *634*, 012043. [CrossRef]
104. Duretek, I.; Schuschnigg, S.; Gooneie, A.; Langecker, G.R.; Holzer, C. Rheological properties of wood polymer composites and their role in extrusion. *J. Phys. Conf. Ser.* **2015**, *602*, 012014. [CrossRef]
105. Murayama, K.; Suzuki, S.; Kojima, Y.; Kobori, H.; Ito, H.; Ogoe, S.; Okamoto, M. The effects of different types of maleic anhydride-modified polypropylene on the physical and mechanical properties of polypropylene-based wood/plastic composites. *J. Wood Chem. Technol.* **2018**, *38*, 224–232. [CrossRef]
106. Hristov, V.; Vlachopoulos, J. Thermoplastic silicone elastomer lubricant in extrusion of polypropylene wood flour composites. *Adv. Polym. Technol.* **2007**, *26*, 100–108. [CrossRef]
107. Sapsrithong, P.; Puksattee, K.; Saewjaidee, K.; Pensuk, N.; Rattanapan, A. The influence of fiber surface treatment and SBR as impact modifier on rheological behavior and mechanical properties of wood plastic composite from acrylate-styrene-acrylonitrile and bagasse. *Key Eng. Mater.* **2017**, *737*, 281–286. [CrossRef]

108. Kaseem, M.; Hamad, K.; Deri, F.; Ko, Y.G. Effect of wood fibers on the rheological and mechanical properties of polystyrene/wood composites. *J. Wood Chem. Technol.* **2017**, *37*, 251–260. [CrossRef]
109. He, P.; Bai, S.; Wang, Q. Structure and performance of poly(vinyl alcohol)/wood powder composite prepared by thermal processing and solid state shear milling technology. *Compos. Part B* **2016**, *99*, 373–380. [CrossRef]
110. Scaffaro, R.; Morreale, M.; Lo Re, G.; La Mantia, F.P. Effect of the processing techniques on the properties of ecocomposites based on vegetable oil-derived Mater-Bi and wood flour. *J. Appl. Polym. Sci.* **2009**, *114*, 2855–2863. [CrossRef]
111. Bi, H.; Ren, Z.; Guo, R.; Xu, M.; Song, Y. Fabrication of flexible wood flour/thermoplastic polyurethane elastomer composites using fused deposition molding. *Ind. Crop. Prod.* **2018**, *122*, 76–84. [CrossRef]
112. Mazzanti, V.; Mollica, F. Rheology of wood flour filled poly(lactic acid). *Proc. Eng.* **2017**, *200*, 61–67. [CrossRef]
113. Durmus, A.; Ozcan, M.; Aydin, I. Quantifying effects of compositional variations on microstructural properties of polypropylene-wood fiber composites by melt rheology and tensile test data. *J. Compos. Mater.* **2018**, *53*, 503–514. [CrossRef]
114. Prado Bettini, S.H.; Pereira de Miranda Josefovich, M.P.; Riveros Munoz, P.A.; Lotti, C.; Capparelli Mattoso, L.H. Effect of lubricant on mechanical and rheological properties of compatibilized PP/sawdust composites. *Carbohydr. Polym.* **2013**, *94*, 800–806. [CrossRef]
115. Ghasemi, I.; Azizi, H.; Naeimian, N. Rheological behaviour of polypropylene/kenaf fibre/wood flour hybrid composites. *Iran. Polym. J.* **2008**, *17*, 191–198.
116. Dai, L.; Wang, X.; Zhang, J.; Wang, F.; Ou, R.; Song, Y. Effects of lubricants on the rheological and mechanical properties of wood flour/polypropylene composites. *J. Appl. Polym. Sci.* **2019**, *136*, 47667. [CrossRef]
117. Soury, E.; Behravesh, A.H.; Rizvi, G.M.; Jam, N.J. Rheological investigation of wood–polypropylene composites in rotational plate rheometer. *J. Polym. Environ.* **2012**, *20*, 998–1006. [CrossRef]
118. Wang, P.; Liu, J.; Yu, W.; Zhou, C. Dynamic rheological properties of wood polymer composites: From linear to nonlinear behaviors. *Polym. Bull.* **2011**, *66*, 683–701. [CrossRef]
119. Gao, H.; Xie, Y.; Ou, R.; Wang, Q. Grafting effects of polypropylene/polyethylene blends with maleic anhydride on the properties of the resulting wood-plastic composites. *Compos. Part A* **2012**, *43*, 150–157. [CrossRef]
120. Habibi, M.; Najafi, S.K.; Ghasemi, I. Rheological and mechanical properties of composites made from wood flour and recycled LDPE/HDPE blend. *Iran. Polym. J.* **2017**, *26*, 949–956. [CrossRef]
121. Twite-Kabamba, E.; Fassi Fehri, Z.; Rodrigue, D. Properties of recycled LDPE/birch fibre composites. *Prog. Rubber Plast. Recycl. Technol.* **2011**, *27*, 1–20. [CrossRef]
122. Hong, H.; Liao, H.; Zhang, H.; He, H.; Liu, T.; Jia, D. Significant improvement in performance of recycled polyethylene/wood flour composites by synergistic compatibilization at multi-scale interfaces. *Compos. Part A* **2014**, *64*, 90–98. [CrossRef]
123. Koohestani, B.; Ganetri, I.; Yilmaz, E. Effects of silane modified minerals on mechanical, microstrucutral, thermal and rheological properties of wood plastic composites. *Compos. Part B* **2017**, *111*, 103–111. [CrossRef]
124. Yang, B.; Liang, C.W.; Lu, F.X.; Chen, P.; Chen, Q.T.; Chen, J.; Hu, L.; Xia, R.; Miao, J.B.; Qian, J.S.; et al. Effect of bamboo flour (BF) content on the dynamic rheological characteristics of BF-filled high-density polyethylene (HDPE). *J. Macromol. Sci. Part B* **2019**, *58*, 341–354. [CrossRef]
125. Li, T.Q.; Wolcott, M.P. Rheology of wood plastics melt, Part 3: Nonlinear nature of the flow. *Polym. Eng. Sci.* **2006**, *46*, 114–121. [CrossRef]
126. Tazi, M.; Erchiqui, F.; Godard, F.; Kaddami, H.; Ajji, A. Characterization of rheological and thermo-physical properties of HDPE-wood composite. *J. Appl. Polym. Sci.* **2014**, *131*, 40495. [CrossRef]
127. Adebayo, G.O.; Hassan, A.; Yahya, R.; Okieimen, F.; Sarih, N.M. Dynamic rheological properties of spotted mangrove/high density polyethylene composites. *J. Thermoplast. Compos. Mater.* **2019**, *34*, 1273–1285. [CrossRef]
128. Hong, H.; Guo, Q.; Zhang, H.; He, H. Effect of interfacial modifiers and wood flour treatment on the rheological properties of recycled polyethylene/wood flour composites. *Prog. Rubber Plast. Recycl. Technol.* **2019**, *36*, 31–46. [CrossRef]
129. Funk, M.; Wimmer, R.; Adamopoulos, S. Diatomaceous earth as an inorganic additive to reduce formaldehyde emissions from particleboards. *Wood Mater. Sci. Eng.* **2017**, *12*, 92–97. [CrossRef]
130. Song, Y.H.; Seo, J.H.; Choi, Y.S.; Kim, D.H.; Choi, B.H.; Cha, H.J. Mussel adhesive protein as an environmentally-friendly harmless wood furniture adhesive. *Int. J. Adhes. Adhes.* **2016**, *70*, 260–264. [CrossRef]
131. Osemeahon, S.A.; Maitera, O.N.; Hotton, A.J.; Dimas, B.J. Influence of starch addition on properties of urea formaldehyde/starch copolymer blends for application as a binder in the coating industry. *J. Environ. Chem. Ecotoxicol.* **2013**, *5*, 181–189.
132. Park, B.D.; Chang Kang, E.; Yong Park, J. Effects of formaldehyde to urea mole ratio on thermal curing behavior of urea–formaldehyde resin and properties of particleboard. *J. Appl. Polym. Sci.* **2006**, *101*, 1787–1792. [CrossRef]
133. Wieland, S.; Pizzi, A.; Grigsby, W.; Warnes, J.; Pichelin, F. Microcrystallinity and colloidal peculiarities of UF/isocyanate hybrid resins. *J. Appl. Polym. Sci.* **2007**, *104*, 2633–2636. [CrossRef]
134. Gu, J.Y.; Zuo, Y.F.; Zhang, Y.H.; Tan, H.Y.; Bin Zhu, L.; Shen, J. Preparation of plywood using starch adhesives modified with isocyanate. *Appl. Mech. Mater.* **2010**, *26-28*, 1065–1068. [CrossRef]
135. Li, H.L.; Liu, G.J.; Zhang, G.X.; Hu, B.; Yu, D.P.; Chen, S.W. The latest progress in study on starch-based adhesives. *Chem. Adhes.* **2008**, *5*.
136. Park, B.D.; Causin, V. Crystallinity and domain size of cured urea–formaldehyde resin adhesives with different formaldehyde/urea mole ratios. *Eur. Polym. J.* **2013**, *49*, 532–537. [CrossRef]

137. Qiao, Z.; Gu, J.; Lv, S.; Cao, J.; Tan, H.; Zhang, Y. Preparation and properties of isocyanate prepolymer/corn starch adhesive. *J. Adhes. Sci. Technol.* **2015**, *29*, 1368–1381. [CrossRef]
138. Shi, J.Y.; Tang, Y.Y. Study on the Rice Straw-Particleboard by Starch-Based API Adhesive. *Adv. Mater. Res.* **2010**, *113*, 1017–1020. [CrossRef]
139. Imam, S.H.; Gordon, S.H.; Mao, L.; Chen, L. Environmentally friendly wood adhesive from a renewable plant polymer: Characteristics and optimization. *Polym. Degrad. Stab.* **2001**, *73*, 529–533. [CrossRef]
140. Basta, A.; El Saied, H.; Lotfy, V.F. Performance of rice straw-based composites using environmentally friendly polyalcoholic polymers-based adhesive system. *Pigment. Resin Technol.* **2013**, *42*, 24–33. [CrossRef]
141. Wang, Z.; Li, Z.; Gu, Z.; Hong, Y.; Cheng, L. Preparation, characterization and properties of starch-based wood adhesive. *Carbohydr. Polym.* **2012**, *88*, 699–706. [CrossRef]
142. Park, B.D.; Kim, Y.S.; Riedl, B. Effect of wood-fiber characteristics on medium density fiberboard (MDF) performance. *J. Korean Wood Sci. Technol.* **2001**, *29*, 27–35.
143. Watkins, D.; Nuruddin, M.; Hosur, M.; Tcherbi-Narteh, A.; Jeelani, S. Extraction and characterization of lignin from different biomass resources. *J. Mater. Res. Technol.* **2015**, *4*, 26–32. [CrossRef]
144. Alonso, M.V.; Rodríguez, J.J.; Oliet, M.; Rodríguez, F.; Garcia, J.; Gilarranz, M.A. Characterization and structural modification of ammonic lignosulfonate by methylolation. *J. Appl. Polym. Sci.* **2001**, *82*, 2661–2668. [CrossRef]
145. Capanema, E.A.; Balakshin, M.Y.; Chen, C.L.; Gratzl, J.S.; Gracz, H. Structural analysis of residual and technical lignins by 1H-13C correlation 2D NMR-spectroscopy. *Holzforschung* **2001**, *55*, 302–308. [CrossRef]
146. Da Silva, C.G.; Grelier, S.; Pichavant, F.; Frollini, E.; Castellan, A. Adding value to lignins isolated from sugarcane bagasse and Miscanthus. *Ind. Crop. Prod.* **2013**, *42*, 87–95. [CrossRef]
147. Cybulska, I.; Brudecki, G.; Rosentrater, K.; Julson, J.L.; Lei, H. Comparative study of organosolv lignin extracted from prairie cord grass, switch grass and corn stover. *Bioresour. Technol.* **2012**, *118*, 30–36. [CrossRef]
148. Ma, R.; Xu, Y.; Zhang, X. Catalytic oxidation of biorefinery lignin to value-added chemicals to support sustainable biofuel production. *ChemSusChem* **2015**, *8*, 24–51.
149. Imam, S.H.; Bilbao-Sainz, C.; Chiou, B.S.; Glenn, G.M.; Orts, W.J. Biobased adhesives, gums, emulsions, and binders: Current trends and future prospects. *J. Adhes. Sci. Technol.* **2013**, *27*, 1972–1997. [CrossRef]
150. Hu, L.; Pan, H.; Zhou, Y.; Zhang, M. Methods to improve lignin's reactivity as a phenol substitute and as replacement for other phenolic compounds: A brief review. *BioResources* **2011**, *6*, 3515–3525. [CrossRef]
151. Ferdosian, F.; Pan, Z.; Gao, G.; Zhao, B. Bio-based adhesives and evaluation for wood composites application. *Polymers* **2017**, *9*, 70. [CrossRef] [PubMed]
152. Kai, D.; Tan, M.J.; Chee, P.L.; Chua, Y.K.; Yap, Y.L.; Loh, X.J. Towards lignin-based functional materials in a sustainable world. *Green Chem.* **2016**, *18*, 1175–1200. [CrossRef]
153. Zhao, W.; Xiao, L.P.; Song, G.; Sun, R.C.; He, L.; Singh, S.; Cheng, G. From lignin subunits to aggregates: Insights into lignin solubilization. *Green Chem.* **2017**, *19*, 3272–3281. [CrossRef]
154. Pizzi, A.; Salvadó, J. Lignin-based wood panel adhesives without formaldehyde. *Holz Roh Werkst.* **2007**, *65*, 65–70.
155. Navarrete, P.; Pizzi, A.; Tapin-Lingua, S.; Benjelloun-Mlayah, B.; Pasch, H.; Rode, K.; Rigolet, S. Low formaldehyde emitting biobased wood adhesives manufactured from mixtures of tannin and glyoxylated lignin. *J. Adhes. Sci. Technol.* **2012**, *26*, 1667–1684. [CrossRef]
156. Mansouri, H.R.; Navarrete, P.; Pizzi, A.; Tapin-Lingua, S.; Benjelloun-Mlayah, B.; Pasch, H.; Rigolet, S. Synthetic-resin-free wood panel adhesives from mixed low molecular mass lignin and tannin. *Eur. J. Wood Wood Prod.* **2011**, *69*, 221–229. [CrossRef]
157. Zhang, Y.; Wu, J.Q.; Li, H.; Yuan, T.Q.; Wang, Y.Y.; Sun, R.C. Heat treatment of industrial alkaline lignin and its potential application as an adhesive for green wood–lignin composites. *ACS Sustain. Chem. Eng.* **2017**, *5*, 7269–7277. [CrossRef]
158. Kües, U. (Ed.) *Wood Production, Wood Technology, and Biotechnological Impacts*; Universitätsverlag Göttingen: Göttingen, Germany, 2007.
159. Pizzi, A. Wood products and green chemistry. *Ann. For. Sci.* **2016**, *73*, 185–203. [CrossRef]
160. Dunky, M. Adhesives based on formaldehyde condensation resins. In *Macromolecular Symposia*; Wiley-VCH Verlag: Weinheim, Germany, 2004; Volume 217, pp. 417–430.
161. Ghahri, S.; Pizzi, A.; Mohebby, B.; Mirshokraie, A.; Mansouri, H.R. Soy-based, tannin-modified plywood adhesives. *J. Adhes.* **2016**, *94*, 218–237. [CrossRef]
162. Pizzi, A. Recent developments in eco-efficient bio-based adhesives for wood bonding: Opportunities and issues. *J. Adhes. Sci. Technol.* **2006**, *20*, 829–846. [CrossRef]
163. Roffael, E.; Dix, B.; Okum, J. Use of spruce tannin as a binder in particleboards and medium density fiberboards (MDF). *Holz Roh Werkst.* **2000**, *58*, 301–305. [CrossRef]
164. López-Suevos, F.; Riedl, B. Effects of Pinus pinaster bark extracts content on the cure properties of tannin-modified adhesives and on bonding of exterior grade MDF. *J. Adhes. Sci. Technol.* **2003**, *17*, 1507–1522. [CrossRef]
165. Gonultas, O. Properties of pine bark tannin-based adhesive produced with various hardeners. *Bioresources* **2018**, *13*, 9066–9078. [CrossRef]
166. Ucar, M.B.; Ucar, G.; Pizzi, A.; Gonultas, O. Characterization of Pinus brutia bark tannin by MALDI-TOF MS and 13C NMR. *Ind. Crop. Prod.* **2013**, *49*, 697–704. [CrossRef]

167. Özacar, M.; Soykan, C.; ŞengülI, A. Studies on synthesis, characterization, and metal adsorption of mimosa and valonia tannin resins. *J. Appl. Polym. Sci.* **2006**, *102*, 786–797. [CrossRef]
168. Li, Z.; Wang, J.; Li, C.; Gu, Z.; Cheng, L.; Hong, Y. Effects of montmorillonite addition on the performance of starch-based wood adhesive. *Carbohydr. Polym.* **2015**, *115*, 394–400. [CrossRef]
169. Santos, J.; Antorrena, G.; Freire, M.S.; Pizzi, A.; González-Álvarez, J. Environmentally friendly wood adhesives based on chestnut (Castanea sativa) shell tannins. *Holz Roh Werkst.* **2017**, *75*, 89–100. [CrossRef]
170. Nath, S.K.; Islam, M.N.; Rahman, K.S.; Rana, M.N. Tannin-based adhesive from Ceriops decandra (Griff.) bark for the production of particleboard. *J. Ind. Acad. Wood Sci.* **2018**, *15*, 21–27. [CrossRef]
171. Cui, J.; Lu, X.; Zhou, X.; Chrusciel, L.; Deng, Y.; Zhou, H.; Brosse, N. Enhancement of mechanical strength of particleboard using environmentally friendly pine (Pinus pinaster L.) tannin adhesives with cellulose nanofibers. *Ann. For. Sci.* **2015**, *72*, 27–32. [CrossRef]
172. Lee, S.H.; Teramoto, Y.; Shiraishi, N. Resol-type phenolic resin from liquefied phenolated wood and its application to phenolic foam. *J. Appl. Polym. Sci.* **2002**, *84*, 468–472. [CrossRef]
173. Chien, M.-Y.; Yang, C.-M.; Chen, C.-H. Effects of physical properties and processing methods on astragaloside IV and flavonoids content in *Astragali radix*. *Molecules* **2022**, *27*, 575. [CrossRef] [PubMed]
174. Antov, P.; Savov, V.; Neykov, N. Sustainable bio-based adhesives for eco-friendly wood composites. A review. *Wood Res.* **2020**, *65*, 51–62. [CrossRef]
175. Qiao, W.; Li, S.; Guo, G.; Han, S.; Ren, S.; Ma, Y. Synthesis and characterization of phenol-formaldehyde resin using enzymatic hydrolysis lignin. *J. Ind. Eng. Chem.* **2015**, *21*, 1417–1422. [CrossRef]
176. Norström, E.; Demircan, D.; Fogelström, L.; Khabbaz, F.; Malmström, E. Green binders for wood adhesives. *Appl. Adhes. Bond. Sci. Technol.* **2018**, *1*, 13–70.
177. Gu, Y.; Cheng, L.; Gu, Z.; Hong, Y.; Li, Z.; Li, C. Preparation, characterization and properties of starch-based adhesive for wood-based panels. *Int. J. Biol. Macromol.* **2019**, *134*, 247–254. [CrossRef]
178. Zhao, X.F.; Peng, L.Q.; Wang, H.L.; Wang, Y.B.; Zhang, H. Environment-friendly urea-oxidized starch adhesive with zero formaldehyde-emission. *Carbohydr. Polym.* **2018**, *181*, 1112–1118. [CrossRef]
179. Qiao, Z.; Gu, J.; Lv, S.; Cao, J.; Tan, H.; Zhang, Y. Preparation and properties of normal temperature cured starch-based wood adhesive. *BioResources* **2016**, *11*, 4839–4849. [CrossRef]
180. Wang, Z.; Gu, Z.; Hong, Y.; Cheng, L.; Li, Z. Bonding strength and water resistance of starch-based wood adhesive improved by silica nanoparticles. *Carbohydr. Polym.* **2011**, *86*, 72–76. [CrossRef]
181. Wang, P.; Cheng, L.; Gu, Z.; Li, Z.; Hong, Y. Assessment of starch-based wood adhesive quality by confocal Raman microscopic detection of reaction homogeneity. *Carbohydr. Polym.* **2015**, *131*, 75–79. [CrossRef]
182. Tan, H.; Zhang, Y.; Weng, X. Preparation of the plywood using starch-based adhesives modified with blocked isocyanates. *Proc. Eng.* **2011**, *15*, 1171–1175. [CrossRef]
183. Gadhave, R.V.; Mahanwar, P.A.; Gadekar, P.T. Starch-based adhesives for wood/wood composite bonding: Review. *Open J. Polym. Chem.* **2017**, *7*, 19–32. [CrossRef]
184. Chen, Y.F.; Li, J.; Chen, Y.X.; Hou, C.M.; Wang, L. Room temperature curing process of urea-formaldehyde/starch composite and its mechanical properties. *Thermosetting Resin* **2010**, *2010*, 3.
185. Samaržija-Jovanović, S.; Jovanović, V.; Konstantinović, S.; Marković, G.; Marinović-Cincović, M. Thermal behavior of modified urea–formaldehyde resins. *J. Therm. Anal. Calorim.* **2011**, *104*, 1159–1166. [CrossRef]
186. Peng, L.; Xiaohua, W. Study on quick dry water resistant type oxidation modification starch adhesive. *Packag. J.* **2012**, 3.
187. Bloembergen, S.; Kappen, F.; Beelen, B. Environmentally friendly biopolymer adhesives and applications based thereon. U.S. Patent No. 6,921,430, 26 July 2005.
188. Sun, J.; Li, L.; Cheng, H.; Huang, W. Preparation, characterization and properties of an organic siloxane-modified cassava starch-based wood adhesive. *J. Adhes.* **2018**, *94*, 278–293. [CrossRef]
189. Wescott, J.M.; Frihart, C.R.; Traska, A.E. High-soy-containing water-durable adhesives. *J. Adhes. Sci. Technol.* **2006**, *20*, 859–873. [CrossRef]
190. Li, K.; Peshkova, S.; Geng, X. Investigation of soy protein-kymene® adhesive systems for wood composites. *J. Am. Oil Chem. Soc.* **2004**, *81*, 487–491. [CrossRef]
191. Lorenz, L.; Frihart, C.R.; Wescott, J.M. Analysis of soy flour/phenol-formaldehyde adhesives for bonding wood. In *Proceedings of the Wood Adhesives 2005, San Diego, CA, USA, 2–4 November 2005*; Forest Products Society: Madison, WI, USA, 2005; pp. 501–505, ISBN 1892529459.
192. Cheng, H.N.; Ford, C.; Dowd, M.K.; He, Z. Use of additives to enhance the properties of cottonseed protein as wood adhesives. *Int. J. Adhes. Adhes.* **2016**, *68*, 156–160. [CrossRef]
193. Yuan, C.; Chen, M.; Luo, J.; Li, X.; Gao, Q.; Li, J. A novel water-based process produces eco-friendly bio-adhesive made from green cross-linked soybean soluble polysaccharide and soy protein. *Carbohydr. Polym.* **2017**, *169*, 417–425. [CrossRef] [PubMed]
194. Lei, H.; Pizzi, A.; Du, G. Environmentally friendly mixed tannin/lignin wood resins. *J. Appl. Polym. Sci.* **2008**, *107*, 203–209. [CrossRef]
195. Zhang, Y.; Ding, L.; Gu, J.; Tan, H.; Zhu, L. Preparation and properties of a starch-based wood adhesive with high bonding strength and water resistance. *Carbohydr. Polym.* **2015**, *115*, 32–37. [CrossRef] [PubMed]

196. Nwokocha, L.M. Adhesive properties of cyanoethyl starch. *J. Adhes. Sci. Technol.* **2011**, *25*, 893–902. [CrossRef]
197. Pradyawong, S.; Qi, G.; Li, N.; Sun, X.S.; Wang, D. Adhesion properties of soy protein adhesives enhanced by biomass lignin. *Int. J. Adhes. Adhes.* **2017**, *75*, 66–73. [CrossRef]
198. Theng, D.; El Mansouri, N.E.; Arbat, G.; Ngo, B.; Delgado-Aguilar, M.; Pèlach, M.À.; Mutjé, P. Fiberboards made from corn stalk thermomechanical pulp and kraft lignin as a green adhesive. *BioResources* **2017**, *12*, 2379–2393. [CrossRef]
199. Liu, C.; Zhang, Y.; Li, X.; Luo, J.; Gao, Q.; Li, J. A high-performance bio-adhesive derived from soy protein isolate and condensed tannins. *RSC Adv.* **2017**, *7*, 21226–21233. [CrossRef]
200. Santiago-Medina, F.; Foyer, G.; Pizzi, A.; Caillol, S.; Delmotte, L. Lignin-derived non-toxic aldehydes for ecofriendly tannin adhesives for wood panels. *Int. J. Adhes. Adhes.* **2016**, *70*, 239–248. [CrossRef]
201. Xiao, Z.; Li, Y.; Wu, X.; Qi, G.; Li, N.; Zhang, K.; Sun, X.S. Utilization of sorghum lignin to improve adhesion strength of soy protein adhesives on wood veneer. *Ind. Crop. Prod.* **2013**, *50*, 501–509. [CrossRef]
202. Tupciauskas, R.; Gravitis, J.; Abolins, J.; Veveris, A.; Andzs, M.; Liitia, T.; Tamminen, T. Utilization of lignin powder for manufacturing self-binding HDF. *Holzforschung* **2017**, *71*, 555–561. [CrossRef]
203. Abdullah, U.H.B.; Pizzi, A. Tannin-furfuryl alcohol wood panel adhesives without formaldehyde. *Eur. J. Wood Wood Prod.* **2013**, *71*, 131–132. [CrossRef]
204. Yusof, M.F.M.; Hashim, R.; Bauk, S.; Sulaiman, O. Characterization of tannin-added Rhizophora spp. particleboards as phantom materials for photon beams. *Ind. Crop. Prod.* **2017**, *95*, 467–474. [CrossRef]
205. Efhamisisi, D.; Thevenon, M.F.; Hamzeh, Y.; Karimi, A.N.; Pizzi, A.; Pourtahmasi, K. Induced tannin adhesive by boric acid addition and its effect on bonding quality and biological performance of poplar plywood. *ACS Sustain. Chem. Eng.* **2016**, *4*, 2734–2740. [CrossRef]
206. Widyorini, R.; Nugraha, P.A.; Rahman, M.Z.A.; Prayitno, T.A. Bonding ability of a new adhesive composed of citric acid-sucrose for particleboard. *BioResources* **2016**, *11*, 4526–4535. [CrossRef]
207. Fitzken Da Vinci, M.; Niro, J.; Kyriazopoulos, M.; Bianchi, S.; Mayer, I.; Eusebio, D.A.; Arboleda, J.R.; Pichelin, F. Development of medium-and low-density fibre boards made of coconut husk and bound with tannin-based adhesives. *Int. Wood Prod. J.* **2016**, *7*, 208–214. [CrossRef]
208. Podschun, J.; Stücker, A.; Buchholz, R.I.; Heitmann, M.; Schreiber, A.; Saake, B.; Lehnen, R. Phenolated lignins as reactive precursors in wood veneer and particleboard adhesion. *Ind. Eng. Chem. Res.* **2016**, *55*, 5231–5237. [CrossRef]
209. Cheng, H.N.; Ford, C.; Dowd, M.K.; He, Z. Effects of phosphorus-containing additives on soy and cottonseed protein as wood adhesives. *Int. J. Adhes. Adhes.* **2017**, *77*, 51–57. [CrossRef]
210. Zheng, P.; Zeng, Q.; Lin, Q.; Fan, M.; Zhou, J.; Rao, J.; Chen, N. Investigation of an ambient temperature-curable soy-based adhesive for wood composites. *Int. J. Adhes. Adhes.* **2019**, *95*, 102429. [CrossRef]
211. Hemmilä, V.; Stergios, A.; Olov, K.; Anuj, K. Development of sustainable bio-adhesives for engineered wood panels–A Review. *RSC Adv.* **2017**, *7*, 38604–38630. [CrossRef]
212. Fabiyi, J.S.; McDonald, A.G.; Wolcott, M.P.; Griffiths, P.R. Wood plastic composites weathering: Visual appearance and chemical changes. *Polym. Degrad. Stab.* **2008**, *93*, 1405–1414. [CrossRef]
213. Butylina, S.; Hyvärinen, M.; Kärki, T. A study of surface changes of wood-polypropylene composites as the result of exterior weathering. *Polym. Degrad. Stab.* **2012**, *97*, 337–345. [CrossRef]
214. Chaochanchaikul, K.; Rosarpitak, V.; Sombatsompop, N. Photodegradation profiles of PVC compound and wood/PVC composites under UV weathering. *Express Polym. Lett.* **2013**, *7*, 146–160. [CrossRef]
215. Khan, M.A.; Haque, N.; Al-Kafi, A.; Alam, M.N.; Abedin, M.Z. Jute Reinforced Polymer Composite by Gamma Radiation: Effect of Surface Treatment with UV Radiation. *Polym. Technol. Eng.* **2006**, *45*, 607–613. [CrossRef]
216. Peng, Y.; Liu, R.; Cao, J.; Chen, Y. Effects of UV weathering on surface properties of polypropylene composites reinforced with wood flour, lignin, and cellulose. *Appl. Surf. Sci.* **2014**, *317*, 385–392. [CrossRef]
217. Syah, A.S.; Rus, A.Z.M.; Aisyah, A.N.A.; Jais, F.H.M.; Sufian, N.A.; Zulhafiz, Z.M.; Amir, N. Physical and mechanical properties of injection molding for waste polypropylene rice husk composite (WPC) upon water absorption and UV irradiation exposure. *Adv. Nat. Appl. Sci.* **2018**, *12*, 32–37.
218. Shkuro, A.E.; Artyomov, A.V.; Savinovskikh, A.V. Physicochemical WPC modification techniques. *Key Eng. Mater.* **2021**, *887*, 144–150. [CrossRef]
219. Zhang, J.; Deng, H.; Taheri, A.; Ke, B.; Liu, C.; Yang, X. Degradation of physical and mechanical properties of sandstone subjected to freeze-thaw cycles and chemical erosion. *Cold Reg. Sci. Technol.* **2018**, *155*, 37–46. [CrossRef]
220. Yu, Y.; Liu, R.; Huang, Y.; Meng, F.; Yu, W. Preparation, physical, mechanical, and interfacial morphological properties of engineered bamboo scrimber. *Constr. Build. Mater.* **2017**, *157*, 1032–1039. [CrossRef]
221. Composites Made from Cellulose-Based Materials and Thermoplastics (Usually Called Wood-Polymer Composites (WPC) or Natural Fibre Composites (NFC)): Test Methods for Characterisation of Compounds and Products, prEN 15534-1. 2012. Available online: https://shop.bsigroup.com/products/composites-made-from-cellulose-based-materials-and-thermoplastics-usually-called-wood-polymer-composites-wpc-or-natural-fibre-composites-nfc-test-methods-for-characterisation-of-compounds-and-products/standard (accessed on 30 December 2021).
222. Alnajjar, H.; Shaker, N.; Naguib, H.; Kandil, U.F.; Ahmed, H.; Farag, A.A. Enhancement of recycled WPC with epoxy nanocomposite coats. *Egypt. J. Chem.* **2019**, *62*, 555–563. [CrossRef]

223. Mohamed, M.R.; Naguib, H.; El-Ghazawy, R.A.; Shaker, N.O.; Amer, A.A.; Soliman, A.M.; Kandil, U.F. Surface Activation of Wood Plastic Composites (WPC) for Enhanced Adhesion with Epoxy Coating. *Mater. Perform. Charact.* **2019**, *8*, 22–40. [CrossRef]
224. Li, X.; Lei, B.; Lin, Z.; Huang, L.; Tan, S.; Cai, X. The utilization of bamboo charcoal enhances wood plastic composites with excellent mechanical and thermal properties. *Mater. Des.* **2014**, *53*, 419–424. [CrossRef]
225. Srubar III, W.V.; Pilla, S.; Wright, Z.C.; Ryan, C.A.; Greene, J.P.; Frank, C.W.; Billington, S.L. Mechanisms and impact of fiber–matrix compatibilization techniques on the material characterization of PHBV/oak wood flour engineered biobased composites. *Compos. Sci. Technol.* **2012**, *72*, 708–715. [CrossRef]
226. Sreekala, M.S.; George, J.; Kumaran, M.G.; Thomas, S. The mechanical performance of hybrid phenol-formaldehyde-based composites reinforced with glass and oil palm fibres. *Compos. Sci. Technol.* **2002**, *62*, 339–353. [CrossRef]
227. Jiang, H.; Pascal Kamdem, D.; Bezubic, B.; Ruede, P. Mechanical properties of poly (vinyl chloride)/wood flour/glass fiber hybrid composites. *J. Vinyl Addit. Technol.* **2003**, *9*, 138–145. [CrossRef]
228. Rizvi, G.M.; Semeralul, H. Glass-fiber-reinforced wood/plastic composites. *J. Vinyl Addit. Technol.* **2008**, *14*, 39–42. [CrossRef]
229. Al Maadeed, M.A.; Kahraman, R.; Khanam, P.N.; Madi, N. Date palm wood flour/glass fibre reinforced hybrid composites of recycled polypropylene: Mechanical and thermal properties. *Mater. Des.* **2012**, *42*, 289–294. [CrossRef]
230. Marcovich, N.E.; Ostrovsky, A.N.; Aranguren, M.I.; Reboredo, M.M. Resin–Sisal and Wood Flour Composites Made from Unsaturated Polyester Thermosets. *Compos. Interfaces* **2009**, *16*, 639–657. [CrossRef]
231. Valente, M.; Sarasini, F.; Marra, F.; Tirillò, J.; Pulci, G. Hybrid recycled glass fiber/wood flour thermoplastic composites: Manufacturing and mechanical characterization. *Compos. Part A Appl. Sci. Manuf.* **2011**, *42*, 649–657. [CrossRef]
232. Sliwa, F.; El Bounia, N.-E.; Charrier, F.; Marin, G.; Malet, F. Mechanical and interfacial properties of wood and bio-based thermoplastic composite. *Compos. Sci. Technol.* **2012**, *72*, 1733–1740. [CrossRef]
233. Cantero, G.; Arbelaiz, A.; Mugika, F.; Valea, A.; Mondragon, I. Mechanical behavior of wood/polypropylene composites: Effects of fibre treatments and ageing processes. *J. Reinf. Plast. Compos.* **2003**, *22*, 37–50. [CrossRef]
234. Lindman, B.; Medronho, B.; Alves, L.; Costa, C.; Edlund, H.; Norgren, M. The relevance of structural features of cellulose and its interactions to dissolution, regeneration, gelation and plasticization phenomena. *Phys. Chem. Chem. Phys.* **2017**, *19*, 23704–23718. [CrossRef]
235. Japanese Industrial Standards (JIS). Plastics-Determination of Charpy Impact Strength: Non-instrumented Impact Test. JIS K 7111-1. 2006. Available online: https://global.ihs.com/doc_detail.cfm?document_name=JIS%20K%207111%2D1&item_s_key=00488434 (accessed on 30 December 2021).
236. Ghahri, S.; Najafi, S.K.; Mohebby, B.; Tajvidi, M. Impact strength improvement of wood flour–recycled polypropylene composites. *J. Appl. Polym. Sci.* **2012**, *124*, 1074–1080. [CrossRef]
237. Sae-Lim, P.; Aht-Ong, D. Physical and mechanical properties of wood-plastics composites: Effect of types and contents of wood flour. *Adv. Mater. Res.* **2013**, *747*, 379–382. [CrossRef]
238. Friedrich, D. Thermoplastic moulding of Wood-Polymer Composites (WPC): A review on physical and mechanical behaviour under hot-pressing technique. *Compos. Struct.* **2021**, *262*, 113649. [CrossRef]
239. Ramesh, R.S.; Kanakuppi, S.; Sharanaprabhu, L.S. Study of hardness and impact behaviour of phenol formaldehyde-based wood plastic composite. *Int. J. Eng. Res. Technol.* **2015**, *3*, 167.
240. Hosseini, S.B. Effects of Dioctyl phthalate and density changes on the physical and mechanical properties of woodflour/PVC composites. *J. Indian Acad. Wood Sci.* **2013**, *10*, 22–25. [CrossRef]
241. Friedrich, D. Natural fiber-reinforced plastics composites: Long-term physico-structural performance in façades. *Acad. J. Civ. Eng.* **2019**, *37*, 412–419.
242. Benthien, J.T.; Ohlmeyer, M. Thickness swelling and water absorption of WPC after immersion in cold and boiling water. *Eur. J. Wood Prod.* **2013**, *71*, 437–442. [CrossRef]
243. Stokke, D.D.; Gardner, D.J. Fundamental aspects of wood as a component of thermoplastic composites. *J. Vinyl Addit. Technol.* **2003**, *9*, 96–104. [CrossRef]
244. Nabinejad, O.; Sujan, D.; Rahman, M.E.; Davies, I.J. Effect of filler load on the curing behavior and mechanical and thermal performance of wood flour filled thermoset composites. *J. Clean. Prod.* **2017**, *164*, 1145–1156. [CrossRef]
245. Bekhta, P.; Proszyk, S.; Krystofiak, T.; Sedliacik, J.; Novak, I.; Mamonova, M. Effects of short-term thermomechanical densification on the structure and properties of wood veneers. *Wood Mater. Sci. Eng.* **2017**, *12*, 40–54. [CrossRef]
246. Hoong, Y.B.; Paridah, M.T. Development a new method for pilot scale production of high grade oil palm plywood: Effect of hot-pressing time. *Mater. Des.* **2013**, *45*, 142–147. [CrossRef]
247. Stark, N.M.; Matuana, L.M.; Clemons, C.M. Effect of processing method on accelerated weathering of wood-flour/HDPE composites. In *Proceedings of Seventh International Conference on Woodfiber-Plastic Composites (and Other Natural Fibers), 19–20 May 2003, Madison, WI, USA*; Forest Products Society: Madison, WI, USA, 2003; pp. 79–87.
248. Ayrilmis, N. Effect of fire retardants on surface roughness and wettability of wood plastic composite panels. *BioResources* **2011**, *6*, 3178–3187.
249. Gindl, M.; Sinn, G.; Gindl, W.; Reiterer, A.; Tschegg, S. A comparison of different methods to calculate the surface free energy of wood using contact angle measurements. *Colloids Surf. A Physicochem. Eng. Asp.* **2001**, *181*, 279–287. [CrossRef]
250. Kada, D.; Migneault, S.; Tabak, G.; Koubaa, A. Physical and mechanical properties of polypropylene-wood-carbon fiber hybrid composites. *BioResources* **2015**, *11*, 1393–1406. [CrossRef]

251. André, L.; Catto Larissa, S.; Montagna Ruth, M.; Santana, C. Abiotic and biotic degradation of post-consumer polypropylene/ethylene vinyl acetate: Wood flour composites exposed to natural weathering. *Polym. Compos.* **2017**, *38*, 571–582.
252. Gozdecki, C.; Wilczynski, A.; Kociszewski, M. Effect of high temperature on the mechanical properties of wood-polymer composites. *Ann. Wars. Univ. Life Sci. SGGW For. Wood Technol.* **2013**, *82*, 291–294.
253. Oksman, K. Improved interaction between wood and synthetic polymers in wood/polymer composites. *Wood Sci. Technol.* **1996**, *30*, 197–205. [CrossRef]
254. Harper, D.; Wolcott, M. Interaction between coupling agent and lubricants in wood–polypropylene composites. *Compos. Part A Appl. Sci. Manuf.* **2004**, *35*, 385–394. [CrossRef]
255. Ranjbarha, Z.; Aberoomand-Azar, P.; Mokhtari-Aliabad, J.; Mirmohammadi, S.A.; Saber-Tehrani, M. High density polyethylene/wood flour composite: Optimization of processing temperature, processing time and coupling agent concentration. *Polym. Polym. Compos.* **2021**, *29*, S106–S116. [CrossRef]
256. Jeske, H.; Schirp, A.; Cornelius, F. Development of a thermogravimetric analysis (TGA) method for quantitative analysis of wood flour and polypropylene in wood plastic composites (WPC). *Thermochim. Acta* **2012**, *543*, 165–171. [CrossRef]
257. Song, Y.; Wang, Y.; Li, H.; Zong, Q.; Xu, A. Role of wood fibers in tuning dynamic rheology, non-isothermal crystallization, and microcellular structure of polypropylene foams. *Materials* **2019**, *12*, 106. [CrossRef] [PubMed]
258. Soccalingame, L.; Bourmaud, A.; Perrin, D.; Bénézet, J.C.; Bergeret, A. Reprocessing of wood flour reinforced polypropylene composites: Impact of particle size and coupling agent on composite and particle properties. *Polym. Degrad. Stab.* **2015**, *113*, 72–85. [CrossRef]
259. Sozen, E.; Aydemir, D.; Zor, M. The effects of lignocellulosic fillers on mechanical, morphological and thermal properties of wood polymer composites/Ucinci lignoceluloznih punila na mehanicka, morfoloskai toplinska svojstva drvnoplasticnih kompozita. *Drvna Industrija* **2017**, *68*, 195–205. [CrossRef]
260. Furtos, G.; Silaghi-Dumitrescu, L.; Pascuta, P.; Sarosi, C.; Korniejenko, K. Mechanical properties of wood fiber reinforced geopolymer composites with sand addition. *J. Natural Fibers* **2021**, *18*, 285–296. [CrossRef]
261. Bengtsson, M.; Oksman, K. The use of silane technology in crosslinking polyethylene/wood flour composites. *Compos. Part A: Appl. Sci. Manuf.* **2006**, *37*, 752–765. [CrossRef]
262. Bengtsson, M.; Oksman, K. Silane crosslinked wood plastic composites: Processing and properties. *Compos. Sci. Technol.* **2006**, *66*, 2177–2186. [CrossRef]
263. Sain, M.; Park, S.H.; Suhara, F.; Law, S. Flame retardant and mechanical properties of natural fibre–PP composites containing magnesium hydroxide. *Polym. Degrad. Stab.* **2004**, *83*, 363–367. [CrossRef]
264. Garcia, M.; Hidalgo, J.; Garmendia, I.; García-Jaca, J. Wood–plastics composites with better fire retardancy and durability performance. *Compos. Part A Appl. Sci. Manuf.* **2009**, *40*, 1772–1776. [CrossRef]
265. Stark, N.M.; White, R.H.; Mueller, S.A.; Osswald, T.A. Evaluation of various fire retardants for use in wood flour–polyethylene composites. *Polym. Degrad. Stab.* **2010**, *95*, 1903–1910. [CrossRef]
266. Kalali, E.N.; Zhang, L.; Shabestari, M.E.; Croyal, J.; Wang, D.Y. Flame-retardant wood polymer composites (WPCs) as potential fire safe bio-based materials for building products: Preparation, flammability and mechanical properties. *Fire Saf. J.* **2019**, *107*, 210–216. [CrossRef]
267. Bouhamed, N.; Souissi, S.; Marechal, P.; Amar, M.B.; Lenoir, O.; Leger, R.; Bergeret, A. Ultrasound evaluation of the mechanical properties as an investigation tool for the wood-polymer composites including olive wood flour. *Mech. Mater.* **2020**, *148*, 103445. [CrossRef]
268. Ramesh, M.; Deepa, C.; Kumar, L.R.; Sanjay, M.R.; Siengchin, S. Life-cycle and environmental impact assessments on processing of plant fibres and its bio-composites: A critical review. *J. Ind. Text.* **2020**, 1528083720924730. [CrossRef]
269. Sommerhuber, P.F.; Wenker, J.L.; Rüter, S.; Krause, A. Life cycle assessment of wood-plastic composites: Analysing alternative materials and identifying an environmental sound end-of-life option. *Resour. Conserv. Recycl.* **2017**, *117*, 235–248. [CrossRef]
270. Väntsi, O.; Kärki, T. Environmental assessment of recycled mineral wool and polypropylene utilized in wood polymer composites. *Resour. Conserv. Recycl.* **2015**, *104*, 38–48. [CrossRef]
271. Khan, M.; Hussain, M.; Deviatkin, I.; Havukainen, J.; Horttanainen, M. Environmental impacts of wooden, plastic, and wood-polymer composite pallet: A life cycle assessment approach. *Int. J. Life Cycle Assess.* **2021**, *26*, 1607–1622. [CrossRef]
272. Elduque, A.; Elduque, D.; Pina, C.; Clavería, I.; Javierre, C. Electricity consumption estimation of the polymer material injection-molding manufacturing process: Empirical model and application. *Materials* **2018**, *11*, 1740. [CrossRef]
273. Pizzol, M.; Laurent, A.; Sala, S.; Weidema, B.; Verones, F.; Koffler, C. Normalisation and weighting in life cycle assessment: Quo vadis? *Int. J. Life Cycle Assess.* **2017**, *22*, 853–866. [CrossRef]
274. Beigbeder, J.; Soccalingame, L.; Perrin, D.; Bénézet, J.C.; Bergeret, A. How to manage biocomposites wastes end of life? A life cycle assessment approach (LCA) focused on polypropylene (PP)/wood flour and polylactic acid (PLA)/flax fibres biocomposites. *Waste Manag.* **2019**, *83*, 184–193. [CrossRef]
275. Ekvall, T.; Azapagic, A.; Finnveden, G.; Rydberg, T.; Weidema, B.P.; Zamagni, A. Attributional and consequential LCA in the ILCD handbook. *Int. J. Life Cycle Assess.* **2016**, *21*, 293–296. [CrossRef]
276. Dahlbo, H.; Bachér, J.; Lähtinen, K.; Jouttijärvi, T.; Suoheimo, P.; Mattila, T.; Saramäki, K. Construction and demolition waste management–a holistic evaluation of environmental performance. *J. Clean. Prod.* **2015**, *107*, 333–341. [CrossRef]

277. Merkel, K.; Rydarowski, H.; Kazimierczak, J.; Bloda, A. Processing and characterization of reinforced polyethylene composites made with lignocellulosic fibres isolated from waste plant biomass such as hemp. *Compos. Part B Eng.* **2014**, *67*, 138–144. [CrossRef]
278. Cherubini, F.; Huijbregts, M.; Kindermann, G.; Van Zelm, R.; Van der Velde, M.; Stadler, K.; Strømman, A.H. Global spatially explicit CO_2 emission metrics for forest bioenergy. *Sci. Rep.* **2016**, *6*, 20186. [CrossRef] [PubMed]
279. Viksne, A.; Berzina, R.; Andersone, I.; Belkova, L. Study of plastic compounds containing polypropylene and wood derived fillers from waste of different origin. *J. Appl. Polym. Sci.* **2010**, *117*, 368–377. [CrossRef]
280. Turku, I.; Keskisaari, A.; Kärki, T.; Puurtinen, A.; Marttila, P. Characterization of wood plastic composites manufactured from recycled plastic blends. *Compos. Struct.* **2017**, *161*, 469–476. [CrossRef]
281. Taufiq, M.; Mansor, M.R.; Mustafa, Z. Characterisation of wood plastic composite manufactured from kenaf fibre reinforced recycled-unused plastic blend. *Compos. Struct.* **2018**, *189*, 510–515. [CrossRef]
282. Jacob, A. WPC industry focuses on performance and cost. *Reinf. Plast.* **2006**, *50*, 32–33. [CrossRef]
283. Bledzki, A.; Franciszczak, P.; Osman, Z.; Elbadawi, M. Polypropylene biocomposites reinforced with softwood, abaca, jute, and kenaf fibers. *Ind. Crop. Prod.* **2015**, *70*, 91–99. [CrossRef]
284. Zhou, M.; Gu, W.; Wang, G.; Zheng, J.; Pei, C.; Fan, F.; Ji, G. Sustainable wood-based composites for microwave absorption and electromagnetic interference shielding. *J. Mater. Chem. A* **2020**, *8*, 24267–24283. [CrossRef]
285. Kristak, L.; Kubovský, I.; Réh, R. New challenges in wood and wood-based materials. *Polymers* **2021**, *13*, 2538. [CrossRef]
286. Gubana, A.; Melotto, M. Experimental tests on wood-based in-plane strengthening solutions for the seismic retrofit of traditional timber floors. *Constr. Build. Mater.* **2018**, *191*, 290–299. [CrossRef]
287. Sun, J.; Jiang, Z.; Liang, Y.; Liu, J. Research status and prospects of bamboo-wood composite container flooring. *World For. Res.* **2018**, *31*, 36–41.
288. Katsiroumpas, K.; Carels, P.; Masoumi, H.; Salkauskis, J. Lightweight floating floor innovations in gym/sports applications. In Proceedings of the INTER-NOISE and NOISE-CON Congress and Conference Proceedings, Chicago, IL, USA, 26–29 August 2018; 2018; Volume 258, pp. 1075–1084.
289. Chen, C.; Kuang, F.; Tor, O.; Quin, F.; Xiong, Z.; Zhang, J. Static lateral load capacity of extruded wood-plastic composite-to-metal single-bolt connections, considering failure at the ends. *Bioresources* **2019**, *14*, 8987–9000.

polymers

MDPI

Article

Properties of Low-Cost WPCs Made from Alien Invasive Trees and rLDPE for Interior Use in Social Housing

Abubakar Sadiq Mohammed and Martina Meincken *

Department of Forest and Wood Science, University of Stellenbosch, Private Bag X1, Matieland, Stellenbosch 7602, South Africa; 23513187@sun.ac.za
* Correspondence: mmein@sun.ac.za

Abstract: Low-cost wood–plastic composites (WPCs) were developed from invasive trees and recycled low-density polyethylene. The aim was to produce affordable building materials for low-cost social housing in South Africa. Both raw materials are regarded as waste materials, and the subsequent product development adds value to the resources, while simultaneously reducing the waste stream. The production costs were minimised by utilising the entire biomass of *Acacia saligna* salvaged from clearing operations without any prior processing, and low-grade recycled low-density polyethylene to make WPCs without any additives. Different biomass/plastic ratios, particle sizes, and press settings were evaluated to determine the optimum processing parameters to obtain WPCs with adequate properties. The water absorption, dimensional stability, modulus of rupture, modulus of elasticity, tensile strength, and tensile moduli were improved at longer press times and higher temperatures for all blending ratios. This has been attributed to the crystallisation of the lignocellulose and thermally induced cross-linking in the polyethylene. An increased biomass ratio and particle size were positively correlated with water absorption and thickness swelling and inversely related with MOR, tensile strength, and density due to an incomplete encapsulation of the biomass by the plastic matrix. This study demonstrates the feasibility of utilising low-grade recycled polyethylene and the whole-tree biomass of *A. saligna*, without the need for pre-processing and the addition of expensive modifiers, to produce WPCs with properties that satisfy the minimum requirements for interior cladding or ceiling material.

Keywords: eco-friendly wood-based composites; wood polymer composites; bio composites; wall cladding; ceiling materials; RDP housing

Citation: Mohammed, A.S.; Meincken, M. Properties of Low-Cost WPCs Made from Alien Invasive Trees and rLDPE for Interior Use in Social Housing. *Polymers* **2021**, *13*, 2436. https://doi.org/10.3390/polym13152436

Academic Editor: Pavlo Bekhta

Received: 8 July 2021
Accepted: 16 July 2021
Published: 24 July 2021

1. Introduction

Alien invasive plants (AIP) in South Africa have enormous adverse consequences on biodiversity and ecosystem services and have led to direct environmental degradation [1,2]. With their high affinity for water and far-reaching roots and rapid spread, they often deplete water resources and degrade the soil, thereby water-stressing other plants and eventually alienating native flora and fauna in a local ecosystem. This has prompted the South African government to commit to a program aimed at clearing the most invasive plants. The cleared biomass is typically left behind to dry, where it poses a fire risk or, at best, is used as firewood. The issue of what to do with the biomass apart from the utilisation of fractions as firewood has emerged as a matter of concern.

Similarly, the surge in plastic pollution and its impact on the health and safety of ecosystems [3–5] is demanding attention. Plastic disposal in landfills results in leakages into the natural environment and consequently has dire adverse effects on wildlife, while incineration releases poisonous emissions. Plastic is therefore increasingly recycled, although the South African recycling industry is not quite as well developed as in Europe.

Making wood plastic composite (WPC) materials from these materials adds value to two different waste streams and can alleviate the financial burden of the government in managing these wastes and potentially form an employment- and income-generating enterprise by reprocessing them into secondary materials [5,6].

In South Africa, the government is trying to rectify the imbalance of property ownership caused by apartheid. Through the Reconstruction and Development Program (RDP), low-cost social housing units—called RDP houses—are being built across the country to address the enormous housing deficits [7,8]. However, the occupants of these budget houses are often exposed to the hazards of environmental stressors, such as heat and humidity, due to the low-quality of the building materials used [9]. The walls typically consist of a single-layer brick wall, without any additional insulation. Moisture condenses on the walls, resulting in mould, which increases their susceptibility to biodegradation and compromises the indoor air quality and health of occupants. The roofs are also often leaky, dripping water onto mostly wood or gypsum ceiling boards, which become soaking wet, develop discolouration, swell out of proportion, and deteriorate in a short time. These houses typically require maintenance shortly after construction, which the occupants cannot afford, resulting in undesirable living standards. Consequently, the use of interior ceiling and wall cladding materials that are high in hydrophobic properties to resist wetting-induced biodegradation while offering extended service lives is necessary. Interior cladding also offers the additional benefits of thermal and acoustic insulation against extreme weather conditions and community noise, which is typical in densely populated RDP settlements.

Utilising abundantly available recycled waste materials in the form of recycled low-density polyethylene (rLDPE) and AIPs as feedstock to develop low-cost WPC boards as wall cladding and ceiling material in RDP houses would add significantly to the living standard in these houses. The cost of the WPC materials is minimised by avoiding any unnecessary processing steps or additives and by using widely available raw materials that are otherwise regarded as waste materials that need to be removed. Recycled LDPE was chosen because of its abundant availability and low cost, as one of the main objectives was to keep the processing and manufacturing costs as low as possible. The recycled polymer was nominally linear low-density polyethylene (LLDPE), which has approximately the same density as LDPE, but with the linearity of HDPE and fairly short polymer chains. However, the obtained polymer was of the lowest quality. The plastic is sorted based on its polymer code and ranked into three quality grades—A, B, and C—in descending order. A-grade plastic is clean material generally sourced from shopping centres and recycled into pellets used for non-food packaging, such as refuse bags, furniture coverings, etc. C-grade is contaminated plastic that is often used in composite materials. Because of the impurities and the uncertainty of the exact composition, we refer to the recycled plastic material as rLDPE, rather than LLDPE.

After over a century since their first introduction, WPCs have evolved from wood-thermoset resin mixtures [10], through in-situ polymerization of plastic monomers within wood pores [10,11], to various blends of woody particles with thermoplastic polymers, as new thermoplastics with new properties and new technologies are continuously emerging [12]. The adhesion mechanism and bond strength between wood and plastic have been found to largely depend on the interlocking mechanism of a continuous plastic matrix through the wood cell lumen and intercellular pores, rather than the chemical bonds between polar wood cells and non-polar plastic [12–14]. Non-woody biomass, such as bark and leaves, are typically considered contaminants and avoided, because they generally lack the properties necessary for strength and bond formation [12,15,16]. Though leaves, bark, and twigs typically contain high amounts of lignin, which add stability, these are offset by the higher amounts of hemicelluloses, compared to woody biomass. Hemicelluloses are loosely bound non-crystalline fibres that oxidise easily and show a low thermal stability [12,13]. The inclusion of non-woody biomass into composites may thus lower their mechanical and thermal stability. In addition, leaves and bark are known to have a high extractive content [13,14,17]. Extractives are known to be rich in functional groups and affect the surface chemistry of wood [18]. Consequently, the bonds in WPC blends that depend on functional groups on the biomass surface may be enhanced by utilising biomass with a large extractive content. However, since polymers are largely unipolar, the majority of the bonds between wood and polymer rely on mechanical interlocking, rather than on chemical adhesion [12,19,20]. Contaminants may, however, add to the performance of WPCs. The higher lignin content in leaves and bark contributes to the moisture resistance, dimensional stability, and durability of composites [21,22] due to the hydrophobic nature of lignin and therefore provides some resistance against biological decay.

In recent reports, various kinds of non-wood lignocellulosic materials, including corncobs, nut shells [23], banana fibres [24], bamboo [25,26], rice husks [27], and many other unconventional biomass types, have found a use in WPC formulations. Owing to their chemical incompatibility, various protocols of pre-treatments have been applied in order to enhance interfacial bonding between biomass and plastic [13,14,28], which led to the introduction of coupling agents, such as maleic anhydride-grafted polypropylene (MAPP), Poly(ethylene-co-vinyl alcohol) (EVOH), and Polyethylene-graft maleic anhydride (PE-g-MA) to enhance bond strength [10,14,29]. However, coupling agents are costly and immensely increase the production costs of WPCs.

To manufacture WPC boards, temperatures between 140–200 °C [16,21,22,25,30] and press times between 5–30 min [17,23,24,31,32] are typically used. These press parameters significantly affect the properties of the WPCs. The upper and lower bound press temperature and time are usually given by the degradation temperature of the biomass type and the melting temperature of the thermoplastic [10,11,16,18]. Within these parameters are many possible combinations of processing settings, which affect the product properties.

Reports on the effect of biomass particle sizes and geometry on the properties of the resulting WPC boards are conflicting. Whereas some researchers report an improvement in some properties and a decrease in other properties with decreasing particle sizes [12,18,33–35], others conclude that particle size is less important to the final board properties than particle geometry [12,25,34,36,37]. Particle size reportedly affects the viscosity and melt flow rate during board formation [30,37,38] and consequently determines board–moisture relations.

Blending ratios between polymer matrix and wood reinforcements or fillers have been widely investigated [21,34,37–41] Depending on the intended properties and production method, ratios between 10% biomass, where the wood particles act as a filler in a continuous plastic matrix, and 90%, where the plastic acts as a binder, rather than a continuous matrix, have been reported. The biomass content at the higher extremes is usually the sole preserve of compression moulding methods, while the lower extremes may be injection moulded or extruded. A higher wood ratio generally results in an increased strength and stiffness, but it may also result in an increased moisture sorption and risks of microbial invasion [24,42].

In a previous study, the use of biomass obtained from *A. saligna*, *A. mearnsii* and *E. camaldulensis*—all invasive trees in South Africa—to produce WPCs was investigated [43]. The biomass was added in two forms: wood only and particles obtained by chipping and milling the entire tree with leaves, twigs, and bark. The results showed that regardless of the biomass type, WPCs made with *A. saligna* had superior mechanical properties, while composites with wood-only biomass were found to have better mechanical properties than composites with whole-tree biomass. In order to explain the superior properties of *A. saligna*, a chemical analysis was conducted on these species to identify the origin of the different properties of the resulting WPCs. Furthermore, it was essential to determine the optimum processing parameters to make WPC boards from entire trees of *A. saligna* and rLDPE. Consequently, different wood to plastic ratios and particle sizes were analysed, as well as different press temperatures and times. The physical and mechanical properties of the resulting WPC boards were analysed, and the effect of processing parameters on the performance of the final composites was determined.

2. Materials and Methods

2.1. Materials

Acacia saligna (Port Jackson) was obtained from clearing operations in the Western Cape of South Africa. Without any further processing, the trees were chipped, including wood, bark, leaves, seeds, and twigs, and air-dried (Figure 1a). The resulting biomass was passed through a 2 mm screen in a hammer mill and further air-dried to about 12% MC. Low-grade recycled low density polyethylene (rLDPE) consisting mostly of recycled shopping bags conglomerates, shown in Figure 1b, with a density of 0.915–0.950 g/cm^3, was sourced from Atlantic Plastic Recycling (APR) CC, a plastic waste recycling plant in Cape Town, South Africa.

(a) (b)

Figure 1. (**a**) Milled whole-tree biomass of *A. saligna* and (**b**) rLDPE.

2.2. Composite Preparation

The biomass was chipped in an OC1 knife chipper (Figure 2a) from Heemaf, Netherlands and milled in a S1 hammermill from Drotsky, South Africa (Figure 2b) to a 2 mm particle size. The raw materials were mixed at three different biomass to plastic ratios—50:50, 60:40, and 70:30—and subsequently compounded in a custom built blender, shown in Figure 2c. The blending process resulted in a frictional breakdown of the WPC granules (Figure 3a), which were size separated into two fractions and characterised according to the procedure described in Section 2.3, from which the WPC boards were pressed.

(a) (b) (c) (d)

Figure 2. (**a**) Knife chipper, (**b**) hammermill, (**c**) custom-built blender, and (**d**) hydraulic press.

The WPC granules were hot pressed at 150 and 180 °C for 10 and 30 min alternately in a hydraulic press from BURKLE, Germany in a 25 × 25 × 4 mm mould, shown in Figure 2d). The resulting boards shown in Figure 3b) were labelled with a three-digit sample code, where the first digit denotes the wood content (5, 6 and 7 = 50, 60 and 70 wt.%), the second letter identifies the particle size (S = 0.31 mm and L = 0.47), and the third letter (A, B, C, or D) identifies the time–temperature combination, where A: t = 10 min and T = 150 °C; B: t

= 30 min and T = 150 °C; C: t = 10 min and T = 180 °C; and D: t = 30 min and T = 180 °C. Table 1 lists the processing parameters.

(a) (b)

Figure 3. (**a**) Dry-compounded feedstock and (**b**) sample board, after hot-pressing at 180 °C for 30 min.

Table 1. Blending ratios and time–temperature-mix treatments of WPCs at ***S*** (0.31 mm) and ***L*** (0.47) particle sizes.

Blending Ratio (wt. %)	Particle size *S*			Pressure (kg/cm^2)	Particle Size *L*		
	Press Temperature (°C)	Press Time (min)	Sample Code		Press Temperature (°C)	Press Time (min)	Sample Code
50:50	150	10	5SA	2500	150	10	5LA
		30	5SB			30	5LB
	180	10	5SC		180	10	5LC
		30	5SD			30	5LD
60:40	150	10	6SA		150	10	6LA
		30	6SB			30	6LB
	180	10	6SC		180	10	6LC
		30	6SD			30	6LD
70:30	150	10	7SA		150	10	7LA
		30	7SB			30	7LB
	180	10	7SC		180	10	7LC
		30	7SD			30	7LD

2.3. Size Analysis of Dry-Compounded Feedstock

Blending of the rLDPE and biomass in a rotary drum compounder resulted in a further frictional fractionation into particle sizes, which are a function of the blending time. The raw materials were compounded for 45 and 90 min, resulting in two different particle fractions designated as '*L*' for larger particles from the 45 min compounding time and '*S*' for the smaller graded particles obtained from the 90 min blending cycle. For a better description of the particle size distribution, samples of 150 g from the compounded feedstock were characterised through sieve analysis in an AS 200 shaker from Retsch, Germany the details of which are presented in Figure 4, indicating D50 as the particle diameter below which 50% of the particles were located.

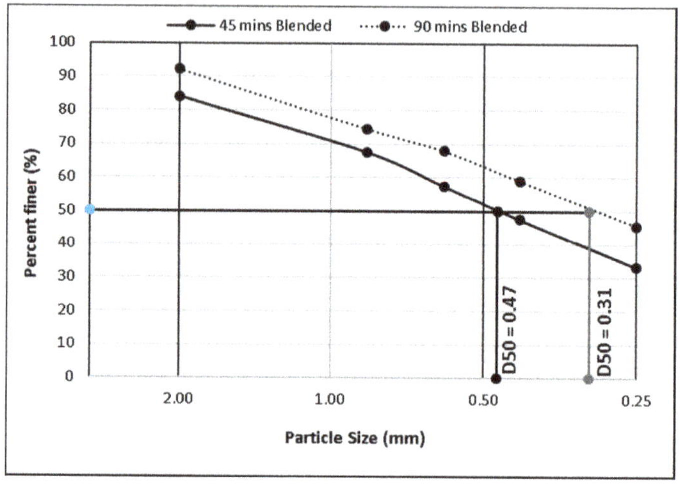

Figure 4. Particle size distribution, showing the variation in D50 as a function of the compounding time.

2.4. Physical Properties

The specimens were conditioned at 20 °C and a 65% ± 5% relative humidity for 4 weeks, prior to testing. Two boards were produced for each treatment, and eight samples per treatment were tested to obtain a mean and a standard deviation. The moisture content (*MC*) was determined in accordance with ASTM D4442-07, using Equation (1):

$$MC\ (\%) = \frac{Original\ mass(g) - Ovendry\ mass(g)}{Ovendry\ mass(g) \times 100} \tag{1}$$

The water absorption (*WA*) and thickness swell (*TS*) of 16 samples per treatment were carried out in accordance with ASTM D1037-12. The specimens were immersed in distilled water for 24 h, 7-, 14-, 21-, and 28-day immersion periods. At the end of each immersion period, the specimens were removed from the water and weighed to the nearest 0.01 g, and their thickness was measured with a digital calliper to the nearest 0.01 mm. The values for *WA* were calculated according to Equation (2):

$$WA_{\rm t}\ (\%) = \frac{W_t - W_0}{W_0} \times 100 \tag{2}$$

where $WA_{\rm t}$ is the water absorption (%) at time t. W_0 and W_t are the respective weights of the specimens prior to immersion and after a given immersion time t.

TS was determined according to Equation (3):

$$TS_{\rm t}\ (\%) = \frac{D_t - D_0}{D_0} \times 100 \tag{3}$$

where $TS_{\rm t}$ is the thickness swelling (%) at time t. D_0 and $D_{\rm t}$ are the respective thicknesses of the specimens prior to immersion and after a given immersion time t.

The board densities were determined according to Equation (4), as prescribed by ASTM standard D2395 (2014):

$$\text{Density}\ (\rho) = \frac{Ovendry\ mass}{airdried\ volume} \tag{4}$$

2.5. Chemical Properties

The compositional analysis of the biomass from *Acacia saligna*, *Acacia mearnsii*, and *Eucalyptus camaldulensis* was conducted in accordance with the standards, NREL (LAP) TP-510-42620, TP-510-42622, TP-510-42621, TP-510-42618, and TP-510-42623 (2008) for ash content, hot water- and ethanol-soluble extractives (setup shown in Figure 5), lignin, sugar, and bulk density in order to determine how the composition affects the physical properties and performance of the WPCs.

Figure 5. Soxhlet extraction setup for biomass compositional analysis.

2.6. Mechanical Properties

2.6.1. Static Bending

Four samples per treatment with a size of 195 × 50 × 4 mm were tested in a three-point bending test in accordance with ASTM D 1037-12 using a universal testing machine from Hottinger Baldwin Messtechnik (HBM), Germany (Figure 6a). The modulus of rupture (*MOR*) and modulus of elasticity (*MOE*) were determined according to Equations (5) and (6). All the calculated values are reported as the averages with the standard deviation.

$$MOR = \frac{3P_{max}L}{2bd^2} \quad (5)$$

$$MOR = \frac{L^3}{4bd^3}\frac{\Delta P}{\Delta y} \quad (6)$$

(**a**) (**b**)

Figure 6. Instron setup for (**a**) the 3-point bending test, and (**b**) tensile test.

2.6.2. Tensile Tests

Five dumbbell shaped samples with a size of 165 × 20 × 4 mm per treatment were tested in the tensile mode in accordance with ASTM D 638-14 using an Universal testing machine 4411 from Instron, Maine, USA equipped with a 5 kN load cell (Figure 6b), at a test speed of 50 mm/min. The tensile strength and tensile modulus were calculated based on Equations (7) and (8), and the average values are reported with the standard deviation.

$$T_S = \frac{P_{max}}{bd} \quad (7)$$

$$E_t = \frac{lg}{bd}\frac{\Delta P}{\Delta y} \quad (8)$$

2.7. Statistical Analysis

A factorial ANOVA with a replicates test in Statistica software 14.0.0.15 was used to determine significant differences within and between treatments at a significance level of $\alpha = 0.05$. Pearson's correlation analysis was conducted in R to quantify the contribution of all the independent variables on the physical properties of the boards.

3. Results

The findings from a previous study by Acheampong et al. [43], analysing WPCs made of biomass obtained from the whole tree or only the wood from *A. saligna*, *A. mearnsii* and *E. camaldulensis*, showed that the addition of bark, twigs, and leaves reduces the mechanical properties somewhat, but all boards still met the minimum requirements for interior use. A surprising finding, however, was that regardless of the biomass type, boards made with *A. saligna* showed significantly better mechanical properties than those made with the other two wood species. This was an unexpected result, especially the large difference between the two—otherwise very similar—Acacia species. To better understand the differences in performance, the chemical composition of the three different wood species was determined.

3.1. Chemical Composition

The chemical composition of the wood from *A. saligna*, *A. mearnsii*, and *E. camaldulensis* is presented in Table 2. *A. saligna* had the highest cellulose content of the three species, but with a low hemicelluloses content and average lignin content. The superior mechanical properties of WPCs made with *A. saligna* can be attributed to the high cellulose content, as the crystalline cellulose fibres are known to impart strength. The total extractive content of *A. saligna* was higher than that of the other two species and appeared to have no negative effect on the strength properties. On the contrary, the increased extractive content seemed to have aided the bonding between the biomass and the polymer matrix.

Table 2. Compositional analysis of the wood of *A. saligna*, *A. mearnsii*, and *E. camaldulensis*.

Parameters (%)	Biomass Type		
	A. saligna	***A. mearnsii***	***E. camaldulensis***
Lignin	19.69 (0.48)	18.92 (0.57)	21.14 (1.96)
Hemicelluloses	15.18 (0.69)	16.24 (0.10)	17.36 (1.23)
Cellulose	41.97 (2.54)	35.55 (0.49)	28.74 (2.34)
Water Extractives	7.62 (0.41)	6.02 (1.22)	4.85 (0.78)
Ethanol Extractives	0.18 (0.19)	0.15 (0.09)	0.18 (0.04)
Total Extractives	7.80 (0.19)	6.17 (0.92)	5.03 (0.64)
Ash	1.07 (0.47)	1.30 (0.76)	0.93 (0.08)

Values in parenthesis are the standard deviations.

Based on the chemical composition and the results from Acheampong et al. [43], boards for further analysis were made from the entire tree of *A. saligna*, and the processing parameters were optimised to obtain the best physical and mechanical properties.

3.2. Water Absorption (WA) and Thickness Swelling (TS)

Observations at 24 h WA of all WPCs (Figure 7) indicate that regardless of the compounding ratio and particle size, the water absorption of all treatment '*A*' samples pressed at T = 150 °C and t = 10 min was the highest, while the WA of treatment '*D*' samples pressed at T = 180 °C and t = 30 min was the lowest. This is because at 180 °C, the biomass acquires a hydrophobic character due to thermal modification, which leads to the bonding of amorphous hydroxyl groups and a reduction of the available bonding sites for water [44,45]. The higher temperature also leads to a reduction in the melt–flow viscosity of rLDPE [14,46]. With a melt temperature of maximum 140 °C, the melt flow rates of the rLDPE at 180 °C would have been high. An increased flow rate and extended pressing time will increase the potential of the polymer to infiltrate the cell wall micro pores, impregnate the cell lumen and encrust entire particles, therefore decreasing the water absorption. At 150 °C, however, which is barely above the melting temperature of the rLDPE, the plastic remained more viscous and left more hydroxyl groups on the wood fibre surface exposed as potential binding sites for water [16,46,47]. While an increase in either the press time (treatment B) or temperature (treatment C) resulted in a reduced *WA*, there was no clear trend as to which of the two factors has more impact in reducing *WA*. However, increasing both factors resulted in the lowest *WA*.

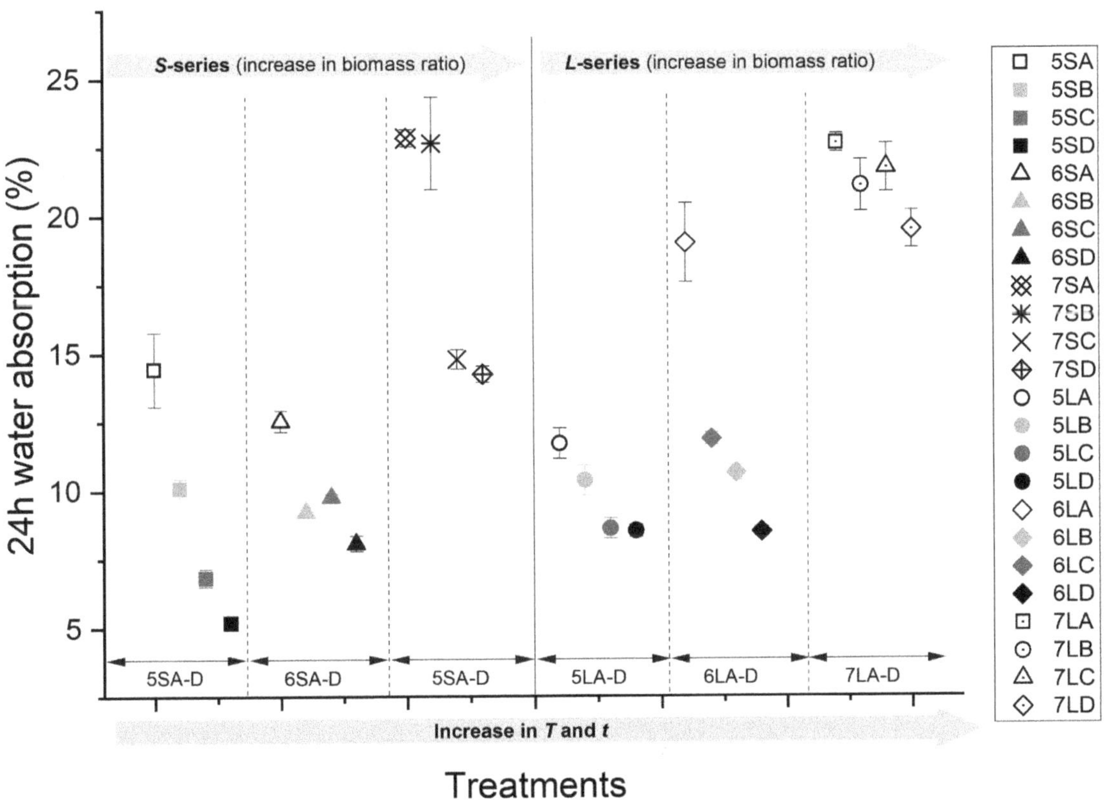

Figure 7. A 24 h WA of WPCs pressed at different temperatures and times (A–D) with 50, 60, and 70% biomass at two particle sizes (*S* and *L*).

Figure 7 shows the *WA* for small (*S*-series) and large particles (*L*-series) at increasing biomass ratios. The trend indicates that an increase in the biomass ratio from 50–70% resulted in a linear increase in *WA* from below 15% to over 20%. As the biomass is hydrophilic and the plastic is hydrophobic, an increase in the biomass to plastic ratio means more available potential binding sites for water, which will increase the *WA*.

Similar to the observed trends in WA, the TS (Figure 8) was independent of the particle size, and the longest pressing time and highest temperature resulted in the lowest TS across all blending ratios. While the increase in the press temperature or time alone resulted in an increased dimensional stability, none of the two factors showed any superiority over the other; however, the effect of the combined increase in press temperature and time resulted in the lowest TS (below 2%).

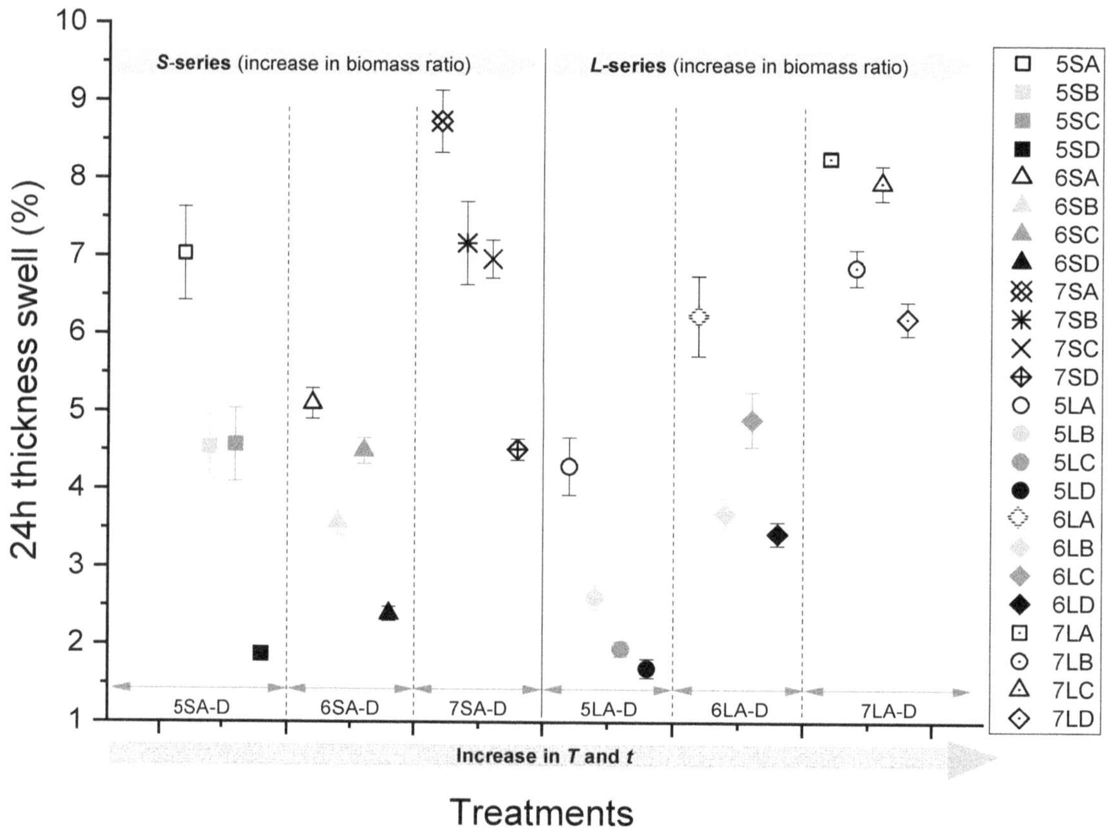

Figure 8. A 24 h TS of WPCs pressed at different temperatures and times (A–D) with 50, 60, and 70% biomass at two particle sizes (*S* and *L*).

The short-term (2 h) and long-term (672 h) *WA* and *TS* for the various press settings (A–D) are presented in Table 2 for all blending ratios and particles sizes. As can be seen, the moisture absorption for press temperatures of T = 150 °C and t = 10 min (A) was the largest, particularly for WPC samples with a 70% biomass, which exceeded 30% MC after water immersion for 672 h. At press settings of T = 180 °C and t = 30 min (D), the lowest WA was recorded. WPCs made with larger particles generally absorbed more water than those made with smaller particles.

Board densities (Table 3) are a direct function of the blending ratio affected by the press temperature and time combination. Samples with a 50% biomass were consistently denser for all press settings (A–D) than WPCs made with 60% or 70% biomass, which had the lowest density. The presence of non-woody biomass in the form of leaves, twigs, and bark significantly reduced the density of the entire biomass below that of the rLDPE. The leaves with their characteristically thin cell walls, numerous intercellular spaces, and loosely packed spongy cells may not have contributed much to the weight but contributed enormously to the bulk volume of the biomass. Consequently, an increase in the biomass ratio resulted in an increase in the filler volume, without a proportionate increase in the mass. The combined effect of a high temperature at 180 °C and longer press time at 30 min under a 250 kg/cm^2 pressure created a suitable condition for a reduced shear viscosity and improved the melt flow rate of the rLDPE, which generally has a melting temperature of 140 °C. The resulting boards were thus more compressed due to the better plastic melt-flow kinetics, and the smaller sized biomass particles allowed for a better compaction and improved the densification [15,35]. At a 70% biomass content, however, the plastic played the role of a binder, rather than a matrix. Since the biomass has a lower density than the plastic, a plastic weight ratio of 30% resulted in a much lower volume ratio, compared to the biomass. This disproportionate volume ratio implied that longer pressing times and higher press temperature were required for the plastic to melt and wet the biomass fibre surfaces, before successfully binding the particles together.

Table 3. Board density (ρ) and WA and TS of the WPCs made with small (*S*) and large (*L*) particle sizes at different blending ratios after 2 h, 24 h, and 672 h.

Treatment	Sample Code [a]	ρ (g/cm^3)	WA (h)			TS (h)			Sample Code [a]	ρ (g/cm^3)	WA (h)			TS (h)		
			2	24	672	2	24	672			2	24	672	2	24	672
A	5SA	1.61	3.27 (0.23)	14.43 (1.34)	30.76 (2.44)	4.39 (0.39)	7.03 (0.60)	9.60 (0.55)	5LA	1.49	3.68 (0.17)	11.71 (0.56)	29.64 (0.83)	2.56 (0.27)	4.30 (0.37)	8.40 (0.68)
	6SA	0.93	5.27 (0.12)	12.54 (0.38)	33.96 (0.62)	1.42 (0.06)	5.10 (0.20)	9.30 (0.30)	6LA	0.94	5.70 (0.27)	19.01 (1.44)	37.46 (1.86)	3.79 (0.25)	6.23 (0.52)	8.94 (0.58)
	7SA	0.96	10.69 (0.19)	22.81 (0.28)	37.05 (0.45)	5.70 (0.15)	8.74 (0.40)	11.46 (0.32)	7LA	0.76	11.40 (0.18)	22.63 (0.34)	37.40 (1.44)	4.76 (0.05)	8.27 (0.07)	12.06 (0.12)
B	5SB	1.52	2.84 (0.15)	10.10 (0.33)	30.53 (0.83)	3.99 (0.14)	4.54 (0.42)	6.91 (0.60)	5LB	1.54	2.50 (0.08)	10.37 (0.54)	28.18 (1.09)	0.71 (0.05)	2.61 (0.15)	6.26 (0.35)
	6SB	1.01	2.65 (0.06)	9.23 (0.20)	30.30 (0.59)	0.65 (0.03)	3.57 (0.16)	6.49 (0.31)	6LB	0.97	2.95 (0.08)	10.63 (0.28)	35.58 (0.85)	0.43 (0.02)	3.70 (0.20)	7.19 (0.32)
	7SB	0.93	8.72 (0.44)	22.65 (1.68)	34.09 (2.39)	3.89 (0.22)	7.17 (0.53)	9.61 (0.71)	7LB	0.88	8.30 (0.22)	21.07 (0.95)	34.87 (1.76)	2.53 (0.05)	6.87 (0.23)	10.79 (0.31)
C	5SC	1.46	2.79 (0.10)	6.83 (0.31)	27.39 (2.28)	3.15 (0.34)	4.57 (0.47)	6.86 (0.63)	5LC	1.39	2.97 (0.12)	8.62 (0.36)	27.36 (0.74)	0.65 (0.04)	1.95 (0.10)	5.90 (0.27)
	6SC	1.12	2.73 (0.05)	9.77 (0.11)	31.02 (0.25)	1.30 (0.06)	4.49 (0.17)	6.95 (0.26)	6LC	0.97	3.82 (0.09)	11.86 (0.23)	32.67 (0.86)	2.30 (0.16)	4.91 (0.35)	7.55 (0.47)
	7SC	0.97	5.24 (0.11)	14.79 (0.34)	29.76 (0.51)	3.51 (0.18)	6.97 (0.25)	9.65 (0.34)	7LC	0.89	8.91 (0.29)	21.72 (0.88)	35.87 (1.71)	4.12 (0.17)	7.95 (0.22)	11.46 (0.33)
D	5SD	1.41	1.32 (0.04)	5.20 (0.17)	19.85 (0.82)	0.58 (0.03)	1.87 (0.09)	6.39 (0.29)	5LD	1.47	2.43 (0.07)	8.53 (0.23)	24.45 (0.85)	0.50 (0.04)	1.70 (0.12)	5.64 (0.37)
	6SD	0.94	2.32 (0.07)	8.09 (0.28)	27.28 (0.81)	1.03 (0.04)	2.39 (0.09)	6.25 (0.16)	6LD	1.06	2.26 (0.04)	8.50 (0.17)	27.88 (0.62)	1.14 (0.05)	3.43 (0.15)	5.84 (0.21)
	7SD	0.99	4.89 (0.12)	14.24 (0.30)	27.96 (0.67)	2.43 (0.06)	4.51 (0.14)	9.04 (0.25)	7LD	0.89	6.62 (0.23)	19.48 (0.68)	30.18 (1.07)	2.39 (0.07)	6.21 (0.21)	10.23 (0.26)

[a] refer to Table 1 for the WPC treatment composition and corresponding codes. (Values in parenthesis are the standard deviations).

From the Pearson's correlation matrices shown in Figure 9, the blending ratio (*B. ratio*) has the highest correlation (0.78) for WA among all the independent input variables and is therefore the most important determinant of water absorption. Since TS is a direct function of WA, as seen from Figure 9, the dimensional stability of the boards is also largely impacted by the blending ratio. As the press time and temperature are increased, WA and TS are decreased. Figures 7 and 8 do not show clearly which of these two variables has a more significant impact on WA and TS. However, in Figure 9, it can be seen that temperature has a larger effect on the reduction of WA and TS than time. While press time is negatively correlated with TS, the correlation is not statistically significant. The effect of particle size on both WA and TS is also significant and positively correlated. Press temperature and time bear no significant correlation with density on their own, but the combined effect is significant. However, the blending ratio correlates positively with density. The board density shows a linear relation with WA and an inverse relation with TS.

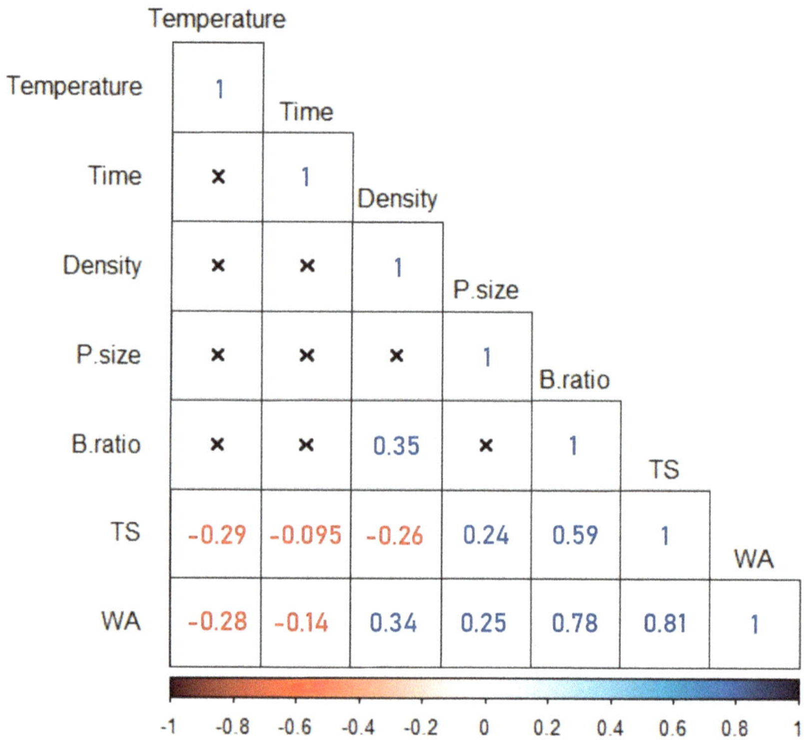

Figure 9. Pearson's R correlations of the impact of independent press parameters on the physical properties (WA, TS and density) of WPCs. The correlation matrices marked (x) are not significant or inapplicable.

3.3. *Mechanical Properties*

3.3.1. Static Bending Strength

The results of the strength modulus of 3-point bending tests for all treatments are presented in Figure 10. For each blending ratio across particle sizes, the highest MOR is obtained for treatment 'D' samples pressed at T =180 °C and t = 30 min, while the lowest strength values were obtained for treatment 'A' samples pressed at T = 150 °C and t = 10 min. The effect of increasing the time (B) or temperature (C) alone resulted in an increase in strength, but there was no clear trend in terms of which of the two variables has a greater impact on strength. However, the increased press time and temperature together resulted in the highest board strength. This can be explained by the formation of a continuous polymer matrix around the biomass fibres and occupation of intercellular voids. The densified composite will have fewer micro pores and fewer stress concentration points and therefore an increased strength.

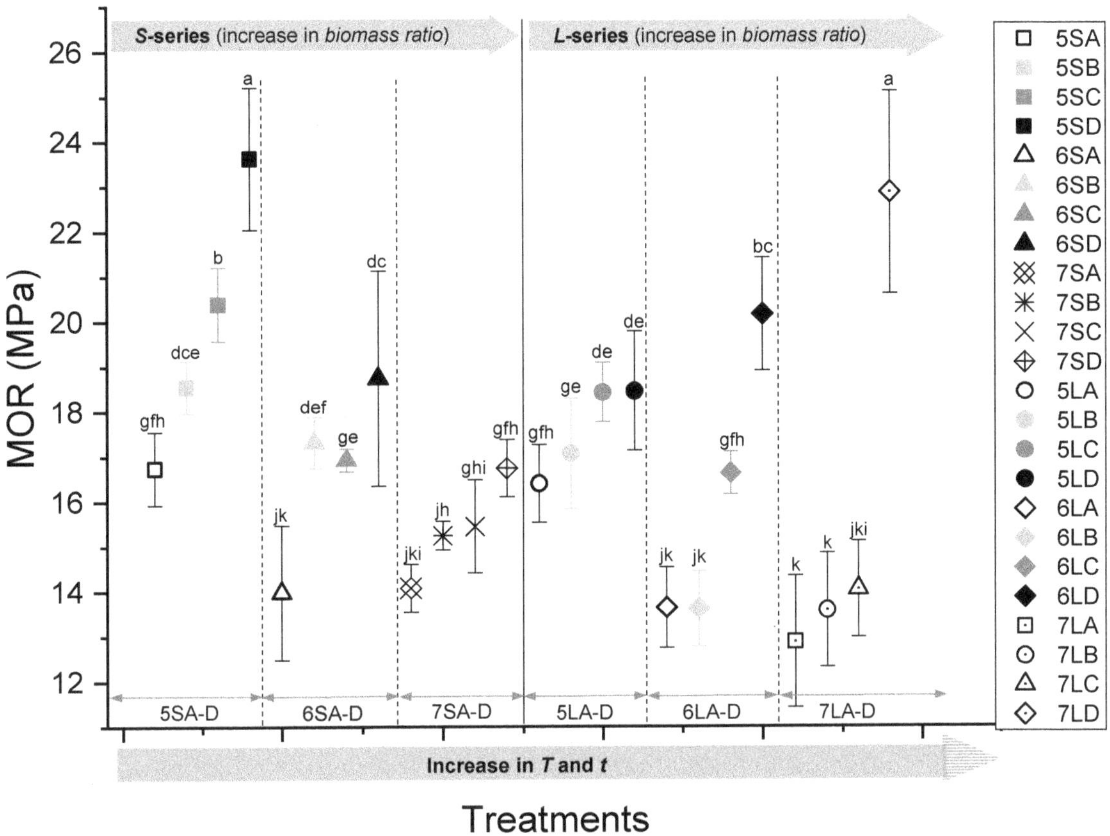

Figure 10. Flexural strength (MOR) of the WPCs pressed at different temperatures and times (A–D) with 50, 60, and 70% biomass at two particle sizes (S and L). Symbols with the same letters are not significantly different.

Boards with a 50% biomass had the highest bending strength properties, with a maximum MOR of about 23 MPa. As the biomass content increased, the MOR was reduced and showed the lowest value of 13 MPa at 70%. This can be explained by the inhomogeneity of the biomass, which contained fibres from leaves, twigs, and bark, which have a lower content of crystalline cellulose and higher portions of short fibres containing amorphous hemicelluloses and lignin. On the whole, boards made with a smaller particle size biomass (*S*) show slightly higher strengths than larger particles (*L*), although the difference is not statistically significant. This can be explained by the better flow of the polymer around smaller particles, which leads to a continuous matrix.

3.3.2. Flexural MOE

Figure 11 illustrates the bending stiffness of the WPC boards. Similar to the MOR, samples 'D' pressed at T = 180 °C and t = 30 min showed a superior bending stiffness for all blending ratios and particle sizes, while treatment samples 'A' pressed at T = 150 °C and t = 10 min recorded the lowest stiffness. As observed before, an increase in either the temperature or time of pressing resulted in an increase in stiffness, but without a clear trend regarding which of the two variables had a greater impact. However, unlike the MOR, composites with a higher biomass content at 70% show the highest bending stiffness, exceeding 900 MPa, and boards with a 50% biomass recorded the lowest stiffness. This indicates that the main contributor to stiffness is the biomass, which acts as a reinforcement in the composite. The impact of particle size on bending stiffness did not follow a clear trend, even though comparative observations of treatment 'A' and 'D' samples across particle sizes indicate that smaller particle sizes generally had better bending moduli.

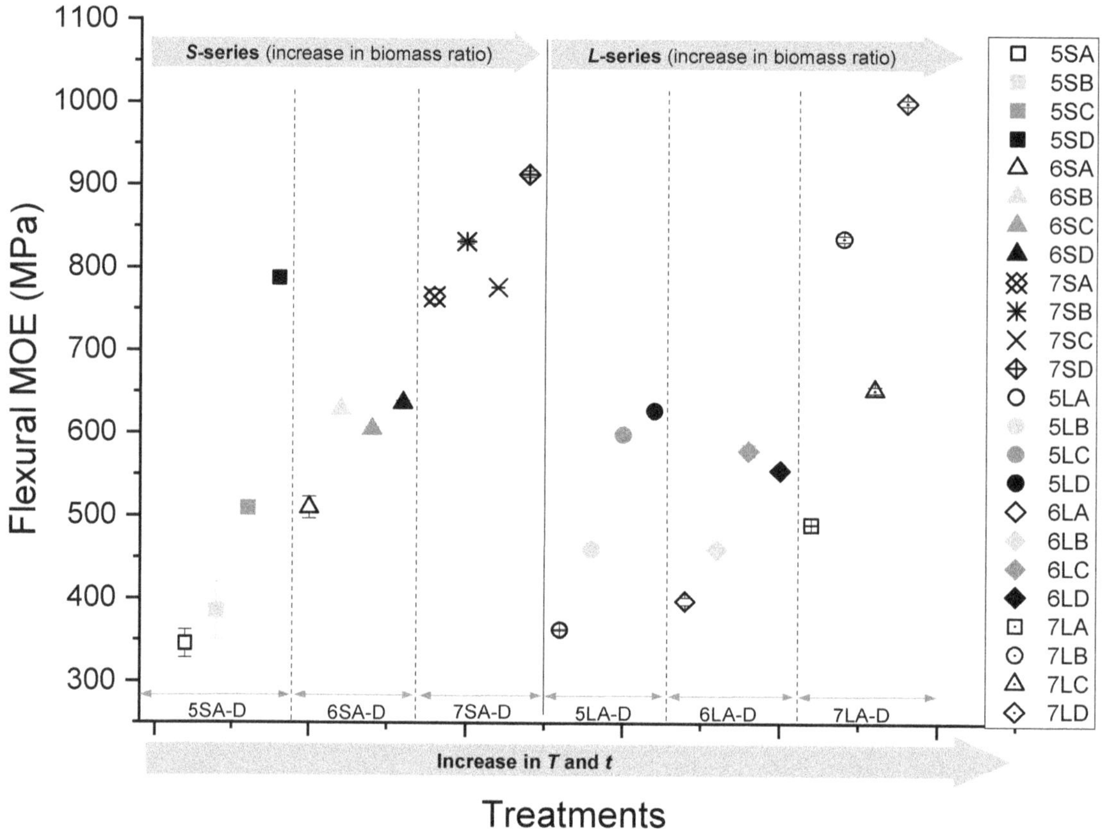

Figure 11. Flexural MOE of the WPCs pressed at different temperatures and times (A–D) with 50, 60, and 70% biomass at two particle sizes (*S* and *L*).

3.3.3. Tensile Strength

The tensile strength of the WPC boards is illustrated in Figure 12 and shows a positive linear relationship with increasing press temperature and/or time. This trend can be explained by the more efficient matrix formation of the rLDPE around the biomass particles. The sustained elevated temperatures above the melt temperature of the rLDPE rendered it less viscous and improved its flow and diffusion into the cell wall micro pores to form continuous matrices, thus imparting a higher tensile strength. An inverse relationship between the tensile strength and biomass ratio was observed across particle sizes. At a higher biomass loading, the composite samples had a higher stiffness due to the reinforcing impact of the biomass. Consequently, an increase in the tensile load broke the samples with a higher biomass ratio at smaller strain levels. With a reduction in the biomass content, the WPCs gained elasticity, which resulted in a higher tensile strength. This phenomenon was observed across particle sizes. However, no clear trend regarding the independent effect of particle size on tensile strength could be statistically established.

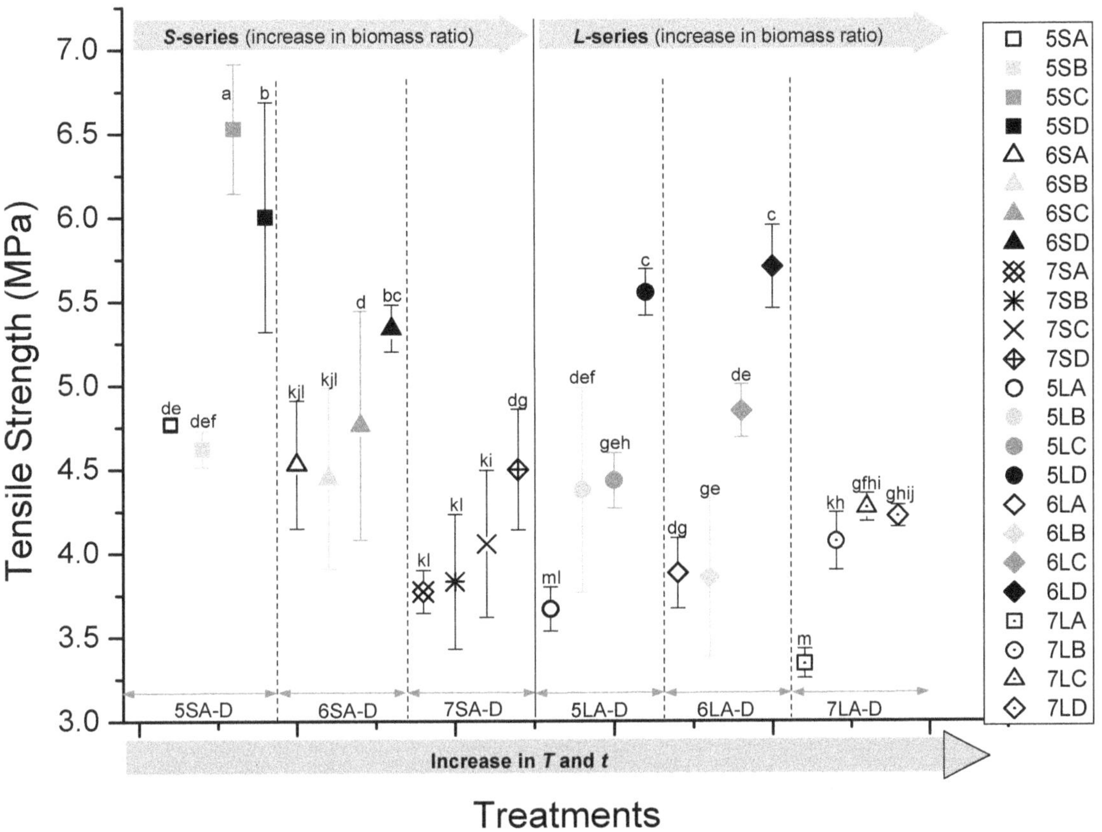

Figure 12. Tensile strength of the WPCs pressed at different temperatures and times (A–D) with 50, 60, and 70% biomass at two particle sizes (*S* and *L*). Symbols with the same letters are not significantly different.

3.3.4. Tensile MOE

The observed tensile moduli were not very different from the tensile strength. From Figure 13, it is observed that treatment 'D' samples pressed at $T = 180$ °C and $t = 30$ min had a significantly higher tensile modulus than treatment 'A' samples pressed at $T = 150$ °C and $t = 10$ min for all blending ratios. This may be attributed to the re-crystallisation of cellulose, in addition to the realignment of lignin at a higher temperature and extended pressing time. However, an increase in the biomass ratio resulted in a rise in tensile stiffness, with the lowest stiffness recorded at approximately 230 MPa for a biomass ratio of 50% wt. and the highest stiffness of nearly 400 MPa at a 70% biomass ratio. The tensile modulus of the WPC is therefore largely determined by the blending ratio and press time and temperature. The effect of particle size generally showed no clear trend. However, the comparative analysis of treatment 'A' and 'D' samples across particle sizes suggests that the WPCs made with larger particles had higher tensile moduli.

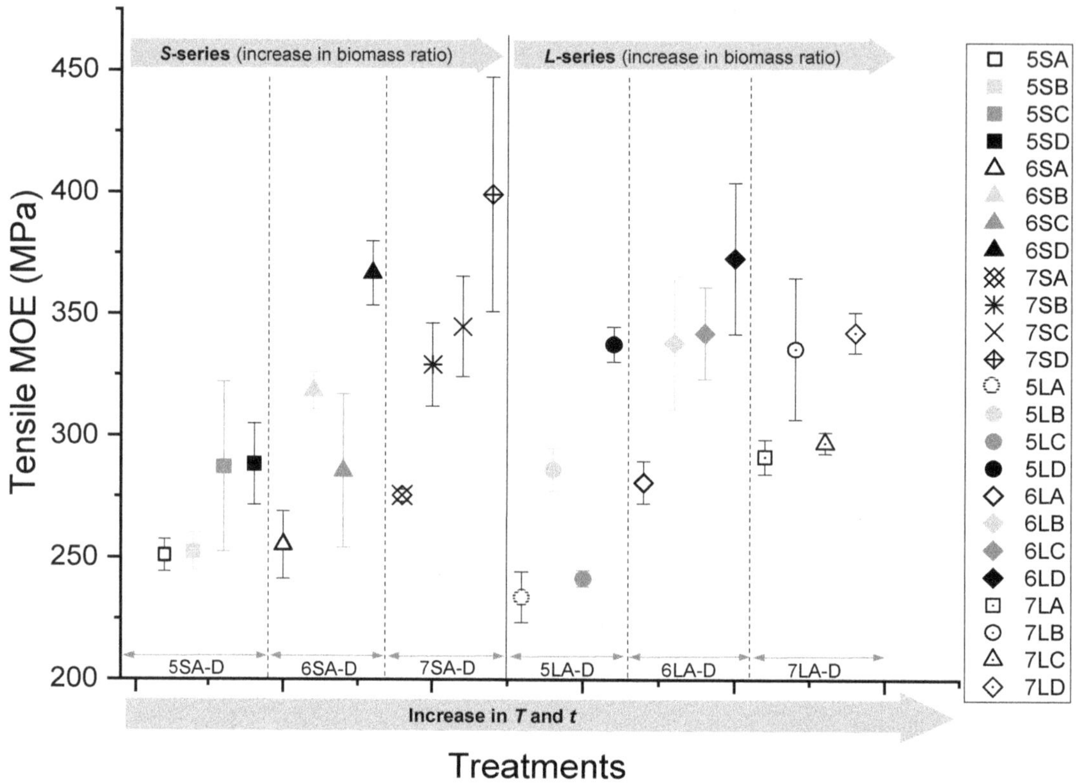

Figure 13. Tensile MOE of the WPCs pressed at different temperatures and times (A-D) with 50, 60, and 70% biomass at two particle sizes (S and L).

4. Discussion

Given that cellulose plays an important role in bond formation [17,37,48], the higher content of cellulose in *A. saligna* suggests its higher potential for wood-polymer bond formation. However, this also implies a more hygroscopic character of the composite. This correlation between cellulose content and a high moisture uptake is confirmed by several researchers [12,15,18,31,47,49,50]. In order to maximise the full benefits of a high cellulose content, while simultaneously limiting the disadvantages, the press conditions that promote an efficient biomass encapsulation by the hydrophobic plastic are required. This is achieved at temperatures above the melting point of the plastic and a sufficient press time to allow for an effective flow, wetting, and diffusion into biomass. This can be observed in Figures 7–13 and Table 3, which show that the elevated temperature and longer press time create better conditions to lower the viscosity of the plastic, therefore allowing it to reach into inter-particle crevices and cell lumina. The resulting formation of a continuous plastic matrix enhances the biomass particle encapsulation, which inhibits the bonding of water molecules to the free hydroxyl groups of holocellulose [15,47,50]. Subsequently, a reduction in water absorption and improvement in dimensional stability are achieved. This is further facilitated by the plasticisation of lignin and the conformational reorganisation of holocellulose due to the dehydration reactions of residual water resulting from the exposure of the biomass to high press temperature conditions [44,45,51]. These findings have been confirmed by several other researchers [14,52], who noted that the choice of press temperature and time affect the melt temperature and melt flow rates of the plastic and the thermal degradation temperature of the biomass [16,29]. The combination of the press temperature and blending ratio ultimately determines the extent of the particle encapsulation within the plastic matrix.

Extractives play a crucial role in wood–plastic bond formation. Polyphenolics and terpenes, including suberin and resin compounds, have hydrophobic functional groups with a considerable influence on the surface chemistry of wood [12,53–55]. Since a higher surface energy is required for a good bonding between biomass and polyethylene, a lower extractive content reduces the potential of wood–plastic adhesion via chemical bonds, while promoting mechanical interlocking. Some components of wood extractives oxidise and form volatiles upon exposure to air and light during drying and subsequent heat exposure during processing. This potentially lowers the surface energy of the biomass as extractives, particularly tannins, which are known to be present in considerable quantities in *A. saligna*, contribute largely to bond formation and water repellence. The reduced surface energy translates into a reduction in the adhesion potential [14,15,52]. At optimised press temperatures, however, extractives do not entirely escape, but migrate to the surfaces of the biomass, where they increase the adhesion between the plastic and biomass. As extractives migrate to the surface, they leave behind micro-voids, which, given sufficient press time will be filled by the plastic matrix through diffusion, which is a slow process [17,50,51,56]. The eventual wood–plastic bonds are both chemical at the interface and mechanical at the cell wall level. The high amount of extractives contained in the leaves and bark of the biomass also act as a plasticiser and can considerably reduce the viscosity of the plastic and enhance the particle dispersion in WPC processing [51,55,57,58]. These properties collectively enhance the moisture resistance of the composite products. Consequently, forming WPCs with biomass high in extractives requires the optimisation of the press settings to maximise both the benefits of the chemical adhesion via a surface energy boost, as well as mechanical interlocking.

At higher temperatures and longer press times, the hemicellulose concentration of the biomass, particularly of the leaves and twigs, may have either partially or wholly degraded. Boonstra et al. [59], Ates et al. [60] and Phuong et al. [61] have reported that the degradation of hemicelluloses starts below 200 °C. Inari et al. [62] found through X-ray photoelectron spectroscopy that after heat treatment at around 200 °C, the hydroxyl concentration of the wood surface decreased due to the thermal degradation of the hemicelluloses. Since the hydrophilic properties of wood depend on an amount of available hydroxyl groups, a reduction in the concentration of these functional groups results in an increased hydrophobic character of the wood [63]. These findings are corroborated by Hakkou et al. [45], Gérardin et al. [64], and Chaouch et al. [65], who reported that the oxygen to carbon ratio of wood at the surface decreases with increasing temperatures above 120 °C. This reduction in the O/C concentration ratio is accompanied by a reduction in the concentration of polar wood components due to dehydration reactions, the depolymerisation of amorphous holocellulose, and the selective degradation of hydroxyl groups. These reactions induce moisture resistance; however, they also lower the surface free energy of biomass and negatively affect wood polymer adhesion [17,59,66], while aiding apolar bonds [64].

Composites pressed at a higher temperature and time exhibited better mechanical properties. Thermal modification is known to induce crosslinks within wood and plastic, which explains the improvement of the mechanical properties. The increase in the stiffness of the WPCs with increasing temperature is in good agreement with the findings of other researchers, such as Lyutyy et al. [67] and Santos [68]. Kubojima et al. [69] noted that the MOE and MOR of wood increases for the first 30 min of heat treatment at 160 °C, after which it begins to decline. This has been attributed to lignin relocation and the recrystallisation of cellulose [61]. However, the positive correlation between the increased press temperature and mechanical strength observed in this study was not confirmed by Hakkou et al. [45], Korkut et al. [70], or Bal and Bektaş [71], who reported that heating wood above 120 °C hindered the mechanical properties.

It was also observed that the dimensional stability and mechanical properties of the WPCs were better in samples made with smaller particles, irrespective of the compounding ratio. This can be explained by the more continuous matrix formation around the smaller particles. This observation is confirmed by various other studies [16,33,35,47,52], where the effect of the biomass particle size on the WPC properties was analysed. Smaller particles also result in better compressibility during hot pressing, which eliminates interparticle voids and reduces the porosity of the composite. The enhanced compressibility also impacts directly on the board density, as better compaction results in a higher board density [30,35,72,73].

The blending ratio had the most significant effect on the board properties. The presence of lightweight non-woody particles, such as leaves and porous bark, lowers the overall density of the biomass, resulting in composites with lower densities as the biomass content increases. The lower biomass density results in a disproportionately higher biomass volume per weight, which reduces the potential of the plastic to wet, infiltrate, and encapsulate the proportionately larger volume of the biomass. Since the resistance of WPCs to moisture sorption, and an effective stress transfer is reliant on the interaction between the stiffer hydrophilic biomass and the ductile hydrophobic polymer, the pathways for moisture movement into the composites and stress concentration points are higher in an incomplete biomass particle enclosure. Consequently, the blending ratio between wood and plastic needs to be adjusted according to the intended application. At a 50% biomass ratio, there is enough plastic to provide moisture resistance, but the boards are less rigid and have a higher density, which is not desirable for insulation boards. Boards with a 60% biomass fraction display a good balance between moisture resistance, strength, and medium density, with thermo-acoustic insulation benefits for interior use [74].

5. Conclusions

WPC boards were made in a cost-effective way, without prior processing of the biomass, such as debarking, separating non-woody and woody parts, or the use of additives, such as compatibilisers. The recycled plastic was the lowest grade—and therefore cheapest—rLDPE. The resulting boards had physical and mechanical properties comparable to commercial products used for insulation purposes and met all the requirements for non-structural interior applications in buildings. The long-term WA and TS results suggest that the panels may be used in high-humidity interior environments, such as kitchens, without much risk of biodegradation and deformation, while the density and mechanical properties make them suitable for interior wall cladding and ceiling boards. All the investigated processing factors—blending ratio, particle size, press time, and temperature—were found to significantly affect the board properties, albeit to varying degrees. The blending ratio is the most important determinant of physical and mechanical properties. A higher press temperature and longer press times lower the viscosity of the plastic and improve the melt-flow rates, while also re-crystallising the cellulose fibres. This gives rise to stronger bonds between the plastic and biomass and a better encapsulation of the biomass particles. However, the maximum temperature is limited to prevent thermal degradation of the biomass. WPC boards made with smaller particle fractions showed better physical and mechanical properties due to a better encapsulation of the biomass particles by the plastic matrix, which leads to improved hydrophobic properties. However, the improvement in the mechanical properties was not as notable.

Ultimately, boards made with a 60% biomass ratio and pressed at 180 °C for 30 min showed the best balance between mechanical and physical properties for interior applications. The obtained density is sufficient but not too high to allow for a good thermo-acoustic insulation.

Author Contributions: Conceptualization: A.S.M. and M.M.; methodology: A.S.M.; formal analysis: A.S.M.; resources: M.M.; writing—original draft preparation: A.S.M.; writing—review and editing: M.M. Both authors have read and agreed to the published version of the manuscript.

Funding: The project on which this work is based was funded by the German Federal Ministry of Education and research, under the grant No. 01DG17007A and in the DAAD project 57359374. The authors are responsible for the content of this publication.

Institutional Review Board Statement: Not applicable.

Informed Consent Statement: Not applicable.

Data Availability Statement: All data is available upon request.

Acknowledgments: The authors wish to thank TKC Amandla for the harvesting and transport of the invasive trees and Atlantic Recycling for the provision of rLDPE.

Conflicts of Interest: The authors declare no conflict of interest. The funders had no role in the design of the study; in the collection, analyses, or interpretation of data; in the writing of the manuscript, or in the decision to publish the results.

References

1. Wilson, J.R.U.; Ivey, P.; Manyama, P.; Nänni, I. A new national unit for invasive species detection, assessment and eradication planning. *S. Afr. J. Sci.* **2013**, *109*, 1–13. [CrossRef]
2. Van Wilgen, B.W.; Reyers, B.; Le Maitre, D.C.; Richardson, D.M.; Schonegevel, L. A biome-scale assessment of the impact of invasive alien plants on ecosystem services in South Africa. *J. Environ. Manag.* **2008**, *89*, 336–349. [CrossRef] [PubMed]
3. Wilcox, C.; Van Sebille, E.; Hardesty, B.D.; Estes, J.A. Threat of plastic pollution to seabirds is global, pervasive, and increasing. *Proc. Natl. Acad. Sci. USA* **2015**, *112*, 11899–11904. [CrossRef] [PubMed]
4. Jambeck, J.; Geyer, R.; Wilcox, C.; Siegler, T.R.; Perryman, M.; Andrady, A.; Narayan, R.; Law, K.L. Plastic waste inputs from land into the ocean. *Mar. Pollut.* **2015**, *347*, 768–771. [CrossRef] [PubMed]
5. WEF. *The New Plastics Economy: Rethinking the Future of Plastics*; WEF: Geneva, Swizerland, 2016. Available online: https://www.ellenmacarthurfoundation.org/assets/downloads/publications/NPEC-Hybrid_English_22-11-17_Digital.pdf (accessed on 2 May 2020).
6. Geyer, R.; Jambeck, J.R.; Law, K.L. Production, use, and fate of all plastics ever made. *Sci Adv.* **2017**, *3*, 25–29. [CrossRef]
7. Department of Human Settlement. *National Housing Policy and Subsidy Programmes*; Department of Human Settlement: Pretoria, South Africa, 2010.
8. Krickova, A. Reconstruction and Development Programme as a tool of socio-economic transformation. *Modern Africa Polit. Hist. Soc.* **2015**, *3*, 57–93.
9. Manomano, T.; Tanga, P.T. Housing needs: The quality and quantity of housing provided by the government for the poor in the Eastern Cape province in South Africa. *Soc. Work* **2018**, *54*, 19–36. [CrossRef]
10. Clemons, C. Interfacing wood-plastic composites industries in the U.S. *For. Prod. J.* **2002**, *52*, 10–18.
11. Li, Y. Wood-polymer composites. In *Advances in Composite Materials—Analysis of Natural and Man-Made Materials*; Tesinova, P., Ed.; IntechOpen: Shanghai, China, 2011; pp. 229–284. Available online: http://www.intechopen.com/books/advances-in-composite-materials-analysis-of-natural-and-man-madematerials/%0Awood-polymer-composites (accessed on 9 September 2020).
12. Stokke, D.D.; Gardner, D.J. Fundamental aspects of wood as a component of thermoplastic composites. *J. Vinyl Addit. Technol.* **2003**, *9*, 96–104. [CrossRef]
13. Cho, J.D.; Kim, S.G.; Hong, J.W. Surface modification of polypropylene sheets by UV-radiation grafting polymerization. *J. Appl. Polym. Sci.* **2006**, *99*, 1446–1461. [CrossRef]
14. Van Der Leeden, M.C.; Frens, G. Surface properties of plastic materials in relation to their adhering performance. *Adv. Eng. Mater.* **2002**, *4*, 280–289. [CrossRef]
15. Schneider, M.H. Wood Polymer Composites. *Wood Fiber Sci.* **1994**, *26*, 142–151.
16. Clemons, C.M.; Caufield, D.F. Wood flour. In *Functional Fillers for Plastics*; Xanthos, M., Ed.; Wiley-VCH Verlag GmbH & Co KgaA: Weinheim, Germany, 2005; pp. 249–270.
17. Vassilev, S.V.; Baxter, D.; Andersen, L.K.; Vassileva, C.G.; Morgan, T.J. An overview of the organic and inorganic phase composition of biomass. *Fuel* **2012**, *94*, 1–33. [CrossRef]
18. Farmer, R.H. *Chemistry in the Utilization of Wood*, 1st ed.; Pergamon Press Ltd.: London, UK, 1967; pp. 1–183. [CrossRef]
19. Migneault, S.; Koubaa, A.; Perré, P.; Riedl, B. Effects of wood fiber surface chemistry on strength of wood-plastic composites. *Appl. Surf. Sci.* **2015**, *343*, 11–18. [CrossRef]
20. Frihart, C.R.; Hunt, C.G. Adhesives with wood materials: Bond formation and performance. In *Wood Handbook—Wood as an Engineering Material*; General Technical Report FPL-GTR-190; Centennial Ross, R.J., Ed.; Forest Products Laboratory: Madison, WI, USA, 2010; pp. 228–251.

21. Irle, M.; Barbu, M.C. Wood-based panel technology. In *Wood-Based Panels—An Introduction for Specialists*; Thoemen, H., Irle, M., Sernek, M., Eds.; Brunel University Press: London, UK, 2010; pp. 1–94.
22. Reddy, K.O.; Uma Maheswari, C.; Muzenda, E.; Shukla, M.; Rajulu, A.V. Extraction and Characterization of Cellulose from Pretreated Ficus (Peepal Tree) Leaf Fibers. *J. Nat. Fibers* **2016**, *13*, 54–64. [CrossRef]
23. Ogah, A.O.; Elom, N.I.; Ngele, S.O.; Nwofe, P.A.; Agbo, P.E.; Englund, K.R. Water Absorption, Thickness Swelling and Rheological Properties of Agro Fibers/HDPE Composites. *IOSR J. Polym Text. Eng.* **2015**, *2*, 66–73. [CrossRef]
24. Liu, H.; Wu, Q.; Zhang, Q. Preparation and properties of banana fiber-reinforced composites based on high density polyethylene (HDPE)/Nylon-6 blends. *Bioresour. Technol.* **2009**, *100*, 6088–6097. [CrossRef] [PubMed]
25. Hung, K.C.; Yeh, H.; Yang, T.C.; Wu, T.L.; Xu, J.W.; Wu, J.H. Characterization of wood-plastic composites made with different lignocellulosic materials that vary in their morphology, chemical composition and thermal stability. *Polymers* **2017**, *9*, 726. [CrossRef]
26. Dehghan, M.; Faezipour, M.; Azizi, M.; Hosseinabadi, H.Z.; Bari, E.; Nicholas, D.D. Assessment of physical, mechanical, and biological properties of bamboo plastic composite made with polylactic acid. *Maderas Cienc. Tecnol.* **2019**, *21*, 599–610. [CrossRef]
27. Ayrilmis, N.; Kaymakci, A.; Akbulut, T.; Elmas, G.M. Mechanical performance of composites based on wastes of polyethylene aluminum and lignocellulosics. *Compos. Part B Eng.* **2013**, *47*, 150–154. [CrossRef]
28. Effah, B.; Van Reenen, A.; Meincken, M. Mechanical properties of wood-plastic composites made from various wood species with different compatibilisers. *Eur. J. Wood Wood Prod.* **2018**, *76*, 57–68. [CrossRef]
29. Frihart, C.R. Wood adhesion and adhesives. In *Handbook of Wood Chemistry and Wood Composites*; Rowell, R.M., Ed.; CRC Press: Boca Raton, FL, USA, 2005; pp. 219–282.
30. Hosseinihashemi, S.K.; Modirzare, M.; Safdari, V.; Kord, B. Decay resistance, hardness, water absorption, and thickness swelling of a bagasse fiber/plastic composite. *BioResources* **2011**, *6*, 3289–3299. [CrossRef]
31. Zabihzadeh, S.M. Water uptake and flexural properties of natural filler/HDPE composites. *BioResources* **2010**, *5*, 316–323. [CrossRef]
32. Jayamani, E.; Hamdan, S.; Ezhumalai, P.; Soon, K.H. Acoustic and thermal properties of polymer composites reinforced with lignocellulosic fibers. *Appl. Mech. Mater.* **2014**, *624*, 25–29. [CrossRef]
33. Bouafif, H.; Koubaa, A.; Perré, P.; Cloutier, A. Effects of fiber characteristics on the physical and mechanical properties of wood plastic composites. *Compos. Part A Appl. Sci. Manuf.* **2009**, *40*, 1975–1981. [CrossRef]
34. AlMaadeed, M.A.; Nógellová, Z.; Janigová, I.; Krupa, I. Improved mechanical properties of recycled linear low-density polyethylene composites filled with date palm wood powder. *Mater. Des.* **2014**, *58*, 209–216. [CrossRef]
35. Adefisan, O.O.; McDonald, A.G. Evaluation of the strength, sorption and thermal properties of bamboo plastic composites. *Maderas Cienc. Tecnol.* **2019**, *21*, 3–14. [CrossRef]
36. Krause, K.C.; Sauerbier, P.; Koddenberg, T.; Krause, A. Utilization of recycled material sources for wood-polypropylene composites: Effect on internal composite structure, particle characteristics and physico-mechanical properties. *Fibers* **2018**, *6*, 86. [CrossRef]
37. Migneault, S.; Koubaa, A.; Erchiqui, F.; Chaala, A.; Englund, K.; Wolcott, M.P. Effects of processing method and fiber size on the structure and properties of wood-plastic composites. *Compos. Part A Appl. Sci. Manuf.* **2009**, *40*, 80–85. [CrossRef]
38. Ozdemir, F.; Ayrilmis, N.; Kaymakci, A.; Kwon, J.H. Improving dimensional stability of injection molded wood plastic composites using cold and hot water extraction methods. *Maderas Cienc. Tecnol.* **2014**, *16*, 365–372. [CrossRef]
39. Pelaez-Samaniego, M.R.; Yadama, V.; Lowell, E.; Amidon, T.E.; Chaffee, T.L. Hot water extracted wood fiber for production of wood plastic composites (WPCs). *Holzforschung* **2013**, *67*, 193–200. [CrossRef]
40. Adhikary, K.B.; Pang, S.; Staiger, M.P. Dimensional stability and mechanical behaviour of wood-plastic composites based on recycled and virgin high-density polyethylene (HDPE). *Compos. Part B Eng.* **2008**, *39*, 807–815. [CrossRef]
41. Stark, N.M.; Cai, Z.; Carll, C. Wood-Based Composite Materials Panel Products, Glued-Laminated Timber, Structural Materials. In *Wood Handbook—Wood as an Engineering Material*; General Technical Report, FPL-GTR-190; Centennial Ross, R.J., Ed.; Wood Products Laboratory: Madison, WI, USA, 2010; pp. 252–279.
42. Luo, Z.; Li, P.; Cai, D.; Chen, Q.; Qin, P.; Tan, T.; Cao, H. Comparison of performances of corn fiber plastic composites made from different parts of corn stalk. *Ind. Crops Prod.* **2017**, *95*, 521–527. [CrossRef]
43. Acheampong, J.B.; de Angelis, M.; Krause, A.; Meincken, M. The effect of raw material selection on physical and mechanical properties of wood plastic composites made from recycled LDPE and wood from invasive trees in South Africa. *Wood Mater. Sci. Eng.* **2021**, *16*, 118–123. [CrossRef]
44. Rowell, R.M. Moisture Properties. In *Handbook of Wood Chemistry and Wood Composites*; Rowell, R.M., Ed.; CRC Press: Boca Raton, FL, USA, 2005; pp. 84–106.
45. Hakkou, M.; Pétrissans, M.; Zoulalian, A.; Gérardin, P. Investigation of wood wettability changes during heat treatment on the basis of chemical analysis. *Polym. Degrad. Stab.* **2005**, *89*, 1–5. [CrossRef]
46. Wang, M.-L.; Chang, R.-Y.; Hsu, C.-H. Material properties of plastics. In *Laser Welding of Plastics*, 1st ed.; Klein, R., Ed.; Wiley-VCH Verlag GmbH & Co. KGaA: Weinheim, Germany, 2011; pp. 3–69.
47. Kallakas, H.; Poltimäe, T.; Süld, T.M.; Kers, J.; Krumme, A. The influence of accelerated weathering on the mechanical and physical properties of wood-plastic composites. In *Proceedings of the Estonian Academy of Sciences, 2015*; Estonian Academy Publishers: Tallinn, Estonia, 2015; Volume 64, pp. 94–104. [CrossRef]

48. Gulitah, V.; Liew, K.C. Three different recycle codes of plastic/Acacia fibre composites: Physical and morphological properties. *Int. J. Biobased Plast.* **2019**, *1*, 1–7. [CrossRef]
49. Wechsler, A.; Hiziroglu, S. Some of the properties of wood-plastic composites. *Build. Environ.* **2007**, *42*, 2637–2644. [CrossRef]
50. Tabarsa, T.; Khanjanzadeh, H.; Pirayesh, H. Manufacturing of wood-plastic composite from completely recycled materials. *Key Eng. Mater.* **2011**, *471–472*, 62–66. [CrossRef]
51. Wiedenhoeft, A. Structure and function of wood. In *Wood Handbook: Wood as an Engineering Material*; General Technical Report, FPL-GTR-19; Centennial Ross, R.J., Ed.; Forest Products Laboratory: Madison, WI, USA, 2010; pp. 62–79. Available online: www.fpl.fs.fed.us (accessed on 30 May 2021).
52. Gallagher, L.W.; McDonald, A.G. The effect of micron sized wood fibers in wood plastic composites. *Maderas Cienc Tecnol.* **2013**, *15*, 357–374. [CrossRef]
53. Clemons, C. Raw materials for wood-polymer composites. In *Composites Science and Engineering, Wood-Polymer Composites*; Niska, K.O., Sain, M., Eds.; Woodhead Publishing: Sawston, UK, 2008; pp. 1–22. [CrossRef]
54. Leskinen, T.; Salas, C.; Kelley, S.S.; Argyropoulos, D.S. Wood Extractives Promote Cellulase Activity on Cellulosic Substrates. *Biomacromolecules* **2015**, *16*, 3226–3234. [CrossRef]
55. Jansson, M.B.; Nilvebrant, N.-O. Wood extractives. In *Pulp and Paper Chemistry and Technology: Wood Chemistry and Wood Biotechnology*; Ek, M., Gellerstedt, G., Henriksson, G., Eds.; Walter de Gruyter GmbH & Co.: Berlin, Germany, 2009; Volume 1, pp. 147–171.
56. Klyosov, A.A. *Wood-Plastic Composites*; John Wiley & Sons, Inc.: Hoboken, NJ, USA, 2007; pp. 1–696.
57. Rowell, R.M.; Pettersen, R.; Han, J.S.; Rowell, J.S.; Tshabalala, M.A. Cell Wall Chemistry. In *Handbook of Wood Chemistry and Wood Composites*; Rowell, R.M., Ed.; CRC Press: Boca Raton, FL, USA, 2005; pp. 43–82.
58. Clemons, C.M.; Stark, N.M. Feasibility of using saltcedar as a filler in injection-molded polyethylene composites. *Wood Fiber Sci.* **2009**, *41*, 2–12.
59. Boonstra, M.J.; Van Acker, J.; Kegel, E.; Stevens, M. Optimisation of a two-stage heat treatment process: Durability aspects. *Wood Sci. Technol.* **2007**, *41*, 31–57. [CrossRef]
60. Ates, S.; Hakan, A.M.; Ozdemir, H. Heat-Treated Calabrian Pine Wood. *BioResources* **2009**, *4*, 1032–1043.
61. Phuong, L.X.; Shida, S.; Saito, Y. Effects of heat treatment on brittleness of Styrax tonkinensis wood. *J. Wood Sci.* **2007**, *53*, 181–186. [CrossRef]
62. Inari, G.N.; Petrissans, M.; Lambert, J.; Ehrhardt, J.J.; Rardin, P.G. XPS characterization of wood chemical composition after heat-treatment. *Surf. Interface Anal.* **2006**, *38*, 1380–1385. [CrossRef]
63. Pétrissans, M.; Gérardin, P.; El Bakali, I.; Serraj, M. Wettability of heat-treated wood. *Holzforschung* **2003**, *57*, 301–307. [CrossRef]
64. Gérardin, P.; Petrič, M.; Petrissans, M.; Lambert, J.; Ehrhrardt, J.J. Evolution of wood surface free energy after heat treatment. *Polym. Degrad. Stab.* **2007**, *92*, 653–657. [CrossRef]
65. Chaouch, M.; Pétrissans, M.; Pétrissans, A.; Gérardin, P. Use of wood elemental composition to predict heat treatment intensity and decay resistance of different softwood and hardwood species. *Polym. Degrad. Stab.* **2010**, *95*, 2255–2259. [CrossRef]
66. Mitsui, K.; Inagaki, T.; Tsuchikawa, S. Monitoring of hydroxyl groups in wood during heat treatment using NIR spectroscopy. *Biomacromolecules* **2008**, *9*, 286–288. [CrossRef]
67. Lyutyy, P.; Bekhta, P.; Sedliacik, J.; Ortynska, G. Properties of flat-pressed wood-polymer composites made using secondary polyethylene. *Acta Fac Xylologiae* **2014**, *56*, 39–50.
68. Santos, J.A. Mechanical behaviour of Eucalyptus wood modified by heat. *Wood Sci. Technol.* **2000**, *34*, 39–43. [CrossRef]
69. Kubojima, Y.; Okano, T.; Ohta, M. Bending strength and toughness of heat-treated wood. *J. Wood Sci.* **2000**, *46*, 8–15. [CrossRef]
70. Korkut, S.; Akgül, M.; Dündar, T. The effects of heat treatment on some technological properties of Scots pine (*Pinus sylvestris* L.) wood. *Bioresour. Technol.* **2008**, *99*, 1861–1868. [CrossRef] [PubMed]
71. Bal, B.C.; Bektaş, I. The Effects of Heat Treatment on Some Mechanical Properties of Juvenile Wood and Mature Wood of Eucalyptus grandis. *Dry Technol.* **2013**, *31*, 479–485. [CrossRef]
72. Rahman, K.S.; Islam, M.N.; Rahman, M.M.; Hannan, M.O.; Dungani, R.; Khalil, H.P.S.A. Flat-pressed wood plastic composites from sawdust and recycled polyethylene terephthalate (PET): Physical and mechanical properties. *Springerplus* **2013**, *2*, 1–7. [CrossRef]
73. Chen, H.C.; Chen, T.Y.; Hsu, C.H. Effects of wood particle size and mixing ratios of HDPE on the properties of the composites. *Holz Roh Werkst* **2006**, *64*, 172–177. [CrossRef]
74. Wang, W.; Morrell, J.J. Water sorption characteristics of two wood-plastic composites. *For. Prod. J.* **2004**, *54*, 209–212.

polymers

MDPI

Review

Recent Progress in Modification Strategies of Nanocellulose-Based Aerogels for Oil Absorption Application

M. A. Iskandar [1], Esam Bashir Yahya [2], H. P. S. Abdul Khalil [2,*], A. A. Rahman [1] and M. A. Ismail [3]

[1] School of Physics, Universiti Sains Malaysia, Penang 11800, Malaysia; iskandar@greentagro.com (M.A.I.); arazhar@usm.my (A.A.R.)
[2] School of Industrial Technology, Universiti Sains Malaysia, Penang 11800, Malaysia; essam912013@gmail.com
[3] Teraju Saga Sdn. Bhd. MP813, Jalan Melaka Perdana 2, Taman Melaka Perdana, Alor Gajah, Melaka 78000, Malaysia; ikrammki00@gmail.com
* Correspondence: akhalilhps@gmail.com

Abstract: Oil spills and oily wastewater have become a major environmental problem in recent years, directly impacting the environment and biodiversity. Several techniques have been developed to solve this problem, including biological degradation, chemicals, controlled burning, physical absorption and membrane separation. Recently, biopolymeric aerogels have been proposed as a green solution for this problem, and they possess superior selective oil absorption capacity compared with other approaches. Several modification strategies have been applied to nanocellulose-based aerogel to enhance its poor hydrophobicity, increase its oil absorption capacity, improve its selectivity of oils and make it a compressible and elastic magnetically responsive aerogel, which will ease its recovery after use. This review presents an introduction to nanocellulose-based aerogel and its fabrication approaches. Different applications of nanocellulose aerogel in environmental, medical and industrial fields are presented. Different strategies for the modification of nanocellulose-based aerogel are critically discussed in this review, presenting the most recent works in terms of enhancing the aerogel performance in oil absorption in addition to the potential of these materials in near future.

Keywords: nanocellulose aerogels; modification techniques; environmental wastes; oil absorption

Citation: Iskandar, M.A.; Yahya, E.B.; Abdul Khalil, H.P.S.; Rahman, A.A.; Ismail, M.A. Recent Progress in Modification Strategies of Nanocellulose-Based Aerogels for Oil Absorption Application. *Polymers* **2022**, *14*, 849. https://doi.org/10.3390/polym14050849

Academic Editor: Pavlo Bekhta

Received: 24 January 2022
Accepted: 16 February 2022
Published: 22 February 2022

1. Introduction

Water pollution from various organic pollutants including frequent oil spills and other synthetic chemical leaks has recently caused significant environmental issues, which become worse every year due to the accumulation of such pollutants in the environment [1]. Conventional approaches that depend on synthetic and petroleum-based polymers, which are nonbiodegradable materials, could introduce another source of environmental pollution [2]. These concerns have encouraged scientists and catalyzed research into green alternatives with improved property profiles. Oily wastewater, as one of the most challenging environmental issues that require many treatment efforts, can be caused by either industrial or domestic factors [3]. Crude oil production is the main cause of industrial oily wastewater, in addition to oil refineries, compressor condensates, petrochemical industry, car washing, metal processing and the use of lubricants and cooling agents in other industries [4]. On the other hand, domestic oil and grease mainly come from total organic matter [5]. In this context, many approaches have been developed for oily water treatment and oil/water separation, including biological degradation by using specific microorganisms to degrade pollutants dissolved in effluents [6]. However, the efficiency of microorganisms in oil degradation is very limited compared with that in the degradation of other pollutants. Other researchers suggested in situ burning of oils; however, this method cannot remove the majority of oil, and it will cause environmental pollution [7]. Currently, physical absorption and membrane separation are the most used methods for oil/water separations and have attracted the greatest attention considering the drawbacks of the other approaches as they are effective and reliable with high absorption efficiency and little adverse environmental effect [8–10]. Thus, researchers nowadays are working on developing ecofriendly and cost-effective absorbents that selectively absorb oil with good reusability without significant reduction in absorption capacity.

Aerogels are highly porous functional materials that have been prepared from organic and inorganic materials and applied for numerous applications, including water treatment. Nanocellulose-based aerogel is known for its high porosity, ultralow density and excellent absorption capacity [11]. As a biopolymer, cellulose is a biodegradable, cost-effective and ecofriendly substance, and developing nanocellulose aerogel only consumes a tiny amount of cellulose due to the high porosity of aerogels [12]. Thus, the past few years witnessed a significant increase in the studies that utilize nanocellulose as precursor material in aerogel fabrication for different absorption applications. Figure 1 presents the number of scientific publications in the past 10 years on the use of aerogel as an absorbent functional material and the utilization of cellulose in this field, which are expected to highly increase in the next five years. Although several other biopolymers have also been used in aerogel fabrication and as absorbent materials, cellulose retains the attention of scientists due to its unique properties.

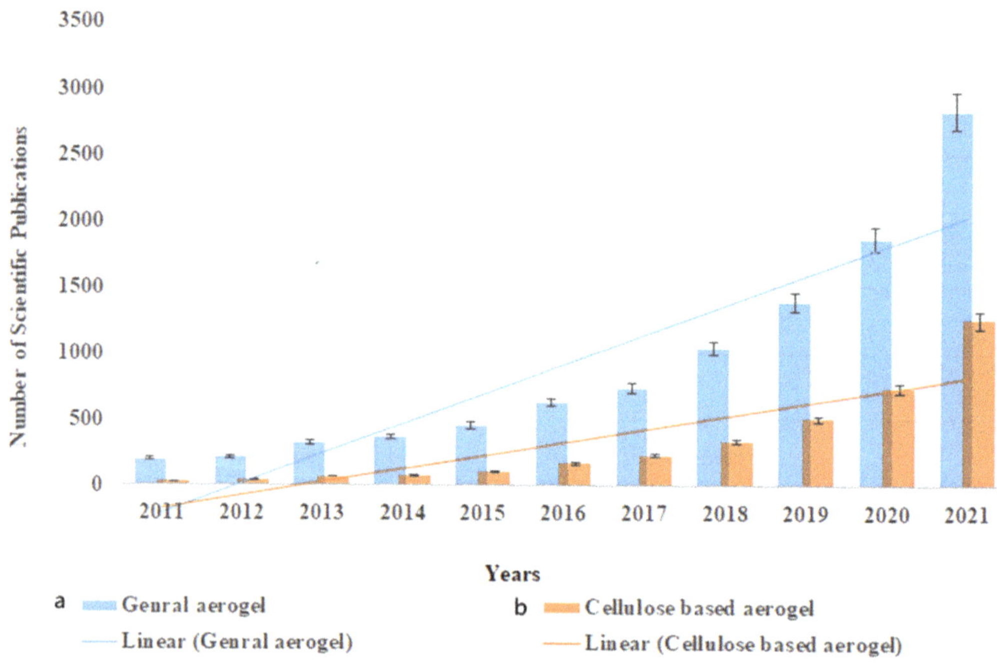

Figure 1. Number of scientific publications in the past 10 years on the use of aerogels as absorbent materials: (**a**) general aerogel; (**b**) cellulose-based aerogel. The research was done using the ScienceDirect database on 30 December 2021, using the keywords (Aerogel absorption) for general aerogel and "cellulose aerogel absorption" for cellulose-based aerogels.

Despite the extraordinary advantages of nanocellulose and aerogels, pure nanocellulose typically suffers from very poor hydrophobicity in addition to inferior mechanical properties [13]. Nanocellulose aerogels are not suitable in their pure form for any water treatment application, due to their high hydrophilicity and poor mechanical stability underwater. Therefore, great effort has been made in the modification of nanocellulose to suit water treatment applications and to even enhance its selective absorption and separation performance. Scientists are convinced that nanocellulose aerogels will play a significant role in the near future in terms of oil absorption and separation. Most of the current studies are only conducted in the lab to develop novel functional materials and investigate their performance compared with conventional ones. Several approaches were developed to change the nature of nanocellulose aerogel to become superhydrophobic, enhance its compressibility and elasticity and make it magnetically responsive for easier recovery and reusability. Several review articles have been published regarding the synthesis and characterization of highly hydrophobic, oil-absorbing aerogels [14] and aerogels for general application and water treatment applications [15]. Zamparas et al. [16] discussed the modification methods for improving the hydrophobicity of natural sorbents without specifying any type of biopolymers. Zaman et al. [17] covered the preparation, properties and general applications of cellulose-based aerogels without focusing on the oil separation. This review presents an overall introduction to nanocellulose-based aerogels, their fabrication and their different applications and covers in detail the most recent techniques for the modification of nanocellulose-based aerogels to suit oil absorption applications. It presents the most recent works in this field and discusses the potential and prospective uses of such materials in the near future.

2. Nanocellulose-Based Aerogels: Properties and Application

The term "nanocellulose" encompasses three major cellulose-based materials, namely cellulose nanofibers, cellulose nanocrystals and cellulose nanoparticles, which all can be isolated by various isolation techniques [18]. The isolation technique has been reported to affect the properties of nanocellulose; using chemical, mechanical and/or enzymatic treatments of the precursor material could affect its surface functional groups, thermal stability and mechanical properties. Apart from the isolation technique, the properties of nanocellulose were also found to depend on the source of precursor material and potential subsequent surface transformations [19]. Many researchers nowadays use surface modification techniques for nanocellulose to enhance its properties in different materials, including aerogels, by introducing charged or hydrophobic moieties [20]. Since the properties of aerogel are mainly determined by the source of precursor material, choosing the appropriate material is important to achieve a high-performance functional material.

2.1. Properties of Nanocellulose-Based Aerogels

As the most abundant biopolymer on earth, cellulose has received the attention of scientists in different fields, including biodegradable plastics, optical films, modern coatings, composites and laminates and controlled release of actives, in addition to membranes and related separation media [2,21]. It has been utilized in the fabrication of different functional materials such as films, membranes, coating agents, hydrogels and aerogels [22]. Many researchers have categorized nanocellulose materials into three main types, namely nanofibrillated cellulose, nanocrystals and nanoparticles, in addition to bacterial cellulose, which is another type of cellulose produced by certain bacteria such as *Acetobacter xylinum* as an extracellular product. Bacterial cellulose is very pure cellulose with narrow size distribution and high crystallinity [23,24]. All the cellulose types have relatively similar chemical compositions but different degrees of crystallinity, particle sizes and morphological properties [25]. Pure nanocellulose aerogel possesses closely similar properties to dried cellulose fibers depending on the effect of fabrication technique on the nanocellulose [26]. As a three-dimensional (3D) fiber network, aerogel is mainly composed of the precursor material(s) after the solvent is removed from the system. Nanocellulose is known for its unique physical, chemical and thermal properties, in addition to its availability, ecofriendliness and nontoxicity. In a recent study, Rizal et al. [26] reported that pure nanocellulose aerogel possessed such a high water absorption ability that it could not remain intact underwater. The authors also added that introducing chitosan into the nanocellulose aerogel enhanced its underwater stability and made the aerogel remain intact underwater. Similarly, Kontturi et al. [27] reported that nanocellulose and its materials are characterized by hydrophilicity and underwater superoleophobicity due to the unique face chemistry and crystallinity of nanocellulose. However, the properties of any aerogel are said to be determined by the precursor materials and their origin, the fabrication approach and the additives [28,29]. The porosity and mechanical strength of pure nanocellulose aerogel can be adjusted by changing the initial concentration of nanocellulose [30]. In a recent study, different concentrations of nanocellulose were used in aerogel preparation using vacuum infusion with biobased epoxy [31]. The study revealed that initial nanocellulose concentration and density were directly proportional with the surface area and porosity were inversely proportional.

Generally, oil/water separation materials based on special wettability are divided into two kinds: hydrophobic oleophilic and hydrophilic oleophobic materials [32]. Hydrophobic oleophilic materials characterized by the rapid spread of oil droplets upon an oil/water mixture touching the surface of that material. Water does not wet the surface of hydrophobic oleophilic materials and is intercepted on the impermeable membrane, in the so-called oil-removing method [33]. However, hydrophilic oleophobic materials are the opposite, as they form a water isolation layer on their surfaces. Nanocellulose must be cross-linked with another material to modify the hydrophilicity of nanocellulose and to improve its mechanical stability. Other researchers have incorporated different materials such as titanium dioxide (TiO_2) [34], nano-silica [35], hydroxyapatite [36] and graphene [37] into nanocellulose to enhance its properties in water treatment applications. Tang et al. [38] recently modified pure cellulose aerogel by directly mixing the raw nanocellulose with poly(propylene glycol adipate), which is a polyester plasticizer. The authors were able to enhance the mechanical strength of the aerogel in addition to its adsorption capacity for dyes. The property enhancement of nanocellulose aerogel continues to satisfy sophisticated requirements for different water treatment applications [39]. Loading of nanocellulose aerogel with a functional material that could enhance both the mechanical properties and the adsorption capacity at the same time is attracting great attention for its potential in easing the fabrication and minimizing the cost of production. The morphology and pore size of nanocellulose aerogels are two important characteristics that typically influence the absorption capacity of the material; they are determined by several factors, including precursor material(s), preparation approach, additives and/or modifiers, cooling rate and physical conditions such as drying [40]. In a recent study by Sakai et al. [41], the authors reported that supercritical drying of nanocellulose aerogel led to the generation of smaller and open pores compared with other drying approaches such as freeze-drying. Nanocellulose-based aerogels have been also utilized in medical applications due to their biocompatibility, noncytotoxicity, nonimmunogenicity and biodegradability; the past few years witnessed the use of such aerogels in drug delivery, tissue engineering, biosensing and wound dressing applications [42–44]. Nanocellulose-derived carbon aerogels are another form of aerogels that exhibit high adsorption capacity and are usually prepared by pyrolyzing the precursor materials [45]. Spongelike carbon aerogel from nanocellulose, with high porosity (99%), ultralow density (0.01 g/cm^3), hydrophobic properties (149° static contact angle) and reusability, has been reported by Meng et al. [46]. The authors reported that the carbonization process removed hydrophilic functional groups from the cellulose, making the resulting aerogel highly hydrophobic with extraordinary ability to absorb oil.

2.2. Fabrication of Nanocellulose-Based Aerogels

As a 3D structured material, the main factor of fabricating aerogels with high porosity and desired stability is to remove the solvent from the polymeric network without any structural disorder. To achieve this, various fabrication techniques have been developed (extensively reviewed by Abdul Khalil et al. [47]), which all initially were established on the formation of polymeric gel and the drying of that gel using supercritical conditions. Supercritical conditions allow the solid form to directly transfer to gas without passing through the liquid state, ensuring the shape of the precursor network structure remains. Depending on the precursor material(s), this process consists of several steps, which can be summarized as dissolving the precursor polymer(s) in water or any suitable solvent, followed by gelation, aging and finally drying (Figure 2) [48]. It has been reported that each step is highly affected by related parameters; chitosan does not dissolve in distilled water and requires weak acid, compared with carrageenan and agar. Pure cellulose does not form gel without cross-linkers, and thus for pure cellulose aerogel, immediate freezing of a homogeneous suspension is required before the freeze-drying process [40].

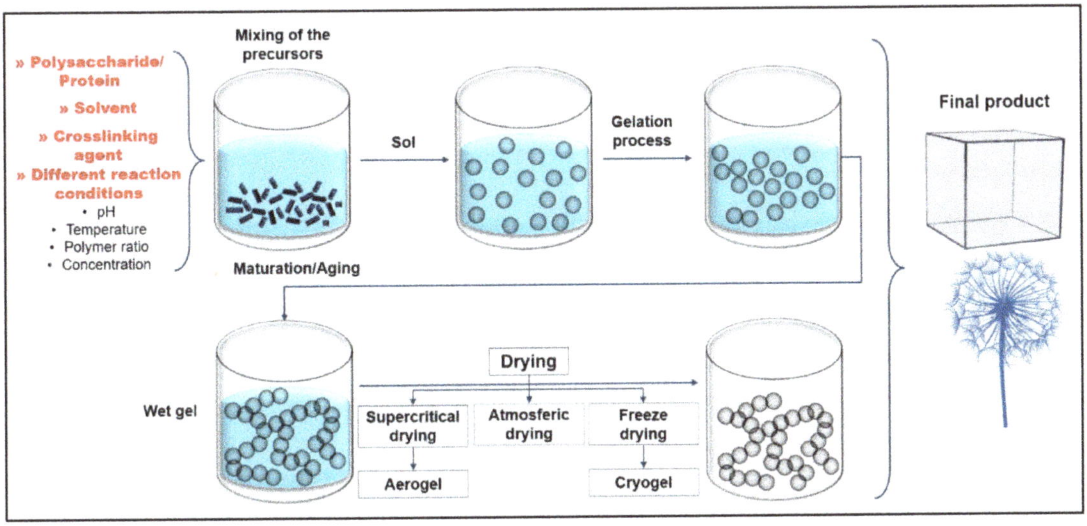

Figure 2. Illustration of the conventional fabrication process of aerogels: dissolving and mixing of biopolymer(s), hydrogel formation, aging and drying process. Adapted from [49], with permission from MDPI, 2022.

Various recent investigations found that type of precursor, its concentration, the type and concentration of solvent(s), pH, temperature and the drying method play a key role in tuning the characteristics and properties of the prepared aerogel structure [48,50]. Among all these parameters, selecting a suitable drying approach is the most important parameter determining the aerogel characteristics. High-temperature and low-temperature supercritical drying are the most used techniques for drying polymeric gels, in addition to freeze-drying and ambient-conditioned evaporation. Although using supercritical carbon dioxide drying for the fabrication of nanocellulose aerogel was effective, it has the drawbacks of being dangerous (due to the use of high pressure), time-consuming and relatively expensive [51]. The sol–gel route of aerogel fabrication was developed in the 1960s by replacing water glass with tetramethoxysilane (TMOS), which makes the removal of supercritical fluid from the gel system easier. Pure nanocellulose does not form thick gel without the use of a chemical cross-linker; thus, the freeze-drying technique is highly suitable for its preparation. Upon freezing the nanocellulose suspension, this approach removes water and ice crystals without the need for high temperature or any other organic solvents. The porosity, pore size and pore shapes can be controlled by changing the concentration of the nanocellulose material. Venkatesan et al. [52] reported that energy consumption can be a drawback associated with the freeze-drying technique, in addition to irregular pore size and shape. In a different study, Kanimozhi et al. [53] compared this technique with salt leaching in the nanocellulose composite aerogel fabrication and revealed a better control of pores by using the salt leaching technique. Such conventional techniques for aerogel preparation require the use of organic solvents and salts, may affect the properties of the material and could lead to chemical pollution. The past two decades witnessed great advances in aerogel preparation techniques using advances in technology aided by computers, known as rapid prototyping techniques. Selective laser sintering, stereolithography, injection molding and 3D printing techniques have been extensively used to fabricate aerogels in desired combinations and shapes [54,55]. Using such techniques, it was possible to study the effect of each combination on the material performance and to control the inner pore structures precisely with high reproducibility [56]. Although these fabrication processes produce the desired shape, pore size and pore structure, due to the limited resources and limited funding, most of the studies at present are conducted by using freeze-drying or supercritical approaches as convenient techniques for lab-based and research evaluations.

2.3. Application of Nanocellulose-Based Aerogels

Nanocellulose-based aerogels have been utilized in a variety of applications depending on their type of cellulose, combined material(s) and modification to suit the desired application. The application of nanocellulose aerogels was further extended to biomedical and medical fields such as drug delivery and tissue engineering due to the biocompatibility and nontoxicity of nanocellulose [40]. Table 2 illustrates the different applications of nanocellulose-based aerogels for different fields.

Table 1. The role of nanocellulose-based aerogels in different applications.

Field of Application	Application	Type of Aerogel	Remark	Ref.
Environmental	Absorption of oils and organic solvents	Cellulose/chitosan aerogel	Highly hydrophobic aerogel from cross-linking oxidized cellulose with chitosan and cold plasma modification	[56]
	Removal of chemicals	Cellulose-based aerogel	Polyaniline used as interface layers; the aerogel had high absorption capacity (409.55 mg/g) for tetracycline	[57]
	Removal of heavy metals	Robust cellulose aerogel	Polyethylenimine used as cross-linker; the aerogel showed high removal capacity (163.4 mg/g), fast adsorption rate and high shape recovery	[58]
	Water treatment	Cellulose nanocrystals/poly(methyl vinyl ether-co-maleic acid)/poly(ethylene glycol) aerogel	The chemically cross-linking aerogel was able to absorb all the cationic dyes from water (116.2 mg/g)	[59]
	Air purification	Robust micro-honeycomb-like nanofibrous aerogels	The aerogel showed extraordinary filtration performance even for 0.3 µm sized particles	[60]
Industrial	Enzyme immobilization	Bacterial cellulose/poly(glycidyl methacrylate) aerogel	The bacterial cellulose was modified with PGMA through atom transfer radical polymerization approach for catalase enzyme immobilization	[61]
	Protein separation	Dendrimer-assisted boronate affinity cellulose	Rapid adsorption rate was achieved with outstanding adsorption capacity for proteins (537.4 mg/g)	[62]
	Thermal insulation	Gelatin/hydroxyethyl cellulose–SiO_2	Hydrogen bonding and chemical cross-linking led the thermal conductivity of aerogel being lowered to 0.035 W/m K	[63]
	Packaging	Arundo donax cellulose aerogel	The superabsorbent bioactive aerogel reduced oxidation processes in red meat, leading to the extension of its shelf life	[64]

Table 2. *Cont.*

Field of Application	Application	Type of Aerogel	Remark	Ref.
Biomedical	Drug delivery	Bamboo shoot cellulose/sodium alginate aerogels	Curcumin initially encapsulated in the aerogel, which then released in a sustained manner	[65]
	Biosensing	Triarylmethane-loaded cellulose acetate aerogel	Urease enzyme was used as a catalytic agent in the aerogel for colorimetric detection of urea	[66]
	Wound healing	Cellulose/konjac glucomannan	The biocompatible aerogel enabled faster wound recovery by enhancing cell proliferation	[67]
	Tissue scaffolding	Cellulose nanofiber–gelatin aerogel	Epichlorohydrin was used as a cross-linker; the aerogel showed adequate cytocompatibility and cell viability	[68]
	Antimicrobial	Nanocellulose aerogel loaded with thymol	The aerogel possessed a high effect against Gram-positive and -negative bacteria in addition to yeasts	[69]
Others	Electric conductivity	Bacterial cellulose/graphene oxide aerogels	The addition of dimethyl sulfoxide enhanced the electric conductivity of the aerogel	[70]
	Catalyst	Gold nanoparticles supported on cellulose aerogel	Cellulose aerogel was loaded with gold nanoparticles and possessed high yield and selectivity for styrene epoxidation	[71]
	Supercapacitor	Carbonized cellulose nanofibril aerogel	The aerogel had excellent electrochemical stability even after 5000 cycles and still kept 89.43% of its specific capacitance	[72]
	Flame retardancy	Bulk Al-doped carboxymethyl cellulose aerogels	The aerogel had better weight-bearing capacity than conventional silica aerogel and could be rated V-0 in terms of UL-94 testing for flame retardancy	[73]

3. Strategies for the Modification of Nanocellulose Aerogels to Suit Oil Absorption Application

Aerogels based on nanocellulose possess great absorption capacity, giving this biopolymer promising potential in oil absorption applications. Despite the modification of the hydrophilicity of nanocellulose aerogels and the enhancement of their underwater stability, the removal of oil absorbents is yet another challenge [74]. Nanocellulose aerogels are known to have high porosity, ultralow density and high water absorption. Thus, modification of these materials is required for water treatment applications. Several approaches have been used to modify nanocellulose aerogels either by chemical treatments or the use of cross-linkers to enhance their hydrophobicity, induce their oil absorption selectivity and improve their shape retention and magnetic or pressure sensitivity (Figure 3).

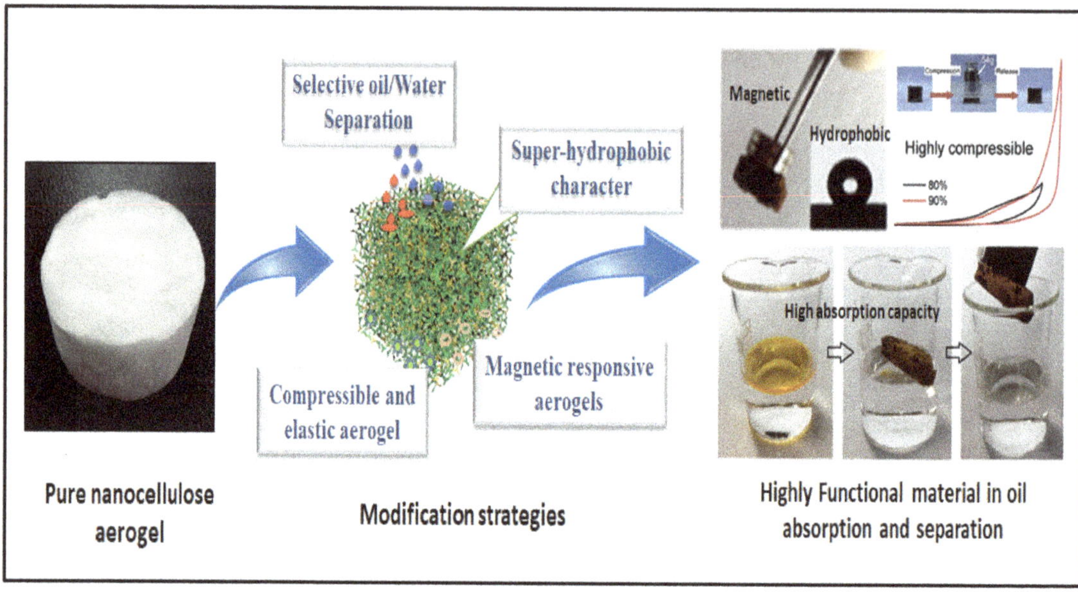

Figure 3. Illustration of commonly used modification strategies for nanocellulose aerogels to suit oil absorption and separation applications.

3.1. *Superhydrophobic Character*

Biopolymers are known for their hydrophilic nature and high water absorbability, which can be serious issues in terms of water absorption applications. Several efforts have been made to develop hydrophobic biopolymer-based materials, including aerogels, for oil absorption applications. Although commercial materials such as polypropylene fiber mats have excellent hydrophobicity and good oil absorption capacities (around 15 g/g), they have the drawback of degradation, which is considered a remaining major environmental challenge [75]. Hydrophobic biopolymer-based aerogel, which can serve the same function with better performance, was developed to overcome this problem. Nanocellulose-based aerogels have been treated with octyl-trichlorosilane to obtain hydrophobic surfaces using vapor phase deposition [76]. The authors obtained superhydrophobic cellulose-based aerogel with absorption capacity comparable to that of commercial absorption materials. The combination of unique aerogel properties of high porosity and surface area and the use of abundant precursors such as plant wastes make the cellulose-based aerogels a highly promising candidate for these applications. In a recent study, nanocellulose aerogel was functionalized with a titanium dioxide layer by using atomic layer deposition [77]. Using the deposition approach produces highly effective absorbents, but this approach is complicated, expensive and requires the use of sophisticated equipment, which may increase the costs of production. To facilitate this and to minimize the costs, Feng et al. [78] developed a facile and cost-effective approach for the fabrication of a very stable superhydrophobic cellulose-based aerogel. The authors used paper waste as the source of cellulose and Kymene as a cross-linker. To modify the hydrophilicity of the cellulose aerogel, the authors coated the aerogel with methyltrimethoxysilane by using chemical vapor deposition, and they were able to achieve excellent oil (but not water) absorption capacities of more than 95 g/g by using only 0.25 wt.% of cellulose in the aerogel. Such recycled cellulose aerogels can be also developed from other wastes following the same approach, gaining the advantages of utilization of such wastes and the benefits of oil absorption. Zhang et al. [79] fabricated hydrophobic aerogel (140.1°) from kapok/microfibrillated cellulose possessing dual-scale hierarchically porous structure. In order to modify the hydrophilicity of cellulose, the authors dispersed the chopped kapok fiber into the aqueous solution of cellulose under magnetic stirring until uniform dispersion. Adding a few drops of vinyltrimethoxysilane and adjusting the pH to 4 was found to promote the hydrolyzation of the vinyltrimethoxysilane in addition to the formation of silanol groups that react with cellulose molecules and make them hydrophobic [79]. Figure 4 presents fabrication steps of hydrophobic cellulose aerogel with ultrahigh absorption ability for oils ranging from 104 to 190.1 g/g, which is comparable to and even higher than that of the conventional environmentally unfriendly aerogels.

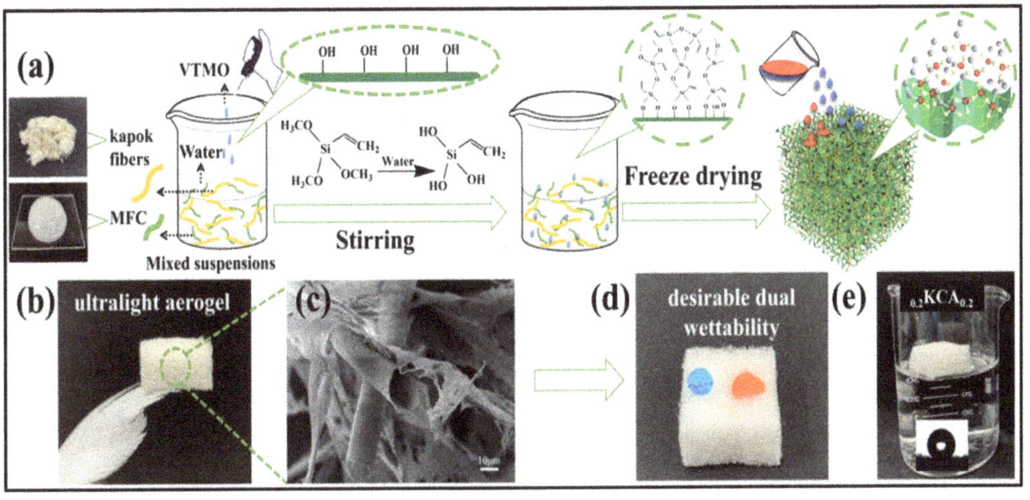

Figure 4. Illustration of hydrophobic cellulose-based aerogel properties: (**a**) schematic drawing of the fabrication approach using freeze-drying; (**b**) realistic image of prepared aerogel; (**c**) SEM image of the aerogel; (**d**) wettability characteristic of the aerogel; (**e**) water contact angle. Adapted from [79], with permission from Elsevier, 2022.

Although numerous researchers have proposed reversing the hydrophilicity of a material intended for use in water treatment applications, others proposed using hydrophilic/oleophobic surface materials such as nanocellulose and chitosan for oil/water separation [80]. However, these materials lose their elaborate topological topography once they are placed underwater and thus cannot achieve hydrophilic/oleophobic properties [81]. Coating suitable-wettability biopolymers onto certain metal meshes or even textiles was found to be an effective approach in the preparation of hydrophilic/oleophobic materials [82]. Zhou et al. [83] developed a different approach for the fabrication of highly porous cellulose aerogel from softwood kraft pulp by using a facile silanization reaction for coating the aerogel with polysiloxane. The authors were able to produce superhydrophobic aerogel with a water contact angle as high as 151.8°. Polysiloxane enhanced the mechanical stability of the aerogel in addition to resulting in a huge oil absorption ability of up to 159 g/g. In different study, Dilamian and Noroozi [84] prepared cellulose-based aerogel with hydrophobicity of 151° by cross-linking rice straw cellulose with polyamideamine-epichlorohydrin followed by freeze-drying the mixture and then oven-heating it at 120 °C for 3 h to achieve covalent cross-linking and finally hydrophobic coating of the aerogel by using methyltrimethoxysilane under silanization reaction. The authors were able to achieve superhydrophobic aerogel with a water contact angle of 151°, dual-scale porous structure and excellent crude oil absorption performance (112 g/g). Such highly functional materials could be even more developed for better performance; the high porosity of aerogels promotes the high adsorption capacity, which is accelerated by the aerogel being modified to become superhydrophilic. This strategy is very important and has the advantage of making the aerogel reusable, as it can only absorb the oil but not the water. Although most of the previous works used chemical modifications to change the hydrophilicity of nanocellulose, tiny quantities are used and the material has the potential of being reusable several times. Using such modification techniques to further enhance the unique properties of nanocellulose aerogels could promote them to become the future replacements for commercial oil absorbents, as ecofriendly and biodegradable oil absorption materials.

3.2. Selective and Versatile Oil/Water Separation

Oil selectivity is an important characteristic that increases the adsorption capacity of the aerogel and enhances its performance. Numerous materials have been incorporated with nanocellulose to increase its oil selectivity and thus enhance its hydrophobicity. Among them, graphene has attracted great attention as a hydrophobic surface material in the field of oil/water separation due to its excellent bonding with nanocellulose, good hydrophobicity, high specific surface area and unique chemical and mechanical stability [85]. Nguyen et al. [86] fabricated a facile approach to develop a superhydrophobic oil absorbent aerogel by coating melamine aerogel with graphene sheets using the dip-coating method. The aerogel exhibits great absorption capacity with excellent selectivity. The same principle was used in a different study to enhance the oil selectivity of nanocellulose, and surprisingly, the aerogel possessed even higher oil absorption capacity with excellent compressibility and recoverability [87]. In another investigation, Chatterjee et al. [88] developed a new approach to synthesize nanocellulose-based oil absorbent aerogel using polyethyleneimine and epoxy as chemical cross-linkers to enhance interaction and elastic recovery of the aerogel. The authors also dip-coated their aerogel with graphene nanosheets to maximize selective oil absorption, and they were able to achieve high selectivity with an absorption capacity reaching 25 to 58 g/g. The surface modification of the aerogel significantly improved the material selectivity in organic solvent absorption with the ability to rapidly retain shape after compression. Thus, mechanical squeezing can be applied for collecting the absorbed oil with over 100 squeezing–swelling cycles. In nanocellulose/graphene aerogels, polyethyleneimine is introduced as a cation part, which integrates with the nanocellulose chain and combines with the anion sheets of graphene [89]. Lu et al. [90] used this modification strategy and was able to fabricate nanostructured aerogel with a specific surface area of 441 m^2/g. The aerogel showed superhydrophobicity with a water contact angle of 155.5°, which can be attributed to the increase in C–C and –NH groups in addition to the pyrolysis of hydrophilic groups. High selectivity toward the oil was observed in the aerogel; the authors reported extraordinary performance in terms of oil/water separation (Figure 5) [90].

Nanocellulose is known for its excellent absorption capacity, and such modification could be used to develop material in large size to address many environmental issues such as oil spills. A different technique was used by Li et al. [91], who used a simple approach to fabricate ecofriendly and highly functionalized cellulose aerogel. The authors coated the aerogel with copper nanoparticles to enhance its oil selectivity, and they were able to produce porous aerogel that can selectively collect oily contaminants in a few minutes with excellent absorption capacity and good recyclability. This production approach is facile, cost-effective and uses sustainable precursors, thus being is suitable for large-scale production without using any organic hydrophobic modification. Recent work has concluded that blending nanocellulose with graphene oxide leads to extensive hydrogen-bonding interactions between the two materials, which leads to a directionally aligned porous structure resulting from the typical porosity of nanocellulose aerogel and sheet type of graphene [92]. Such a blend gave the resulting aerogel prominent compression, elastic and high recoverability properties, resulting from the superoleophilicity of graphene, making the aerogel selectively collect oils with excellent absorption capacities [92]. In contrast, polycondensation of nanocellulose with three silane-coupling agents has been used to fabricate a hydrophilic–oleophobic aerogel with a robust compressive ability and selective water absorption from oil/water mixtures with high reusability [93]. Blending nanocellulose with a particular material and the fabrication approach determine the properties of the resulting aerogel in addition to its absorption capacity and/or hydrophobicity. Although the interaction between nanocellulose and one particular material may look the same in several works, different findings may be explained by the origin of nanocellulose as well as the type of additive and fabrication situations, which all have a great influence on the aerogel properties [56,94]. Although graphene may not be a green material, blending it with nanocellulose facilitates its degradation. Selective and versatile separation has the advantage of separating the two materials at the same time rather than absorbing one from the other, which makes the process faster, less expensive and suitable for large-scale applications. Combining aerogel with certain materials with specific ligands could give it the potential to be a selective absorbent for specific chemicals or drugs, which could open new doors for this functional material in pharmaceutical applications.

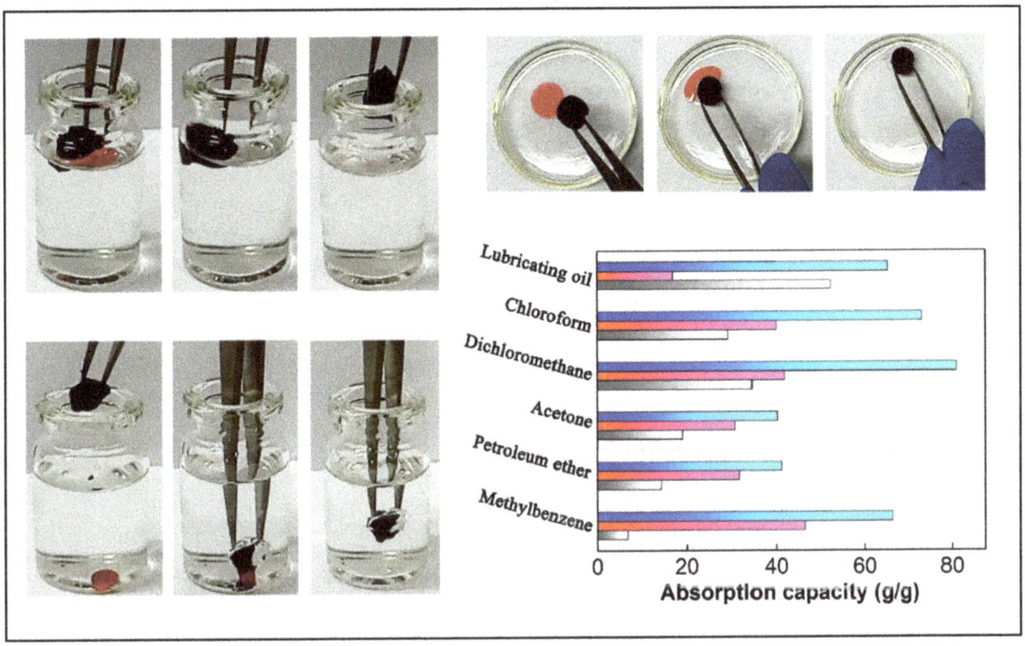

Figure 5. Nanocellulose/graphene aerogel in selective different oil absorption application, showing the high selectivity of the aerogel in oil absorption. Adapted from [90], with permission from Elsevier, 2022.

3.3. Magnetic Aerogels

The incorporation of iron oxide nanoparticles in cellulose-based aerogel to enhance the recovery ability of the aerogel by using a magnetic field has been gaining tremendous attention recently. The first attempt at this principle was made in 1973 by Turbeville, who used ferromagnetic sorbents for oil absorption and recovery [95]. Since then, a significant number of studies have been conducted using different iron compounds and different sorbent materials. Iron oxide nanoparticles have been reported to have excellent oil adsorption capacity besides their magnetic property [96]. However, due to the agglomeration and air oxidization characteristics, as highly chemically active compounds, these nanoparticles are not used in their naked form [97]. Nanocellulose aerogels have been proposed as a protecting shell on the surface of these nanoparticles, which can stabilize and maintain the function of iron oxide nanoparticles. A SiO_2 shell should be used to modify Fe_3O_4 nanoparticles to enhance the compatibility of the two materials with the organosilanes and thus obtain uniform magnetic aerogels. Fe_3O_4 nanoparticles were found to increase the hydrophobicity and oleophilicity of the cellulose aerogel, in addition to enhancing its mechanical stability and increasing the roughness [98]. Chin et al. [99] used a facile approach to fabricate magnetic cellulose-based aerogel by incorporating magnetic Fe_3O_4 nanoparticles into the aerogel and then coating the surfaces with a thin layer of TiO_2 to enhance the hydrophobicity of the aerogel. The authors used the sol–gel process for the coating process, and they were able to produce hydrophobic cellulose aerogel with the ability to easily be recovered from water by application of a magnetic field. Owing to the high porosity and the unique properties of cellulose, this aerogel was able to selectively absorb oil up to 28 times its own weight from a water/oil mixture in a short time. The use of iron oxide nanoparticles made the aerogel recovery quite easy upon the application of an external magnetic field. As an ecofriendly material, this type of biopolymeric aerogel has the potential of being the future solution for oil removal; it is an easy-to-prepare, unexpansive and selective oil absorbent with a great absorption capacity and efficiency. Dip-coating and adsorption approaches have been used to synthesize magnetic aerogels to transfer iron oxide nanoparticles into the aerogel texture [100]. Such an approach could also improve the hydrophobicity of nanocellulose aerogel in addition to its oil adsorption efficiency. However, it could lead to a slight loss of iron oxide nanoparticles during the oil adsorption process due to the weak bonding between nanocellulose and iron oxide, which could impairs its reusability [101]. In order to address this issue, a chemical cross-linker is used to enhance the bonding and immobilize the iron oxide nanoparticles within the aerogel system [99]. Such an aerogel was able to absorb up to approximately 28 times of the aerogel's own weight within 10 min and possessed excellent recovery upon use of an external magnetic field. Lu et al. [102] developed for the first time an ecofriendly magnetic aerogel based on ethyl cellulose (Figure 6); the aerogel was superhydrophobic (θ water > 150°) in several kinds of solutions, with low density and high porosity, and was prepared by silanizing ethyl cellulose with hexadecyl-trimethoxysilane and then incorporating ferroferric oxide nanoparticles into the mixture. The authors were able to achieve 37–51 times absorption capacity for the prepared magnetic aerogel, which was found to be well maintained upon repeated use. This aerogel demonstrates superior recyclability compared with other magnetic aerogels.

Considering the poor selectivity of pure nanocellulose aerogels for oil, many researchers have introduced hydrophobic groups that possess low surface energy into the aerogel system to improve its oil adsorption capacity and introduced magnetic nanoparticles to improve the recovery of aerogel [103]. Lin et al. [104] treated nanocellulose aerogel with trimethyl-chlorosilane to improve its oil adsorption by using the low-temperature plasma technology and achieved an adsorption capacity of more than 34.5 g/g. The authors were able to enhance the adsorption capacity of the aerogel, but its removal and recovery were achieved in a similar study, which used the same approach and introduced Fe_3O_4 into the aerogel [105]. Rapid removal and easy reusability and separation of the adsorbent can be achieved by introducing magnetic property into nanocellulose aerogel under a permanent magnet. Regenerated cellulose was incorporated with manganese iron oxide ($MnFe_2O_4$) nanoparticles by in situ coprecipitation approach, using both sodium chlorite and sodium periodate as oxidants to achieve modified magnetic cellulose aerogels [106]. The addition of $MnFe_2O_4$ nanoparticles promotes the oil adsorption and enhances the mechanical stability of the aerogel. A different study used Fe_3O_4 nanoparticles in nanocellulose/inorganic silica fiber aerogel and reported that the addition of Fe_3O_4 nanoparticles significantly improved the aerogel surface area, water contact angle and compressive property, in addition to the adsorption capacity of the aerogel [107]. However, with the increase in Fe_3O_4 nanoparticles, the bulk density increased, and thus the adsorption capacity decreased. This strategy significantly facilitates the recovery of aerogel after the adsorption process, it can be used for large projects using large pieces of aerogel to address the marine oil spill issues. However, this strategy has the limitation of using iron nanoparticles, which could have adverse effects on ecosystems; currently, studies on the effects of iron oxides on human and environmental ecosystems are limited.

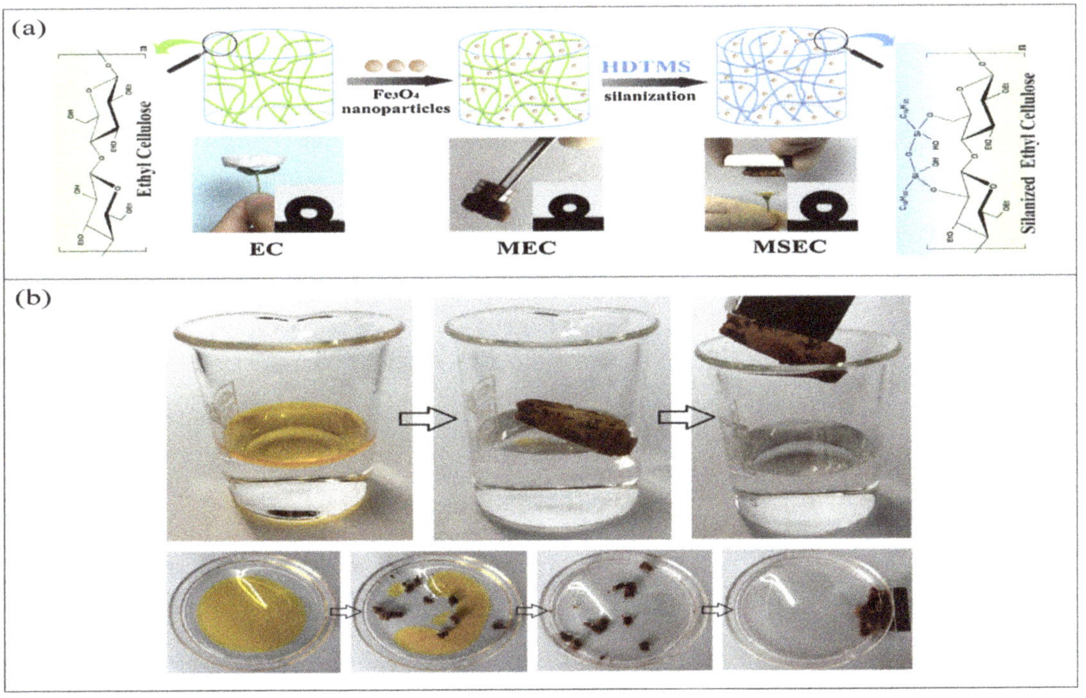

Figure 6. Illustration of modified ethyl cellulose magnetic aerogel: (**a**) schematic drawing of fabrication approach; (**b**) selective oil absorption and magnetic-responsive character. Adapted from [102], with permission from Elsevier, 2022.

3.4. Compressibility and Elasticity (Shape Retention)

Bioaerogels mostly possess low compressibility and elasticity due to their low density and high porosity. Preventing the permanent deformation and collapse of the polymeric network within the aerogel is the key to attaining aerogel with high compressibility and elasticity [108]. Nanocellulose aerogel has an anisotropic macroscopic structure; thus, permanent deformation and collapse of its structures occur upon the application of mechanical force. Designing the anisotropic porous structure of the aerogel was found to be a feasible solution for boosting the compressibility and elasticity of nanocellulose aerogels [109]. To prove this principle, Gao et al. [110] were able to fabricate a carbon–graphene monolith exhibiting an anisotropic lamellar structure with excellent compressibility and recoverability. The introduction of micro- and nanolattice structures into the material could even endow compressibility as well as elasticity to rigid metals and fragile ceramics [111]. Recent investigations showed that cross-linking nanocellulose with graphene oxide increases the elasticity of the resulting aerogel and enhances its compression [112]. Mi et al. [94] modified nanocellulose to produce 3D highly compressible, anisotropic and elastic aerogel using a bidirectional freeze-drying approach. The authors incorporated graphene aerogels with nanocellulose to produce hybrid aerogel and then modified it by using the DDTS approach, which includes using saline with a long carbon chain, and they were able to produce highly compressible superhydrophobic aerogel (Figure 7). The outstanding compression and recoverability properties of the aerogel came as a result of mixing the properties of flexible cellulose and stiff graphene. Upon the compression of aerogel to 60% and 90%, strain the authors were able to recover 99.8% and 96.3%, respectively. Owing to the ultralight weight of the aerogel and high surface area, its absorption capacity was extremely higher than that of conventional aerogel and ranged from 80 to 197 times its weight. Such compressible aerogel also possesses efficient recovery of oil upon simply squeezing the aerogel. The authors claim that the high compressibility of their aerogel and its high oil recovery ability have never been obtained by any other cellulose-based aerogel. This strategy enhances the mechanical stability of the material, making it remain intact and avoiding any possible leaching that may occur in the environment. Future studies should use more green materials to enhance the compressibility of aerogels to avoid any possible long- or short-term adverse effects on humans and/or environment after their use.

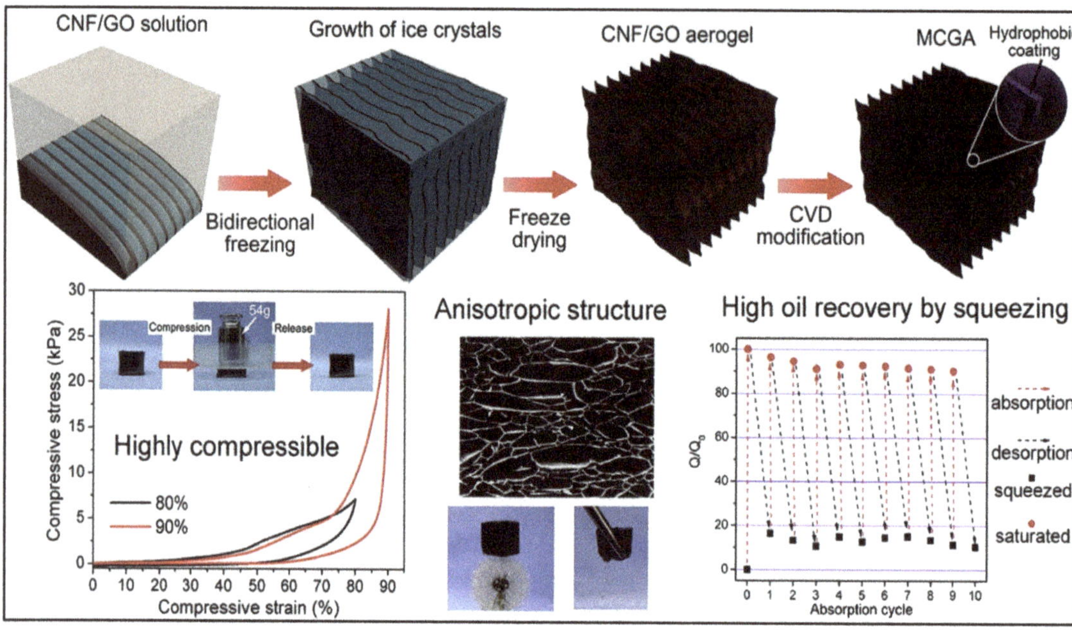

Figure 7. Illustration of the fabrication steps of compressible, anisotropic and elastic nanocellulose/graphene aerogel, showing its high oil recovery by squeezing. Adapted from [94], with permission from Elsevier, 2022.

4. Biopolymer Aerogel Composites for Oil Absorption and Separation Applications

The use of biodegradable materials to address oil spill issues has attracted significant attention in the past few years. Owing to the unique properties of modified nanocellulose-based aerogels, they could serve as the future materials for all oil absorption and separation purposes. In recent work, Korhonen et al. [77] fabricated a modified nanocellulose aerogel by using the atomic layer deposition technique to coat the aerogel with a TiO_2 layer and used it for oil removal. The authors were able to achieve good oil absorption capacity with high selectivity, but the approach requires the use of sophisticated equipment that raises its costs compared to conventional removal and absorption approaches. To solve this problem and reduce the production costs, Chin et al. [99] developed a facile approach to synthesize superhydrophobic and magnetic nanocellulose aerogel by incorporating magnetic nanoparticles within the nanocellulose and coating the surfaces with a thin TiO_2 layer. The resulting aerogel was able to rapidly absorb the oil with an absorption capacity of 28 times the aerogel weight in a few minutes. This unique aerogel was also magnetically sensitive, and the authors were able to recover it without reducing its absorption capacity. In a recent study, Zhang et al. [113] developed a novel aerogel system for oily wastewater treatment, using nanocrystalline cellulose and chitosan as the major precursor materials. The authors used a quaternized N-halamine siloxane monomer to modify the hydrophilicity of the aerogel by hydrolyzing it into polymer form in the nanocellulose/chitosan solution. Owing to the hydrophilic nature of nanocellulose and the porous structure of aerogel that endow the prepared aerogel with underwater oleophobic character, the authors were able to achieve high separation efficiency for several types of oil/water solutions with separation efficiency close to 100% (Figure 8). Furthermore, the reusability of such aerogel did not affect the separation performance; thus, it could be a promising material for different oil/water separation applications.

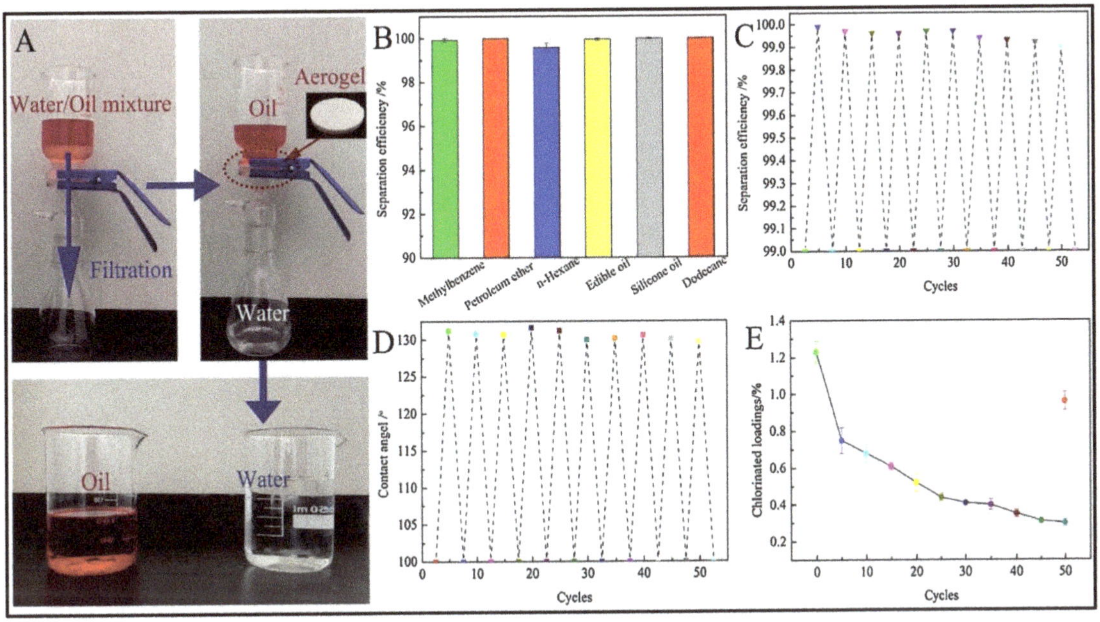

Figure 8. Illustration of the great performance of nanocellulose/chitosan aerogel in oil/water separation, showing the separation of the two materials in high purity; (**A**) oil/water separation experiment, (**B**,**C**) separation efficiency values, (**D**) water contact angle, and (**E**) chlorinated loadings. Adapted from [113], with permission from Elsevier, 2022.

Numerous publications apply multiple modifications to the same aerogel to enhance its absorption performance in terms of superhydrophobicity, compressibility and ease of reusability. Compared with the conventional commercial materials such as polypropylene and its derivatives that possess an absorption capacity of 8.1–24.6 g/g and a water contact angle of about 102.1° [114], most of the fabricated cellulose aerogel possess superior performance. Table 3 presents a summary of the most recent publications on nanocellulose-based aerogels in oil absorption applications compared with polypropylene.

Table 3. Literature summary of nanocellulose-based aerogels for oil absorption applications.

Type of Aerogel	Preparation Method	Absorption Capacity (g/g)	Water Contact Angle	Outcomes	Ref.
Polypropylene-based material (reference)	Commercial material	8.1–24.6	102.1°	The hollow fiber had higher absorption rate than solid fibers	[114]
Budget cotton-based aerogel	CO_2 supercritical drying and vapor deposition	16.0	153.0°	Simple, rapid and effective superhydrophobic material for oil absorption	[115]
Bacterial cellulose and fumed silica aerogel	Freeze-drying of fumed silica and infiltrated bacterial cellulose	28	142.0°	Used 10 times for oil recovery without any reduction in the uptake capacity	[116]
Modified waste cellulose fibers	Freeze-drying	142.9	159.0°	Super oil absorption capacity for at least 30 cycles	[117]
Anisotropic graphene oxide/polyvinyl alcohol/CNF carbon aerogel	Freeze-drying	155–288	140.0°	Outstanding compressibility and recyclability with extremely large oil absorption capacity	[118]
Nanocellulose/ nanochitosan/reduced graphene aerogel	Hydrothermal and freeze-drying technique	120–176	115.2°	Oil/water pump apparatus containing the aerogel continuously removed/collected the oil from wastewater	[119]
Wool waste fiber aerogel	Direct freeze-drying	136.2	138.0°	9.1–15.3 times better than commercial oil absorption materials	[120]
Rice straw nanocellulose aerogel	Freeze-drying	170	151.0°	Superlative adsorbent for remediation of polluted water	[84]
Bagasse-based aerogels	Freeze-drying and thermo-crosslinking	31.65	148.0°	High oil absorption capacity and low thermal conductivity	[121]
Collagen and dialdehyde carboxymethylcellulose	Freeze-drying and surface coating	20.4–57.2	144.4°	Excellent reusability and recyclability for oily liquids	[122]
Kapok/ microfibrillated cellulose aerogels	Vacuum freeze-drying and surface modification	104–190.1	140.1°	Rapid, selective and ultrahigh absorption and recycling ability	[79]
Raw cotton fiber macroporous cellulose aerogel	Sol–gel and freeze-drying techniques	19.8–41.5	154.0°	The aerogel possessed superb oil retention capability	[123]
Nanocellulose/silica fiber/Fe3O4 aerogel	Direct freeze-drying and surface modification	34.2–58.3	150.0°	Superior stability in wide pH range with multiple uses	[107]
Graphene/cellulose/ silica aerogel	Hydrothermal and subsequent freeze-drying	39–68	157.0°	Extraordinary absorption efficiency for various oils	[87]

5. Potential and Future Perspective of Nanocellulose-Based Aerogels

In recent years, several novel adsorbent and separation materials have appeared, such as graphene sponges and aerogels, carbon nanotube frameworks, carbon fiber aerogels and biopolymer-based aerogels [124,125]. Although these materials possess excellent oil/water adsorption and/or separation performance, most of them are nonbiodegradable, which could raise another issue for the environment. Among them, biopolymer-based aerogels have been modified to become highly competitive to inorganic adsorbents, with minimal disadvantages. In comparison, the modified nanocellulose-based aerogels proved to have a superhigh oil-absorbing capacity and can also be reused for several cycles without significant reduction in the adsorption capacity. Although nanocellulose-based aerogels for oil absorption and separation have not yet entered commercialization and are still limited to the research and laboratory sectors, we believe that the following years will witness great advances in the commercialization of these materials.

Great advances have been made in the past few years in terms of fabrication techniques; the traditional approaches that use excessive amounts of organic solvents are no longer applicable, which is significantly reducing the production costs [20]. The use of freeze-drying may consume time and energy, but the following years will witness modified printing techniques that will enhance the fabrication of such functional materials that can maintain the function and achieve the possibility of large-scale production and commercialization of nanocellulose-based aerogels. The further functionalization and multifunctionality of nanocellulose aerogels in different water treatment applications have encountered several challenges, including designable structures, undesirable dispersion of fillers within the aerogel, modification of surface functional groups and the possibility of transition from laboratory and research into industrialization. Future studies should consider the mentioned challenges and comprehensively consider subjective initiation to develop the applications and determine the future role of nanocellulose-based aerogels in future life. However, many functional fillers have been already incorporated with nanocellulose aerogels to endow structurally anisotropic functionalities such as magnetism and electricity, thereby easing their function in water treatment and broadening their application fields. Such studies are still in their initial stage and require further laboratory and economical evaluation before they reach industrialization.

6. Conclusions

Nanocellulose aerogels are a new category of high-efficiency adsorbents for treating oil spills and water pollution. They not only present typical characteristics of high porosity, large specific surface area and light weight but also incorporate excellent inherent properties such as high availability, low cost, easy scale-up, nontoxicity and sustainability. Although most of the strategies for modifying nanocellulose aerogels involve physical blending or chemical modification, nanocellulose is still the major part of the material. Such modifications can endow nanocellulose aerogels with additional special features, including superhydrophobicity, oil absorption selectivity, improved shape retention and magnetic or pressure sensitivity. Therefore, apart from oil/water separation, nanocellulose aerogels have great application prospects in other water treatment applications such as the adsorption of pesticides, herbicides, pharmaceuticals and heavy metals.

Author Contributions: Conceptualization, H.P.S.A.K. and E.B.Y.; data curation, E.B.Y. and M.A.I. (M. A. Ismail); formal analysis, H.P.S.A.K., M.A.I. (M. A. Ismail) and M.A.I. (M. A. Iskandar); funding acquisition, H.P.S.A.K. and A.A.R.; investigation, E.B.Y.; methodology, H.P.S.A.K. and E.B.Y.; project administration, H.P.S.A.K. and A.A.R.; resources, H.P.S.A.K.; software, A.A.R., M.A.I. (M. A. Ismail). and E.B.Y.; supervision, H.P.S.A.K.; writing—original draft, E.B.Y.; writing—review and editing, E.B.Y. and H.P.S.A.K. All authors have read and agreed to the published version of the manuscript.

Funding: This research was funded by the Ministry of Higher Education, Fundamental Research Grant Scheme—Malaysia's Research Star Award (FRGS-MRSA) with Project Code FRGS/1/2019/TK05/USM/01/6 and grant No. R115 Elemental Analysis and Material Characterization of Industrial Wastes, Teraju Saga Sdn. Bhd.

Institutional Review Board Statement: Not applicable.

Informed Consent Statement: Not applicable.

Data Availability Statement: Not applicable.

Acknowledgments: The authors would like to thank the Ministry of Higher Education, Fundamental Research Grant Scheme—Malaysia's Research Star Award (FRGS-MRSA), Project Code FRGS/1/2019/TK05/USM/01/6, for making this work possible.

Conflicts of Interest: The authors declare no conflict of interest.

References

1. Chen, C.; Zhu, X.; Chen, B. Durable superhydrophobic/superoleophilic graphene-based foam for high-efficiency oil spill cleanups and recovery. *Environ. Sci. Technol.* **2019**, *53*, 1509–1517. [CrossRef] [PubMed]
2. Rizal, S.; Olaiya, F.G.; Saharudin, N.; Abdullah, C.; NG, O.; Mohamad Haafiz, M.; Yahya, E.B.; Sabaruddin, F.; Abdul Khalil, H.P.S. Isolation of textile waste cellulose nanofibrillated fibre reinforced in polylactic acid-chitin biodegradable composite for green packaging application. *Polymers* **2021**, *13*, 325. [CrossRef] [PubMed]
3. Al-Anzi, B.S.; Siang, O.C. Recent developments of carbon based nanomaterials and membranes for oily wastewater treatment. *RSC Adv.* **2017**, *7*, 20981–20994. [CrossRef]
4. Zhang, Y.; Yin, M.; Lin, X.; Ren, X.; Huang, T.-S.; Kim, I.S. Functional nanocomposite aerogels based on nanocrystalline cellulose for selective oil/water separation and antibacterial applications. *Chem. Eng. J.* **2019**, *371*, 306–313. [CrossRef]
5. Azamzam, A.A.; Rafatullah, M.; Yahya, E.B.; Ahmad, M.I.; Lalung, J.; Alharthi, S.; Alosaimi, A.M.; Hussein, M.A. Insights into Solar Disinfection Enhancements for Drinking Water Treatment Applications. *Sustainability* **2021**, *13*, 10570. [CrossRef]
6. Rastib, A.; Memariani, M.; Riahi, M. Investigation of enterobacter aerogenes effects on heavy oil from biological degradation aspects by GC* GC technique. *Int. J. Petrochem. Sci. Eng.* **2019**, *4*, 47–52. [CrossRef]
7. Kong, D.; He, X.; Khan, F.; Chen, G.; Ping, P.; Yang, H.; Peng, R. Small scale experiment study on burning characteristics for in-situ burning of crude oil on open water. *J. Loss Prev. Process Ind.* **2019**, *60*, 46–52. [CrossRef]
8. Mi, H.-Y.; Jing, X.; Xie, H.; Huang, H.-X.; Turng, L.-S. Magnetically driven superhydrophobic silica sponge decorated with hierarchical cobalt nanoparticles for selective oil absorption and oil/water separation. *Chem. Eng. J.* **2018**, *337*, 541–551. [CrossRef]
9. Zhong, Q.; Shi, G.; Sun, Q.; Mu, P.; Li, J. Robust PVA-GO-TiO_2 composite membrane for efficient separation oil-in-water emulsions with stable high flux. *J. Membr. Sci.* **2021**, *640*, 119836. [CrossRef]
10. Gupta, R.K.; Dunderdale, G.J.; England, M.W.; Hozumi, A. Oil/water separation techniques: A review of recent progresses and future directions. *J. Mater. Chem. A* **2017**, *5*, 16025–16058. [CrossRef]
11. Zhao, X.-Q.; Wahid, F.; Cui, J.-X.; Wang, Y.-Y.; Zhong, C. Cellulose-based special wetting materials for oil/water separation: A review. *Int. J. Biol. Macromol.* **2021**, *185*, 890–906. [CrossRef] [PubMed]
12. Rizal, S.; Abdul Khalil, H.P.S.; Oyekanmi, A.A.; Gideon, O.N.; Abdullah, C.K.; Yahya, E.B.; Alfatah, T.; Sabaruddin, F.A.; Rahman, A.A. Cotton Wastes Functionalized Biomaterials from Micro to Nano: A Cleaner Approach for a Sustainable Environmental Application. *Polymers* **2021**, *13*, 1006. [CrossRef] [PubMed]
13. Zhang, N.; Qi, Y.; Zhang, Y.; Luo, J.; Cui, P.; Jiang, W. A review on oil/water mixture separation material. *Ind. Eng. Chem. Res.* **2020**, *59*, 14546–14568. [CrossRef]
14. Patil, P.G.; Sharma, R.; Chakrabarty, A. Synthesis and characterization of highly hydrophobic, oil-absorbing aerogels for oil spill applications; A Review. *J. Phys. Conf. Ser.* **2020**, *1644*, 012047. [CrossRef]
15. Ganesamoorthy, R.; Vadivel, V.K.; Kumar, R.; Kushwaha, O.S.; Mamane, H. Aerogels for water treatment: A review. *J. Clean. Prod.* **2021**, *329*, 129713. [CrossRef]
16. Zamparas, M.; Tzivras, D.; Dracopoulos, V.; Ioannides, T. Application of sorbents for oil spill cleanup focusing on natural-based modified materials: A review. *Molecules* **2020**, *25*, 4522. [CrossRef]
17. Zaman, A.; Huang, F.; Jiang, M.; Wei, W.; Zhou, Z. Preparation, properties, and applications of natural cellulosic aerogels: A review. *Energy Built Environ.* **2020**, *1*, 60–76. [CrossRef]
18. Qiao, A.; Cui, M.; Huang, R.; Ding, G.; Qi, W.; He, Z.; Klemeš, J.J.; Su, R. Advances in nanocellulose-based materials as adsorbents of heavy metals and dyes. *Carbohydr. Polym.* **2021**, *272*, 118471. [CrossRef]
19. Thomas, B.; Raj, M.C.; Joy, J.; Moores, A.; Drisko, G.L.; Sanchez, C. Nanocellulose, a versatile green platform: From biosources to materials and their applications. *Chem. Rev.* **2018**, *118*, 11575–11625. [CrossRef]
20. Chen, Y.; Zhang, L.; Yang, Y.; Pang, B.; Xu, W.; Duan, G.; Jiang, S.; Zhang, K. Recent progress on nanocellulose aerogels: Preparation, modification, composite fabrication, applications. *Adv. Mater.* **2021**, *33*, 2005569. [CrossRef]
21. Fingolo, A.C.; de Morais, V.B.; Costa, S.V.; Corrêa, C.t.C.; Lodi, B.; Santhiago, M.; Bernardes, J.S.; Bufon, C.C. Enhanced Hydrophobicity in Nanocellulose-Based Materials: Toward Green Wearable Devices. *ACS Appl. Bio Mater.* **2021**, *4*, 6682–6689. [CrossRef] [PubMed]
22. Giachini, P.; Gupta, S.; Wang, W.; Wood, D.; Yunusa, M.; Baharlou, E.; Sitti, M.; Menges, A. Additive manufacturing of cellulose-based materials with continuous, multidirectional stiffness gradients. *Sci. Adv.* **2020**, *6*, eaay0929. [CrossRef] [PubMed]
23. Huang, Y.; Wang, J.; Yang, F.; Shao, Y.; Zhang, X.; Dai, K. Modification and evaluation of micro-nano structured porous bacterial cellulose scaffold for bone tissue engineering. *Mater. Sci. Eng. C Mater. Biol. Appl.* **2017**, *75*, 1034. [CrossRef] [PubMed]
24. Kim, J.-H.; Shim, B.S.; Kim, H.S.; Lee, Y.-J.; Min, S.-K.; Jang, D.; Abas, Z.; Kim, J. Review of nanocellulose for sustainable future materials. *Int. J. Precis. Eng. Manufacturing-Green Technol.* **2015**, *2*, 197–213. [CrossRef]
25. Abdul Khalil, H.; Banerjee, A.; Saurabh, C.K.; Tye, Y.; Suriani, A.; Mohamed, A.; Karim, A.; Rizal, S.; Paridah, M. Biodegradable films for fruits and vegetables packaging application: Preparation and properties. *Food Eng. Rev.* **2018**, *10*, 139–153. [CrossRef]
26. Rizal, S.; Yahya, E.B.; Abdul Khalil, H.P.S.; Abdullah, C.; Marwan, M.; Ikramullah, I.; Muksin, U. Preparation and Characterization of Nanocellulose/Chitosan Aerogel Scaffolds Using Chemical-Free Approach. *Gels* **2021**, *7*, 246. [CrossRef]

27. Kontturi, E.; Suchy, M.; Penttila, P.; Jean, B.; Pirkkalainen, K.; Torkkeli, M.; Serimaa, R. Amorphous characteristics of an ultrathin cellulose film. *Biomacromolecules* **2011**, *12*, 770–777. [CrossRef]
28. Cervin, N.T.; Johansson, E.; Larsson, P.A.; Wågberg, L. Strong, water-durable, and wet-resilient cellulose nanofibril-stabilized foams from oven drying. *ACS Appl. Mater. Interfaces* **2016**, *8*, 11682–11689. [CrossRef]
29. Chen, W.; Li, Q.; Wang, Y.; Yi, X.; Zeng, J.; Yu, H.; Liu, Y.; Li, J. Comparative study of aerogels obtained from differently prepared nanocellulose fibers. *ChemSusChem* **2014**, *7*, 154–161. [CrossRef]
30. Yahya, E.B.; Amirul, A.; Abdul Khalil, H.P.S.; Olaiya, N.G.; Iqbal, M.O.; Jummaat, F.; AK, A.S.; Adnan, A. Insights into the Role of Biopolymer Aerogel Scaffolds in Tissue Engineering and Regenerative Medicine. *Polymers* **2021**, *13*, 1612. [CrossRef]
31. Nissilä, T.; Karhula, S.S.; Saarakkala, S.; Oksman, K. Cellulose nanofiber aerogels impregnated with bio-based epoxy using vacuum infusion: Structure, orientation and mechanical properties. *Compos. Sci. Technol.* **2018**, *155*, 64–71. [CrossRef]
32. Ding, L.; Wang, Y.; Zhu, P.; Bai, Y. One-step plant-inspired reaction that transform membrane hydrophobicity into high hydrophilicity and underwater super oleophobicity for oil-in-water emulsion separation. *Appl. Surf. Sci.* **2019**, *479*, 423–429. [CrossRef]
33. Li, J.; Kang, R.; Tang, X.; She, H.; Yang, Y.; Zha, F. Superhydrophobic meshes that can repel hot water and strong corrosive liquids used for efficient gravity-driven oil/water separation. *Nanoscale* **2016**, *8*, 7638–7645. [CrossRef] [PubMed]
34. Wang, Q.; Wang, Y.; Chen, L.; Cai, J.; Zhang, L. Facile construction of cellulose nanocomposite aerogel containing TiO_2 nanoparticles with high content and small size and their applications. *Cellulose* **2017**, *24*, 2229–2240. [CrossRef]
35. Yuan, B.; Zhang, J.; Mi, Q.; Yu, J.; Song, R.; Zhang, J. Transparent cellulose–silica composite aerogels with excellent flame retardancy via an in situ sol–gel process. *ACS Sustain. Chem. Eng.* **2017**, *5*, 11117–11123. [CrossRef]
36. Guo, W.; Wang, X.; Zhang, P.; Liu, J.; Song, L.; Hu, Y. Nano-fibrillated cellulose-hydroxyapatite based composite foams with excellent fire resistance. *Carbohydr. Polym.* **2018**, *195*, 71–78. [CrossRef] [PubMed]
37. Sehaqui, H.; Salajková, M.; Zhou, Q.; Berglund, L.A. Mechanical performance tailoring of tough ultra-high porosity foams prepared from cellulose I nanofiber suspensions. *Soft Matter* **2010**, *6*, 1824–1832. [CrossRef]
38. Tang, M.; Jia, R.; Kan, H.; Liu, Z.; Yang, S.; Sun, L.; Yang, Y. Kinetic, isotherm, and thermodynamic studies of the adsorption of dye from aqueous solution by propylene glycol adipate-modified cellulose aerogel. *Colloids Surf. A Physicochem. Eng. Asp.* **2020**, *602*, 125009. [CrossRef]
39. Jiang, J.; Zhang, Q.; Zhan, X.; Chen, F. A multifunctional gelatin-based aerogel with superior pollutants adsorption, oil/water separation and photocatalytic properties. *Chem. Eng. J.* **2019**, *358*, 1539–1551. [CrossRef]
40. Abdul Khalil, H.P.S.; Adnan, A.; Yahya, E.B.; Olaiya, N.; Safrida, S.; Hossain, M.; Balakrishnan, V.; Gopakumar, D.A.; Abdullah, C.; Oyekanmi, A. A review on plant cellulose nanofibre-based aerogels for biomedical applications. *Polymers* **2020**, *12*, 1759. [CrossRef]
41. Sakai, K.; Kobayashi, Y.; Saito, T.; Isogai, A. Partitioned airs at microscale and nanoscale: Thermal diffusivity in ultrahigh porosity solids of nanocellulose. *Sci. Rep.* **2016**, *6*, 20434. [CrossRef] [PubMed]
42. Yahya, E.B.; Jummaat, F.; Amirul, A.; Adnan, A.; Olaiya, N.; Abdullah, C.; Rizal, S.; Mohamad Haafiz, M.; Abdul Khalil, H.P.S. A review on revolutionary natural biopolymer-based aerogels for antibacterial delivery. *Antibiotics* **2020**, *9*, 648. [CrossRef] [PubMed]
43. Jummaat, F.; Yahya, E.B.; Abdul Khalil, H.P.S.; Adnan, A.; Alqadhi, A.M.; Abdullah, C.; AK, A.S.; Olaiya, N.; Abdat, M. The Role of Biopolymer-Based Materials in Obstetrics and Gynecology Applications: A Review. *Polymers* **2021**, *13*, 633. [CrossRef] [PubMed]
44. Yahya, E.B.; Alzalouk, M.M.; Alfallous, K.A.; Abogmaza, A.F. Antibacterial cellulose-based aerogels for wound healing application: A review. *Biomed. Res. Ther.* **2020**, *7*, 4032–4040. [CrossRef]
45. Lee, Y.J.; Jung, J.C.; Yi, J.; Baeck, S.-H.; Yoon, J.R.; Song, I.K. Preparation of carbon aerogel in ambient conditions for electrical double-layer capacitor. *Curr. Appl. Phys.* **2010**, *10*, 682–686. [CrossRef]
46. Meng, Y.; Young, T.M.; Liu, P.; Contescu, C.I.; Huang, B.; Wang, S. Ultralight carbon aerogel from nanocellulose as a highly selective oil absorption material. *Cellulose* **2015**, *22*, 435–447. [CrossRef]
47. Abdul Khalil, H.P.S.; Jummaat, F.; Yahya, E.B.; Olaiya, N.; Adnan, A.; Abdat, M.; NAM, N.; Halim, A.S.; Kumar, U.; Bairwan, R. A review on micro-to nanocellulose biopolymer scaffold forming for tissue engineering applications. *Polymers* **2020**, *12*, 2043. [CrossRef]
48. Soorbaghi, F.P.; Isanejad, M.; Salatin, S.; Ghorbani, M.; Jafari, S.; Derakhshankhah, H. Bioaerogels: Synthesis approaches, cellular uptake, and the biomedical applications. *Biomed. Pharmacother.* **2019**, *111*, 964–975. [CrossRef]
49. Nita, L.E.; Ghilan, A.; Rusu, A.G.; Neamtu, I.; Chiriac, A.P. New trends in bio-based aerogels. *Pharmaceutics* **2020**, *12*, 449. [CrossRef]
50. Jafari, S.; Derakhshankhah, H.; Alaei, L.; Fattahi, A.; Varnamkhasti, B.S.; Saboury, A.A. Mesoporous silica nanoparticles for therapeutic/diagnostic applications. *Biomed. Pharmacother.* **2019**, *109*, 1100–1111. [CrossRef]
51. Khedkar, M.V.; Somvanshi, S.B.; Humbe, A.V.; Jadhav, K. Surface modified sodium silicate based superhydrophobic silica aerogels prepared via ambient pressure drying process. *J. Non-Cryst. Solids* **2019**, *511*, 140–146. [CrossRef]
52. Venkatesan, J.; Bhatnagar, I.; Kim, S.-K. Chitosan-alginate biocomposite containing fucoidan for bone tissue engineering. *Mar. Drugs* **2014**, *12*, 300–316. [CrossRef] [PubMed]

53. Kanimozhi, K.; Basha, S.K.; Kumari, V.S.; Kaviyarasu, K. Development of biomimetic hybrid porous scaffold of chitosan/polyvinyl alcohol/carboxymethyl cellulose by freeze-dried and salt leached technique. *J. Nanosci. Nanotechnol.* **2018**, *18*, 4916–4922. [CrossRef] [PubMed]
54. Tang, A.; Li, J.; Li, J.; Zhao, S.; Liu, W.; Liu, T.; Wang, J.; Liu, Y. Nanocellulose/PEGDA aerogel scaffolds with tunable modulus prepared by stereolithography for three-dimensional cell culture. *J. Biomater. Sci. Polym. Ed.* **2019**, *30*, 797–814. [CrossRef]
55. Feng, J.; Su, B.-L.; Xia, H.; Zhao, S.; Gao, C.; Wang, L.; Ogbeide, O.; Feng, J.; Hasan, T. Printed aerogels: Chemistry, processing, and applications. *Chem. Soc. Rev.* **2021**, *50*, 3842–3888. [CrossRef]
56. Mariana, M.; Abdul Khalil, H.P.S.; Yahya, E.B.; Olaiya, N.; Alfatah, T.; Suriani, A.; Mohamed, A. Recent trends and future prospects of nanostructured aerogels in water treatment applications. *J. Water Process Eng.* **2022**, *45*, 102481. [CrossRef]
57. Liu, Q.; Yu, H.; Zeng, F.; Li, X.; Sun, J.; Hu, X.; Pan, Q.; Li, C.; Lin, H.; Su, Z.M. Polyaniline as interface layers promoting the in-situ growth of zeolite imidazole skeleton on regenerated cellulose aerogel for efficient removal of tetracycline. *J. Colloid Interface Sci.* **2020**, *579*, 119–127. [CrossRef]
58. Tang, C.; Brodie, P.; Li, Y.; Grishkewich, N.J.; Brunsting, M.; Tam, K.C. Shape recoverable and mechanically robust cellulose aerogel beads for efficient removal of copper ions. *Chem. Eng. J.* **2020**, *392*, 124821. [CrossRef]
59. Liang, L.; Zhang, S.; Goenaga, G.A.; Meng, X.; Zawodzinski, T.A.; Ragauskas, A.J. Chemically cross-linked cellulose nanocrystal aerogels for effective removal of cation dye. *Front. Chem.* **2020**, *8*, 570. [CrossRef]
60. Ma, X.Y.D.; Zeng, Z.; Wang, Z.; Xu, L.; Zhang, Y.; Ang, J.M.; Wan, M.P.; Ng, B.F.; Lu, X. Robust microhoneycomb-like nanofibrous aerogels derived from cellulose and lignin as highly efficient, low-resistant and anti-clogging air filters. *J. Membr. Sci.* **2022**, *642*, 119977. [CrossRef]
61. Liu, X.; Zheng, H.; Li, Y.; Wang, L.; Wang, C. A novel bacterial cellulose aerogel modified with PGMA via ARGET ATRP method for catalase immobilization. *Fibers Polym.* **2019**, *20*, 520–526. [CrossRef]
62. Yao, X.; Zhang, S.; Qian, L.; Du, M. Dendrimer-assisted boronate affinity cellulose foams for the efficient and selective separation of glycoproteins. *Carbohydr. Polym.* **2021**, *265*, 118082. [CrossRef] [PubMed]
63. Peng, T.; Zhu, J.; Huang, T.; Jiang, C.; Zhao, F.; Ge, S.; Xie, L. Facile preparation for gelatin/hydroxyethyl cellulose-SiO_2 composite aerogel with good mechanical strength.h, heat insulation, and water resistance. *J. Appl. Polym. Sci.* **2021**, *138*, 50539. [CrossRef]
64. Fontes-Candia, C.; Erboz, E.; Martínez-Abad, A.; López-Rubio, A.; Martínez-Sanz, M. Superabsorbent food packaging bioactive cellulose-based aerogels from Arundo donax waste biomass. *Food Hydrocoll.* **2019**, *96*, 151–160. [CrossRef]
65. Zhang, A.; Zou, Y.; Xi, Y.; Wang, P.; Zhang, Y.; Wu, L.; Zhang, H. Fabrication and characterization of bamboo shoot cellulose/sodium alginate composite aerogels for sustained release of curcumin. *Int. J. Biol. Macromol.* **2021**, *192*, 904–912. [CrossRef] [PubMed]
66. Alharthi, S.; El-Naggar, M.E.; Abu-Saied, M.; Khattab, T.A.; Saleh, D.I. Preparation of biosensor based on triarylmethane loaded cellulose acetate xerogel for the detection of urea. *Mater. Chem. Phys.* **2022**, *276*, 125377. [CrossRef]
67. Yuan, Z.; Cheng, J.; Lan, G.; Lu, F. A cellulose/Konjac glucomannan–based macroporous antibacterial wound dressing with synergistic and complementary effects for accelerated wound healing. *Cellulose* **2021**, *28*, 5591–5609. [CrossRef]
68. Mirtaghavi, A.; Baldwin, A.; Tanideh, N.; Zarei, M.; Muthuraj, R.; Cao, Y.; Zhao, G.; Geng, J.; Jin, H.; Luo, J. Crosslinked porous three-dimensional cellulose nanofibers-gelatine biocomposite scaffolds for tissue regeneration. *Int. J. Biol. Macromol.* **2020**, *164*, 1949–1959. [CrossRef]
69. Darpentigny, C.; Marcoux, P.R.; Menneteau, M.; Michel, B.; Ricoul, F.; Jean, B.; Bras, J.; Nonglaton, G. Antimicrobial cellulose nanofibril porous materials obtained by supercritical impregnation of thymol. *ACS Appl. Bio. Mater.* **2020**, *3*, 2965–2975. [CrossRef]
70. Pinto, S.C.; Gonçalves, G.; Sandoval, S.; López-Periago, A.M.; Borras, A.; Domingo, C.; Tobias, G.; Duarte, I.; Vicente, R.; Marques, P.A. Bacterial cellulose/graphene oxide aerogels with enhanced dimensional and thermal stability. *Carbohydr. Polym.* **2020**, *230*, 115598. [CrossRef]
71. Keshipour, S.; Khezerloo, M. Gold nanoparticles supported on cellulose aerogel as a new efficient catalyst for epoxidation of styrene. *J. Iran. Chem. Soc.* **2017**, *14*, 1107–1112. [CrossRef]
72. Zhang, Z.; Li, L.; Qing, Y.; Lu, X.; Wu, Y.; Yan, N.; Yang, W. Manipulation of nanoplate structures in carbonized cellulose nanofibril aerogel for high-performance supercapacitor. *J. Phys. Chem. C* **2019**, *123*, 23374–23381. [CrossRef]
73. Hu, W.; Lu, L.; Li, Z.; Shao, L. A facile slow-gel method for bulk Al-doped carboxymethyl cellulose aerogels with excellent flame retardancy. *Carbohydr. Polym.* **2019**, *207*, 352–361. [CrossRef] [PubMed]
74. Zhang, S.; He, J.; Xiong, S.; Xiao, Q.; Xiao, Y.; Ding, F.; Ji, H.; Yang, Z.; Li, Z. Construction and nanostructure of chitosan/nanocellulose hybrid aerogels. *Biomacromolecules* **2021**, *22*, 3216–3222. [CrossRef]
75. Bayat, A.; Aghamiri, S.F.; Moheb, A.; Vakili-Nezhaad, G.R. Oil spill cleanup from sea water by sorbent materials. *Chem. Eng. Technol. Ind. Chemistry-Plant Equipment-Process Engineering-Biotechnol.* **2005**, *28*, 1525–1528. [CrossRef]
76. Cervin, N.T.; Aulin, C.; Larsson, P.T.; Wågberg, L. Ultra porous nanocellulose aerogels as separation medium for mixtures of oil/water liquids. *Cellulose* **2012**, *19*, 401–410. [CrossRef]
77. Korhonen, J.T.; Kettunen, M.; Ras, R.H.; Ikkala, O. Hydrophobic nanocellulose aerogels as floating, sustainable, reusable, and recyclable oil absorbents. *ACS Appl. Mater. Interfaces* **2011**, *3*, 1813–1816. [CrossRef]
78. Feng, J.; Nguyen, S.T.; Fan, Z.; Duong, H.M. Advanced fabrication and oil absorption properties of super-hydrophobic recycled cellulose aerogels. *Chem. Eng. J.* **2015**, *270*, 168–175. [CrossRef]

79. Zhang, H.; Wang, J.; Xu, G.; Xu, Y.; Wang, F.; Shen, H. Ultralight, hydrophobic, sustainable, cost-effective and floating kapok/microfibrillated cellulose aerogels as speedy and recyclable oil superabsorbents. *J. Hazard. Mater.* **2021**, *406*, 124758. [CrossRef]
80. Xu, Z.; Zhu, Z.; Li, N.; Tian, Y.; Jiang, L. Continuous in situ extraction toward multiphase complex systems based on superwettable membrane with micro-/nanostructures. *ACS Nano* **2018**, *12*, 10000–10007. [CrossRef]
81. Ge, J.; Zong, D.; Jin, Q.; Yu, J.; Ding, B. Biomimetic and superwettable nanofibrous skins for highly efficient separation of oil-in-water emulsions. *Adv. Funct. Mater.* **2018**, *28*, 1705051. [CrossRef]
82. Lv, J.; Gong, Z.; He, Z.; Yang, J.; Chen, Y.; Tang, C.; Liu, Y.; Fan, M.; Lau, W.-M. 3D printing of a mechanically durable superhydrophobic porous membrane for oil–water separation. *J. Mater. Chem. A* **2017**, *5*, 12435–12444. [CrossRef]
83. Zhou, S.; Liu, P.; Wang, M.; Zhao, H.; Yang, J.; Xu, F. Sustainable, reusable, and superhydrophobic aerogels from microfibrillated cellulose for highly effective oil/water separation. *ACS Sustain. Chem. Eng.* **2016**, *4*, 6409–6416. [CrossRef]
84. Dilamian, M.; Noroozi, B. Rice straw agri-waste for water pollutant adsorption: Relevant mesoporous super hydrophobic cellulose aerogel. *Carbohydr. Polym.* **2021**, *251*, 117016. [CrossRef]
85. Chen, P.-H.; Sie, M.-C.; Jeng, P.-D.; Wu, R.-C.; Wang, C.-B. Graphene sponge as an efficient and recyclable oil sorbent. *AIP Conf. Proc.* **2017**, *1877*, 030005.
86. Nguyen, D.D.; Tai, N.-H.; Lee, S.-B.; Kuo, W.-S. Superhydrophobic and superoleophilic properties of graphene-based sponges fabricated using a facile dip coating method. *Energy Environ. Sci.* **2012**, *5*, 7908–7912. [CrossRef]
87. Mi, H.-Y.; Jing, X.; Huang, H.-X.; Peng, X.-F.; Turng, L.-S. Superhydrophobic graphene/cellulose/silica aerogel with hierarchical structure as superabsorbers for high efficiency selective oil absorption and recovery. *Ind. Eng. Chem. Res.* **2018**, *57*, 1745–1755. [CrossRef]
88. Chatterjee, S.; Ke, W.-T.; Liao, Y.-C. Elastic nanocellulose/graphene aerogel with excellent shape retention and oil absorption selectivity. *J. Taiwan Inst. Chem. Eng.* **2020**, *111*, 261–269. [CrossRef]
89. Syverud, K.; Xhanari, K.; Chinga-Carrasco, G.; Yu, Y.; Stenius, P. Films made of cellulose nanofibrils: Surface modification by adsorption of a cationic surfactant and characterization by computer-assisted electron microscopy. *J. Nanoparticle Res.* **2011**, *13*, 773–782. [CrossRef]
90. Lu, Y.; Wang, H.; Lu, Y. An architectural exfoliated-graphene carbon aerogel with superhydrophobicity and efficient selectivity. *Mater. Des.* **2019**, *184*, 108134. [CrossRef]
91. Li, Z.; Zhong, L.; Zhang, T.; Qiu, F.; Yue, X.; Yang, D. Sustainable, flexible, and superhydrophobic functionalized cellulose aerogel for selective and versatile oil/water separation. *ACS Sustain. Chem. Eng.* **2019**, *7*, 9984–9994. [CrossRef]
92. Wang, J.; Zhang, W.; Zhang, C. Versatile fabrication of anisotropic and superhydrophobic aerogels for highly selective oil absorption. *Carbon* **2019**, *155*, 16–24. [CrossRef]
93. Wu, Z.; Zhang, T.; Zhang, H.; Liu, R.; Chi, H.; Li, X.; Wang, S.; Zhao, Y. One-pot fabrication of hydrophilic-oleophobic cellulose nanofiber-silane composite aerogels for selectively absorbing water from oil–water mixtures. *Cellulose* **2021**, *28*, 1443–1453. [CrossRef]
94. Mi, H.-Y.; Jing, X.; Politowicz, A.L.; Chen, E.; Huang, H.-X.; Turng, L.-S. Highly compressible ultra-light anisotropic cellulose/graphene aerogel fabricated by bidirectional freeze drying for selective oil absorption. *Carbon* **2018**, *132*, 199–209. [CrossRef]
95. Turbeville, J. Ferromagnetic sorbents for oil spill recovery and control. *Environ. Sci. Technol.* **1973**, *7*, 433–438. [CrossRef] [PubMed]
96. Zhang, H.; Zhong, X.; Xu, J.-J.; Chen, H.-Y. Fe_3O_4/polypyrrole/Au nanocomposites with core/shell/shell structure: Synthesis, characterization, and their electrochemical properties. *Langmuir* **2008**, *24*, 13748–13752. [CrossRef]
97. Qiao, K.; Tian, W.; Bai, J.; Wang, L.; Zhao, J.; Du, Z.; Gong, X. Application of magnetic adsorbents based on iron oxide nanoparticles for oil spill remediation: A review. *J. Taiwan Inst. Chem. Eng.* **2019**, *97*, 227–236. [CrossRef]
98. Li, L.; Li, B.; Wu, L.; Zhao, X.; Zhang, J. Magnetic, superhydrophobic and durable silicone sponges and their applications in removal of organic pollutants from water. *Chem. Commun.* **2014**, *50*, 7831–7833. [CrossRef]
99. Chin, S.F.; Romainor, A.N.B.; Pang, S.C. Fabrication of hydrophobic and magnetic cellulose aerogel with high oil absorption capacity. *Mater. Lett.* **2014**, *115*, 241–243. [CrossRef]
100. Pinto, J.; Heredia-Guerrero, J.; Athanassiou, A.; Fragouli, D. Reusable nanocomposite-coated polyurethane foams for the remediation of oil spills. *Int. J. Environ. Sci. Technol.* **2017**, *14*, 2055–2066. [CrossRef]
101. Maurer-Jones, M.A.; Gunsolus, I.L.; Murphy, C.J.; Haynes, C.L. Toxicity of engineered nanoparticles in the environment. *Anal. Chem.* **2013**, *85*, 3036–3049. [CrossRef] [PubMed]
102. Lu, Y.; Wang, Y.; Liu, L.; Yuan, W. Environmental-friendly and magnetic/silanized ethyl cellulose sponges as effective and recyclable oil-absorption materials. *Carbohydr. Polym.* **2017**, *173*, 422–430. [CrossRef] [PubMed]
103. Gu, H.; Zhou, X.; Lyu, S.; Pan, D.; Dong, M.; Wu, S.; Ding, T.; Wei, X.; Seok, I.; Wei, S. Magnetic nanocellulose-magnetite aerogel for easy oil adsorption. *J. Colloid Interface Sci.* **2020**, *560*, 849–856. [CrossRef] [PubMed]
104. Lin, R.; Li, A.; Zheng, T.; Lu, L.; Cao, Y. Hydrophobic and flexible cellulose aerogel as an efficient, green and reusable oil sorbent. *RSC Adv.* **2015**, *5*, 82027–82033. [CrossRef]
105. Li, J.; Zhou, L.; Jiang, X.; Tan, S.; Chen, P.; Zhou, H.; Xu, Z. Directional preparation of superhydrophobic magnetic CNF/PVA/MWCNT carbon aerogel. *IET Nanobiotechnol.* **2019**, *13*, 565–570. [CrossRef] [PubMed]
106. Wang, X.; Jiang, S.; Cui, S.; Tang, Y.; Pei, Z.; Duan, H. Magnetic-controlled aerogels from carboxylated cellulose and $MnFe_2O_4$ as a novel adsorbent for removal of Cu (II). *Cellulose* **2019**, *26*, 5051–5063. [CrossRef]

107. Mi, H.-Y.; Li, H.; Jing, X.; Zhang, Q.; Feng, P.-Y.; He, P.; Liu, Y. Superhydrophobic cellulose nanofibril/silica fiber/Fe_3O_4 nanocomposite aerogel for magnetically driven selective oil absorption. *Cellulose* **2020**, *27*, 8909–8922. [CrossRef]
108. Si, Y.; Yu, J.; Tang, X.; Ge, J.; Ding, B. Ultralight nanofibre-assembled cellular aerogels with superelasticity and multifunctionality. *Nat. Commun.* **2014**, *5*, 5802. [CrossRef]
109. Xu, X.; Zhang, Q.; Yu, Y.; Chen, W.; Hu, H.; Li, H. Naturally dried graphene aerogels with superelasticity and tunable Poisson's ratio. *Adv. Mater.* **2016**, *28*, 9223–9230. [CrossRef]
110. Gao, H.-L.; Zhu, Y.-B.; Mao, L.-B.; Wang, F.-C.; Luo, X.-S.; Liu, Y.-Y.; Lu, Y.; Pan, Z.; Ge, J.; Shen, W. Super-elastic and fatigue resistant carbon material with lamellar multi-arch microstructure. *Nat. Commun.* **2016**, *7*, 12920. [CrossRef]
111. Meza, L.R.; Das, S.; Greer, J.R. Strong, lightweight, and recoverable three-dimensional ceramic nanolattices. *Science* **2014**, *345*, 1322–1326. [CrossRef] [PubMed]
112. Yao, X.; Yu, W.; Xu, X.; Chen, F.; Fu, Q. Amphiphilic, ultralight, and multifunctional graphene/nanofibrillated cellulose aerogel achieved by cation-induced gelation and chemical reduction. *Nanoscale* **2015**, *7*, 3959–3964. [CrossRef] [PubMed]
113. Zhang, Y.; Yin, M.; Li, L.; Fan, B.; Liu, Y.; Li, R.; Ren, X.; Huang, T.-S.; Kim, I.S. Construction of aerogels based on nanocrystalline cellulose and chitosan for high efficient oil/water separation and water disinfection. *Carbohydr. Polym.* **2020**, *243*, 116461. [CrossRef] [PubMed]
114. Raman, P.K.Y.; Jain, A.; Ramkumar, S. Oil absorption and desorption by polypropylene fibers. *Tappi J.* **2017**, *16*, 507–513. [CrossRef]
115. Zhao, Y.; Peng, C.; Cui, S.; Wu, X.; Jiang, S. Microstructure Characterization and Oil Absorption Performance of Superhydrophobic Cotton Cellulose Aerogel. *J. Wuhan Univ. Technology-Mater. Sci. Ed.* **2021**, *36*, 538–545. [CrossRef]
116. Padmanabhan, S.K.; Protopapa, C.; Licciulli, A. Stiff and tough hydrophobic cellulose-silica aerogels from bacterial cellulose and fumed silica. *Process Biochem.* **2021**, *103*, 31–38. [CrossRef]
117. Laitinen, O.; Suopajärvi, T.; Österberg, M.; Liimatainen, H. Hydrophobic, superabsorbing aerogels from choline chloride-based deep eutectic solvent pretreated and silylated cellulose nanofibrils for selective oil removal. *ACS Appl. Mater. Interfaces* **2017**, *9*, 25029–25037. [CrossRef]
118. Zhou, L.; Xu, Z. Ultralight, highly compressible, hydrophobic and anisotropic lamellar carbon aerogels from graphene/polyvinyl alcohol/cellulose nanofiber aerogel as oil removing absorbents. *J. Hazard. Mater.* **2020**, *388*, 121804. [CrossRef]
119. Gu, H.; Gao, C.; Zhou, X.; Du, A.; Naik, N.; Guo, Z. Nanocellulose nanocomposite aerogel towards efficient oil and organic solvent adsorption. *Adv. Compos. Hybrid. Mater.* **2021**, *4*, 459–468. [CrossRef]
120. Loh, J.; Goh, X.Y.; Nguyen, P.T.; Thai, Q.B.; Ong, Z.; Duong, H.M. Advanced aerogels from wool waste fibers for oil spill cleaning applications. *J. Polym. Environ.* **2021**, *30*, 681–694. [CrossRef]
121. Li, W.; Li, Z.; Wang, W.; Li, Z.; Li, Q.; Qin, C.; Cao, F. Green approach to facilely design hydrophobic aerogel directly from bagasse. *Ind. Crops Prod.* **2021**, *172*, 113957. [CrossRef]
122. Zhang, F.; Wang, C.; Mu, C.; Lin, W. A novel hydrophobic all-biomass aerogel reinforced by dialdehyde carboxymethyl cellulose for oil/organic solvent-water separation. *Polymer* **2021**, *238*, 124402. [CrossRef]
123. Wang, J.; Liu, S. Remodeling of raw cotton fiber into flexible, squeezing-resistant macroporous cellulose aerogel with high oil retention capability for oil/water separation. *Sep. Purif. Technol.* **2019**, *221*, 303–310. [CrossRef]
124. He, Y.; Liu, Y.; Wu, T.; Ma, J.; Wang, X.; Gong, Q.; Kong, W.; Xing, F.; Liu, Y.; Gao, J. An environmentally friendly method for the fabrication of reduced graphene oxide foam with a super oil absorption capacity. *J. Hazard. Mater.* **2013**, *260*, 796–805. [CrossRef] [PubMed]
125. Camilli, L.; Pisani, C.; Gautron, E.; Scarselli, M.; Castrucci, P.; D'Orazio, F.; Passacantando, M.; Moscone, D.; De Crescenzi, M. A three-dimensional carbon nanotube network for water treatment. *Nanotechnology* **2014**, *25*, 065701. [CrossRef] [PubMed]

polymers MDPI

Review

Panel Products Made of Oil Palm Trunk: A Review of Potency, Environmental Aspect, and Comparison with Wood-Based Composites

Arif Nuryawan [1,*], Jajang Sutiawan [1], Rahmawaty [2], Nanang Masruchin [3,4] and Pavlo Bekhta [5]

1 Department of Forest Products Technology, Faculty of Forestry, Universitas Sumatera Utara, Medan 20155, North Sumatra, Indonesia; jajangsutiawan@usu.ac.id
2 Department of Forest Management, Faculty of Forestry, Universitas Sumatera Utara, Medan 20155, North Sumatra, Indonesia; rahmawaty@usu.ac.id
3 Research Center for Biomass and Bioproducts, National Research and Innovation Agency (BRIN), Jl. Raya Bogor Km. 46, Cibinong 16911, West Java, Indonesia; nana021@brin.go.id
4 Research Collaboration Center for Biomass and Biorefinery, National Research and Innovation Agency (BRIN), Jl. Raya Bogor Km. 46, Cibinong 16911, West Java, Indonesia
5 Department of Wood-Based Composites, Cellulose and Paper, Ukrainian National Forestry University, 79057 Lviv, Ukraine; bekhta@nltu.edu.ua
* Correspondence: arif5@usu.ac.id

Abstract: Oil palm plantations have expanded rapidly in Southeast Asia, particularly in Indonesia and Malaysia. A lot of products, including food and other edible products, oleo-chemicals, cosmetics, personal and household care, pharmaceutical products, and biodiesels are derived from palm oil, thus making them one of the most economically important plants. After 25–30 years of age, the palms are felled and replaced due to declining oil production. Oil palm trunks (OPT) are considered significant waste products. The trunks remain on the plantation site for nutrient recycling or burning. This increases insect and fungi populations causing environmental problems for the new palm generation or air pollution due to the fire. Up till now, OPT has received less attention in research studies. Therefore, this review summarizes the utilization of OPT into products made of oil palm fibers mainly derived from OPT and its application as the substitution of wood panel products. Some research works have been carried out on oil palm fibers that are derived from OPT for exploiting their potential as raw material of composite panel products, which is the objective of this review. Areas of development are processed into various conventional composite panel products such as plywood and laminated board which are usually predominantly made of wood and bonded by synthetic resins, particleboard with binder, or binderless and cement board which is arranged with wood as a minor component. All of the products have been presented and described technically according to best knowledge of the authors and literature review.

Keywords: composite panel; biomass; oil palm trunk (OPT)

Citation: Nuryawan, A.; Sutiawan, J.; Rahmawaty; Masruchin, N.; Bekhta, P. Panel Products Made of Oil Palm Trunk: A Review of Potency, Environmental Aspect, and Comparison with Wood-Based Composites. *Polymers* **2022**, *14*, 1758. https://doi.org/10.3390/polym14091758

Academic Editor: Antonios N. Papadopoulos

Received: 7 March 2022
Accepted: 23 April 2022
Published: 26 April 2022

1. Introduction

Elaeis guineensis Jacq. is the broadest species in oil palm plantations for palm oil production. According to specimens that were collected in Martinique Island, it was native from the Guinea coast of West Africa and was formally named by Jacquin in 1763. The first commercial crops in estates were established in Sumatra Island-Indonesia and Peninsular Malaysia in 1911 [1]. Thus, this foundation made oil palm an important integrated industry, with both upstream and downstream sectors. Plantation as an upstream agriculture sector provides fresh fruit bunches (FFB) as raw materials for further processing in downstream mills such as vegetable oils and fat industries. They manufacture end-products for food and other edible uses (cooking oil, frying fats, margarine, confectionery/cocoa butter substitute, ice cream, non-dairy creamer, salad oil, shortening, etc.) [2]; oleo-chemical (stearine, fatty acids, fatty alcohols, fatty methyl esters, and glycerin) [3]; cosmetics, personal care, and household cleaner (balms, lipsticks, emulsifiers, soaps, detergents, candles and cleaning products) [4]; skin-based pharmaceutical products (vitamin E supplement, tocopherol, and tocotrienol) [5]; and biodiesel production (lubricant, surfactant, transesterification of triglycerides with alcohol) [6,7].

Since palm oil is the primary yield of oil palm plantations and economically considerable revenue is gained through this commodity, these crops have expanded dramatically. Nowadays, worldwide 43 countries have cultivated this palm [8], and almost 80% of the world oil palm plantation is centered in Southeast Asia [9], with most of it occurring in Indonesia, Malaysia, and Thailand [10]. Both Indonesia and Malaysia are the two major palm oil-producing countries and the two top world producers of oil palm products, having the largest planted oil palm area in the world [11]. Figure 1 illustrates their development in the last decade. The data area harvested and oil palm production in the predominant country in Southeast Asia and the world a decade back showed an increase [12].

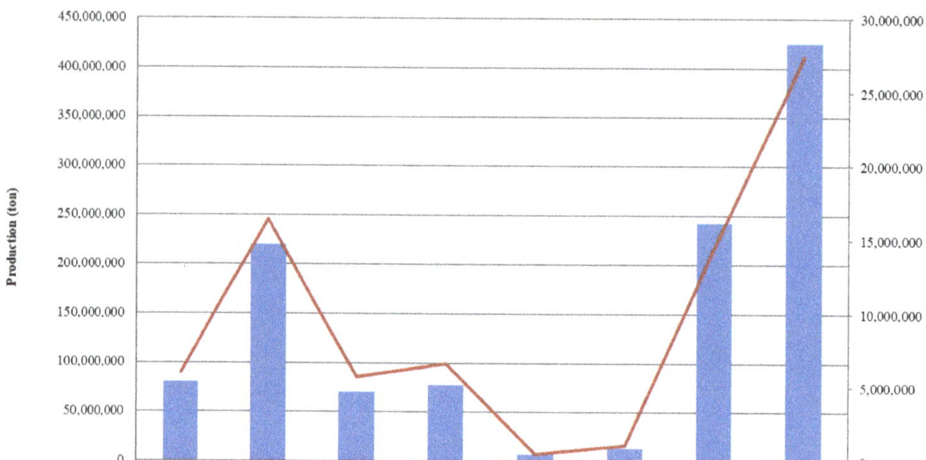

	Indonesia 2009	Indonesia 2019	Malaysia 2009	Malaysia 2019	Thailand 2009	Thailand 2019	World 2009	World 2019
Area (ha)	5,370,000	14,677,560	4,691,160	5,216,822	510,003	896,322	16,158,445	28,312,612
Production (ton)	90,000,000	245,633,087	85,705,499	99,065,364	8,162,703	16,772,430	216,414,772	410,696,692

Figure 1. World area that was harvested and oil palm production within the predominant three countries in Southeast Asia during 2009–2019, Indonesia, Malaysia, and Thailand [12].

Both Indonesia and Malaysia produce 85% of the world's palm oil [13], where oil palm estates occupy 22 of 34 provinces in Indonesia [10] and cover more than 60% of the agricultural land of both the west (the peninsular) and the east (the States of Sabah and Sarawak) in Malaysia [13]. The large extent of oil palm plantations in both countries has generated waste products in large quantities. For instance, the fibrous part can be gained after the removal of oil seeds from FFB for oil extraction which results in crude palm oil (CPO) and crude palm kernel oil (CPKO) that are derived from the mesocarp fiber (MF) and the palm kernel (the seed enclosed in a shell of the endocarp), respectively (Figure 2) [13].

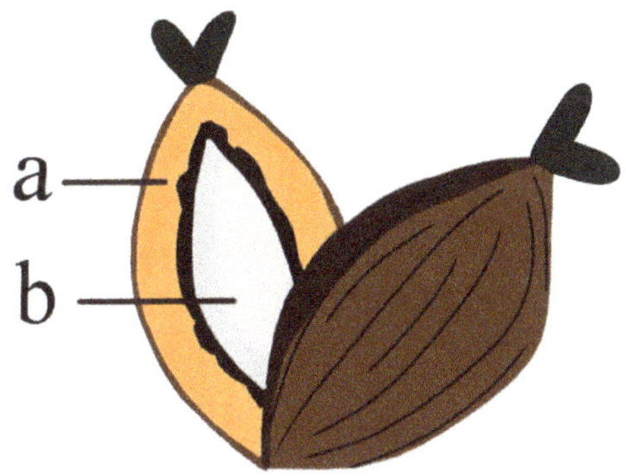

Figure 2. Cross-section of an oil palm fruit that is comprised of (**a**) an outer oily flesh or pericarp (consisted of exocarp, mesocarp, and endocarp) and (**b**) an oil-rich seed/nut or kernel (endosperm).

In addition to oil palm processing, various waste consisting of empty fruit bunches (EFB), palm kernel shells (PKS), and mill effluents (ME) can also be obtained [9]; however, these are not under the scope of the present study and thus will not be discussed here because all of these only credited 10% of biomass [11] and are derived from by-products of the industrial process. This study discusses a large amount of biomass that is produced from oil palm growing on plantations, either during pruning or replanting as shown in Figure 3.

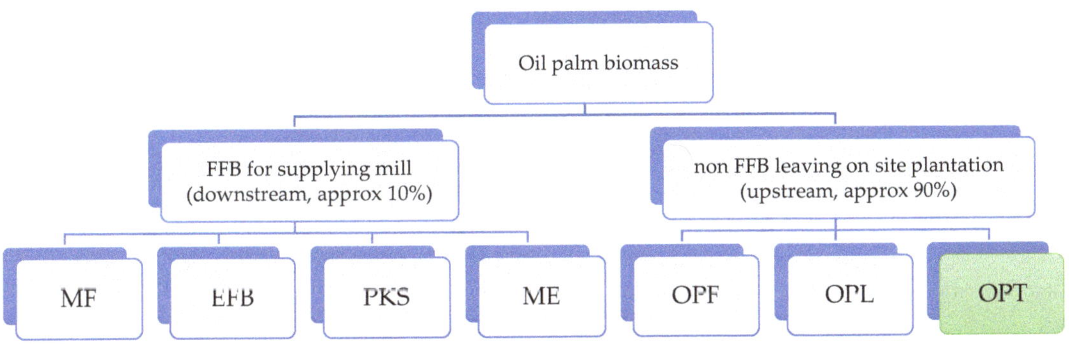

Figure 3. Biomass of oil palm that is generated both from downstream (fruit origin) and upstream (non-fruit). The focus of this study is OPT which shown in green color.

Biomass refers to any organic plant products with general uses [10] categorized as agro-based resources [11]. They have broad characteristics such as being renewable, widely distributed but locally available, voluminous or bulky, moldable, variety in many forms and sizes, anisotropic, hygroscopic, recyclable, versatile, non-abrasive, porous, biodegradable, combustible, compostable, and reactive. They also have a high aspects ratio, high strength to weight ratio, relatively low energy conversion, and good insulation properties (sounds, electrical, and thermal). The general fiber structure is hollow, laminated with molecular layers, and an integrated matrix [14–16].

Even though there are two primary biomass residues, namely oil palm frond (OPF) with oil palm leaves (OPL) and oil palm trunk (OPT) in oil palm growing on plantations [17], only OPT was considered in this review. OPF with the leaves which are generated during harvesting of FFB has been used as a substitute for grass as well as the source of roughage for decades as for ruminant feed such as for goats, beef, and dairy cattle without any report of toxicity [18] where the farm sites are usually located near the oil palm plantations.

Long-term feeding of OPF-based feeds, either fresh or processed as silage or pellet, has produced good quality carcasses, and the meat is safe for consumption [19]. However, OPT has received less attention in research studies [20] even though the potency is abundantly available. OPT is available only when the economic lifespan of the palm is achieved at the time of replanting. The average age of replanting is approximately 25 years when OPT is discarded for re-plantation after 25–30 years of oil production [8]. The main economic criteria for felling are the height of the palm, reaching 13 m or above years, either due to their decreasing yield or because they have grown too tall, which makes harvesting very difficult [21]. During replanting, the diameter of the felled trunk is around 45 cm to 65 cm, measured at breast height [9]. In other words, the OPT is fallen during replanting when the stems have reached the end of their useful life [22]. It is a less expensive lignocelluloses raw material as compared to wood. Using OPT as a raw material to produce value-added products will reduce the overall costs of production and increase economic returns [9]. Further, OPT is a sustainable non-wood source that is available in abundance at replanting operations would have seemed for alternative raw materials for some of the products that are traditionally made from wood such as veneer, sawn lumber, and fiber-based bio-composites with considering position on the trunk [23]. Unfortunately, OPT is monocotyledon and consists of vascular bundles and parenchyma, which are the main obstacle in utilizing OPT as a wood solid. In monocotyledon wood, fibers have less strength and are irregular in characteristics compared to the dicotyledonous wood. OPT wood consists of primary tissue, and it is not comparable to dicotyledons and gymnosperms in terms of wood development because oil palm does not possess cambium. Indeed, OPT is not wood compared to softwood and hardwood, which wood-based industries are used to [24]. Khalil et al. [24] mentioned that fibers within monocotyledons have had less strength and irregular properties compared to wood in dicotyledons. Therefore, in this regard, OPT would be utilized as alternative wood products because so far, many studies of OPT could be engineered into the substitution of traditional wood products such as plywood, laminated veneer lumber (LVL), particleboard, fiberboard, and cement board predominant matrix which will be discussed later.

In this review, the potential for transforming the waste biomass of OPT to an alternative substitution of conventional wood products was discussed, including their environmental aspects. Based on the availability of OPT and the environmental and economic impacts of the fibrous parts from OPT, this study's objective is to evaluate the quality of composite panel products that are made of OPT to be an alternative substitution of wood products. The success of these alternative products will be supported by previously published studies, performed prototypes, and standard specifications.

2. Method

A review of the literature found 599 titles on sciencedirect.com (accessed on 21 December 2020) and 74 titles via a search engine in the target journal Polymers (MDPI). After reviewing the title, abstract, and content of the article to ensure that the raw materials that were used to create the composite products that were derived from OPT were listed, there were only 31 titles on the search engine sciencedirect.com and 2 titles on the journal Target Polymers (MDPI), as presented in Table 1.

From this literature review, OPT or OPS has been used as the main raw material of various composite panel products ranged from tiny size such as fine particles and fibers up to small dimensions such as chips. Moderate size was found as strands of OPT, resulted in a separation between vascular bundles and the parenchyma. Larger dimensions were found in the form of veneer mats and veneer lumber. These materials are used to make composite panel products or bio-composites, which often replace wood-based panel goods such as particleboard, fiberboard, chipboard, oriented strand board (OSB), plywood, and LVL. Sometimes, OPT was found as minor part, such as filler or substitution of aggregate in cement board, gypsum board, or concrete brick. The type of binder that was used ranged from binderless (without adhesive), using bacteria (PHA), thermoplastics (PVA, PVAc), up to conventional formaldehyde-based resins (UF, MF, PF). In mineral boards, cement or gypsum has a role as both a binder and a matrix. Discussions further about these raw materials will be conveyed in next sub chapter.

Table 1. Results of the search engine sciencedirect.com uses the keywords: "oil palm trunk" OR "oil palm stem" OR "oil palm waste".

	Sciencedirect.com (31 Articles)		
No	**Type of Panels**	**Year**	**Ref**
1	Particleboard made of OPT bonded by poly(vinyl)alcohol/PVA	2020	[25]
2	Plywood made of OPT veneer without binder	2020	[26]
3	Cement (gypsum) board with OPT filler	2019	[27]
4	Binderless particleboard made of inner part of OPT	2019	[28]
5	OPT chips particleboard bonded by urea-formaldehyde (UF) resin	2018	[29]
6	OPT chips binderless particleboard with different thickness	2015	[30]
7	OPT fine particleboard with addition of polyhydroxylalkanoates (PHA)	2013	[31]
8	Particleboard made of treated OPT particles bonded by UF resin	2014	[32]
9	Laminated panels made of compressed OPT and poly(vinyl)acetate/PVAc	2013	[33]
10	OPT fine particleboard with addition of PHA	2012	[34]
11	Binderless OPT fine particleboard	2011	[35]
12	Three-layers laminated veneer lumber (LVL) made of OPT and UF resin	2013	[36]
13	Binderless particleboard made of OPT fine particles and strands	2010	[37]
14	Oil palm stem (OPS) plywood treated by phenol-formaldehyde (PF)	2011	[38]
15	Binderless particleboard surface made of OPT and acacia wood particles	2019	[39]
16	LVL made of OPT veneer with different types of formaldehyde resins	2009	[40]
17	Hybrid plywood made of OPT and oil palm empty fruit bunches (OPEFB)	2010	[41]
18	Plywood made of OPS pre-preg by PF	2013	[42]
19	Plywood, particleboard, and fiberboard made of OPT	2012	[43]
20	Properties of binderless particleboard made of OPT and acacia wood	2014	[44]
21	OPS plywood treated by low molecular weight PF	2012	[45]
22	Seasoning of veneer OPT	2017	[46]
23	Plywood made of OPT veneer treated with various concentration of PF	2013	[47]
24	Biocomposite made of oil palm biomass including OPT	2015	[48]
25	Binderless particleboard made of young and old OPT	2014	[49]
26	Binderless particleboard made of fine particles and vascular strands OPT	2014	[50]
27	Binderless fiberboard made of agricultural biomass including OPT	2019	[51]
28	Binderless particleboard made of oil palm biomass including OPT	2011	[52]
29	Laminated composites made of agricultural biomass including OPT	2017	[53]
30	OPT as sandwich panel core bonded by melamine-formaldehyde (MF)	2015	[54]
31	OPS veneer treated by low molecular weight PF	2011	[55]
	mdpi.com (2 articles)		
No	Type of panels	Year	Reference
1	Concrete brick with cement, gypsum, gravel and OPT aggregate	2020	[56]
2	Hybrid plywood made of OPT veneer and OPEFB mat bonded by PF	2020	[57]

3. Environmental Aspect

Currently, Malaysia's oil palm industry generates more than 80 million tons of oil palm waste per year (dry weight basis) [58,59], with the figure expected to increase by at least 40% by 2020 [60]. Annual oil palm production is expected to increase by 50 million tons by 2030 in response to Malaysia's proposed expanding oil palm plantations [61]. The most frequently generated co-products that are derived from fruits include palm kernel shell (PKS) [62], OPEFB, mesocarp fiber (MF), and palm oil mill effluent (POME) while from the non-fruit include oil palm frond (OPF), oil palm leaves (OPL), and OPT [63–65]. Table 2 shows the comparison of oil palm biomass residues that originated from pruning, replanting, and milling activities between the two highest producing countries of oil palm, namely Indonesia [22] and Malaysia [66].

Table 2. Biomass residue of oil palm in two countries, Indonesia and Malaysia, that are derived from the fruit and non-fruit (million tonnes).

	Indonesia [22]	Malaysia [66]
Fruit		
Shell (PKS)	4.83	4.72
OPEFB	15.87	7.78
Fiber (MF)	8.97	8.18
POME	n.a	3.38
Non-fruit		
Frond (OPF)	43.05	23.39
Trunk (OPT/OPS)	13.94	3.74

Remarks: n.a is no data available.

Malaysia is internationally renowned for its palm cultivation, and it is a runner up in palm oil production and exports [67] after first rank of Indonesia. It accounts for more than 30% of the global palm oil production and 37% of global exports in 2016 [68]. Apart from that, the oil palm industry and palm oil mills generate approximately 90% of biowaste, even though the oil accounts for only 10% of the biomass [69]. As a result, large quantities of oil palm biomass are generated each year and are then discarded or burned without further use due to a lack of technology to utilize this material [70]. This situation exacerbates the problem of biomass overload, wastes valuable cellulose-rich resources, and may result in serious environmental problems such as air pollution [71]. Additionally, the improper handling of this residue and biowaste may raise environmental and health concerns. It may contribute to eutrophication, pollution, and other types of disturbances for aquatic and terrestrial life [72].

The biomass data that originated from the extraction of fruit (FFB) was dependent on palm oil mills, their operation, and capacity. In Indonesia there were 421 mills [22] spread out in 18 provinces in five big islands (Sumatra, Java, Kalimantan, Sulawesi, and Papua) while in Malaysia it was slightly higher about 454 mills [66] that were concentrated in Peninsular Malaysia, Sabah, and Sarawak. The data of OPF was predominant in both countries because the fronds (and leaves) were harvested during the regular pruning through the lifetime of the palm. OPT or OPS were collected after the palm exceeded its economical age about 25–30 years. The traditional method for establishing new oil palm plantations or replanting is through mechanical or fire clearing of the land. Mechanical clearing is frequently required in large-scale plantations, resulting in soil compaction and other physical degradation of the soil. When old palm stands are felled, stacked, and burned, a large amount of smoke is released into the environment. Slashing and burning removes the aboveground biomass, understory vegetation, and ground litter, resulting in high environmental costs. The purpose of burning biomass in the land clearing process is to eliminate waste material that obstructs plantation management and eradicate pests and diseases by destroying their breeding medium.

Additionally, burning of the unwanted biomass and another waste material appears to be the most cost-effective and time-efficient method of waste disposal. This activity results in a large amount of CO_2 being released into the atmosphere, contributing to climate change and global warming. Particles that are produced by the burning undoubtedly contribute to the haze problem. Fortunately, all these wastes are classified as organic wastes that degrade naturally. However, because these wastes are generated in large quantities, they can pollute the environment.

The presence of this waste that is not managed correctly promotes the growth of Ganoderma disease. Vascular wilt, bud and spear rot, sudden wilt, red ring disease, and basal stem rot disease are the most common diseases that attack oil palm trees. The basal stem rot (BSR) disease poses the greatest threat to oil palm plantations. BSR disease, also known as Ganoderma disease, is caused by a pathogenic basidiomycete, a white-rot fungus called *Ganoderma boninense*. Indonesia and Malaysia have a high prevalence of BSR, whereas Africa, Papua New Guinea, and Thailand have a low prevalence [73].

According to a Malaysian study, oil palm stands that are planted after coconut trees in coastal areas are more susceptible to this disease. According to reports, up to 50% of yields have been lost, and 80% of oil palm has been killed before their economic age [74]. In comparison, another survey discovered that incidences of BSR infection in Malaysia, Peninsular Malaysia, Peninsular Malaysian Coast, Sarawak, and Sabah were 29, 34, 37, 9%, and 29%, respectively [75]. In Indonesia, a survey that was conducted in an estate in North Sumatra found that 22% or about 30 palms per hectare died after being infected this disease causing a significant loss in yield production as well as threatening palm oil sustainability [76].

Ganoderma disease begins with the pathogenic fungi colonizing the oil palm root, followed by the degradation of the basal stem tissue [77]. The disease then destroys the internal tissue and the palm xylem, disrupting the water and nutrient supply from the root to the plant's upper parts [78]. Infected stands will show signs of wilting, desiccated fronds, and unopened spears. Later, basidiocarps or fruiting bodies in the shape of a conch will develop on the palm trunk. At this point, the fungal infections have spread widely throughout the palm, typically resulting in the plant's death [79]. The fruiting body can then release spores and spread to the soil or other palms. As a result, Ganoderma reduces the useful life of oil palms and results in significant yield losses for the oil palm industry [80]. The Malaysian yield losses due to this disease are estimated to be up to USD 500 million, with an average mortality rate of 3.7% [81].

If no effort is made to utilize oil palm biomass, it will become a problem for the environment. When it is used correctly, it will eliminate disposal issues and generate value-added products. As a result, the oil palm industry must be prepared to capitalize on this situation and maximize the utilization of available oil palm biomass to convert waste to wealth [82]. This biomass residue may be used in place of wood to manufacture wood-based panels [83]. Worldwide, research has been conducted to maximize the economic value of biomass residues through the production of lumber, furniture, and lightweight construction. Malaysia pioneered oil palm biomass to create medium-density fiberboard (MDF) [84]. Globally, an increased awareness of oil palm biomass-based products has increased their demand [85].

4. Overview Production of Wood-Based Panels in the World

Global wood-based panel production totaled 408 million m^3 in 2018, a 1% increase over the previous year (404 million m^3) and a 9% increase over the previous nine years. Wood-based panels were the fastest growing product category in production, owing to the Asia-Pacific region's rapid and consistent growth up to 2016. Global production has stabilized in recent years. In 2018, Asia-Pacific accounted for 61% of the global output (248 million m^3), followed by Europe (90 million m^3, or 22%), Northern America (48 million m^3, or 12%), Latin America and the Caribbean (19 million m^3, or 4%), and Africa (3 million m^3, or 1%) [86].

Since 2014, the global trade in wood-based panels has been steadily increasing. In 2018, it increased by 3% to 91 million m^3, accounting for 22% of the total production. In 2018, Europe and Asia-Pacific dominated the international trade in wood-based panels, accounting for 71% of total imports and 80% of total exports, respectively. In 2018, the top five producers of wood-based panels (China, the United States of America, Russia, Germany, and Canada) accounted for 69% (282 million m^3) of global production. Meanwhile, the top four consumers of wood-based panels are the same as the top four producers, indicating that the products are primarily consumed domestically. Consumption trends are parallel with the production trends [86].

Plywood (which includes blockboard and LVL) produced 163 million m^3 (or 40% of the total wood-based panel production) in 2018, an increase of 11% over 2014. Regional variations in the composition of various wood-based panel products exist. In North America and Europe, reconstituted panels (OSB, particleboard, and fiberboard) dominate other product categories, whereas plywood (including blockboard) is the most widely used wood-based panel product in Asia-Pacific (mainly in China). Each primary wood-based panel product accounts for roughly the same proportion of the total production in Latin America and the Caribbean as the others [86].

Between 2014 and 2018, OSB and particleboard production increased at the fastest rates, by 25% and 13%, respectively. Both products multiplied in Eastern Europe, most notably in the Russian Federation. The global fiberboard production peaked in 2016 (121 million m^3) and then fell 4% to 116 million m^3 in 2018, remaining at 2014 levels. Between 2014 and 2018, both MDF and high-density fiberboard production increased by 5%, accounting for 85% of total fiberboard production. Other types of fiberboards saw a 19% decline during the same time [86].

The development of wood-based panel production in Indonesia and Malaysia is shown in Figure 4. Wood chip and particle production in Indonesia and Malaysia have experienced the same characteristics, which have increased in the last decade (2009–2019). However, there are differences in veneer sheet and plywood production in Indonesia and Malaysia. The production of veneer sheet and plywood in Indonesia has increased, while veneer sheet and plywood in Malaysia has decreased in the last decade. These phenomena are due to changes in industrial development from plywood to particleboard in Malaysia, which has increased. Meanwhile, the development of OSB in the two countries has not been done much. Therefore, this is very interesting to study the development of the OSB industry in Indonesia and Malaysia.

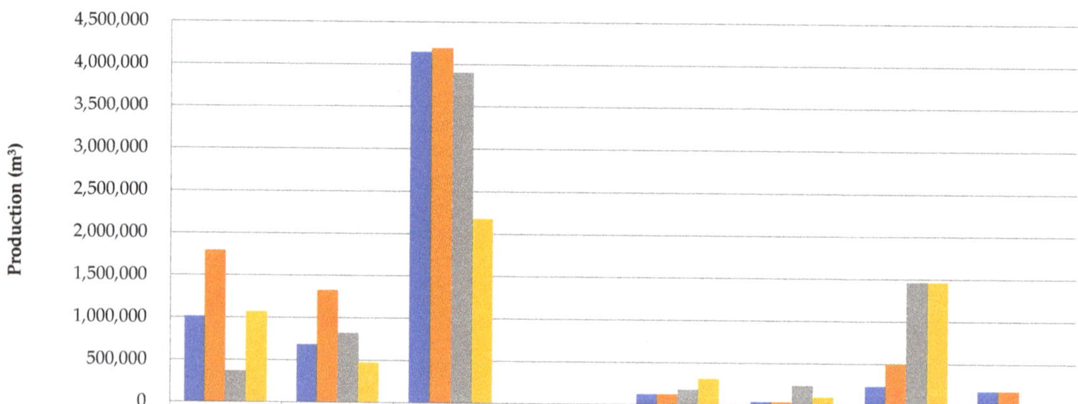

	Wood chips and particles	Veneer sheets	Plywood	OSB	Particle board	Hardboard	MDF/HDF	Other fibreboard
	Raw Material		Structural Panels		Non Structural Panels			
Indonesia 2009	1,013,000	685,000	4,150,000	-	125,000	40,000	229,000	178,000
Indonesia 2019	1,788,000	1,322,000	4,196,000	-	125,000	40,000	500,000	178,000
Malaysia 2009	371,000	821,000	3,901,000	-	177,000	236,000	1,460,000	-
Malaysia 2019	1,069,000	473,000	2,180,298	-	310,000	100,000	1,460,000	-

Figure 4. Data production of wood-based panels of Indonesia and Malaysia in a decade (2009–2019) consisting of raw material (wood chips and particles and veneer sheets), structural panels (plywood & OSB), and non-structural panels (particleboard and fiberboard) [12].

5. Types of Wood-Based Panels

5.1. Particleboard

Particleboard is a broad term that refers to any panel products that are made entirely of wood particles. Naturally, particleboards are composed of a diverse range of particle shapes and sizes. As a result, the particle type is used to define the type of particleboard product. For instance, particleboard is a generic name for products made of particle while particular size of chips, wafer, or strands resulted in chipboards, waferboards, or OSB, respectively. Another characteristic of particleboards is that the wood particles are bonded together using a synthetic resin adhesive and then compressed at high pressures and temperatures. This is critical because the manufacturing process has a significant impact on the properties of the resulting panel products [87].

5.2. OSB

OSB are multilayered panels that are made from pre-shaped strands of wood (typically 15–25 mm wide, 75–150 mm long, and 0.3–0.7 mm thick) that are bonded together under pressure and heat with a binder (often water-resistant). The strands of the outer layers run parallel to the long board edge and the production line. While the strands in the core layer are frequently smaller and can be oriented randomly or at right angles to the strands in the face layers, the strands in the core layer are frequently larger and can be aligned at right angles to the strands in the face layers. OSB is classified into four categories by the European Norms (EN) 300 and EN 13986 standards: OSB-1 is used for general indoor (dry) purposes, OSB-2 is used for load-bearing purposes in dry conditions, OSB-3 is used for load-bearing purposes in humid conditions, and OSB-4 is used for heavy-duty construction in humid conditions. OSB is typically available in thicknesses ranging from 10 to 32 mm. The internal bonding (IB) and modulus of rupture (MOR) tests following cyclic boiling are frequently the most difficult to pass for OSB-3 and 4, which is why manufacturers use moisture-resistant resins such as isocyanates (PMDI), phenolic-based resins (PF, MUPF), or UF resins that are reinforced with melamine (MUF). MOR and modulus of elasticity (MOE) values that are parallel to the long panel edge are typically double those that are observed across the panel. This is due to the strand orientation of the face layer [87].

5.3. Fiberboard

MDF is an engineered wood product that is composed of fine lignocellulosic fibers with or without synthetic resin to form panels. Binderless fiberboard utilizes a wet manufacturing process for activating hydrogen bonding within the panels. While wood is the most common lingo-cellulosic fiber, other plant fibers such as bagasse and cereal straws can also be used [87]. MDF is available in a variety of panel sizes and thicknesses ranging from 2 to 100 mm. These panels have a density of between 500 and 900 kg/m^3. The faces of the panels are smooth and dense (for dry manufacturing process using synthetic resin), and the color ranges from pink-brown to dark brown unless a dye is added during manufacture [87].

5.4. Plywood

This product is a flat panel that is constructed from veneer layers of either soft or hardwood, which are arranged perpendicularly in odd numbers and then glued either using traditional thermosetting resin or using thermoplastics. It is stacked together to form a panel using a clamp or a pressing machine. Each layer is given a unique name; for example, the outer layer is referred to as the face, the inner layer as the middle or core, and the wood grain layer that is perpendicular to both the face and back sides is referred to as the crosswise [88]. The properties of plywood are determined by the quality of the veneer layers, the layering order, the adhesive that is used, and the control of bonding conditions.

6. Comparison Wood versus Oil-Palm Trunk Biomass

6.1. Physical Properties of Oil Palm Trunk

According to Srivaro et al. [89], the water uptake (WU) of the OPT ranges between 73 and 415%. According to Srivaro et al. [90], the WU increases gradually along with the trunk height and the central region. The outer and lower zones have significantly lower values than the other two zones. The increasing trend in WU can be attributed to the distribution of parenchymatous cells, which retain more moisture than the vascular bundles. According to Lamaming et al. [91], between the parenchyma tissues and vascular bundles are comprised of different compositions of extractive, lignin, and holocellulose. By contrast, parenchyma tissues are more abundant near the apex of the OPT and the central region. Recently, Wulandari and Erwinsyah [92] reported that the highest moisture content (MC) was found at the inner and gradually decreased toward centre to the trunk's outer part. These percentages ranged from 36 to 141%. A decrease in the percentages of parenchyma cells with a high capacity for water absorption (WA) increased the number of vascular bundles. Generally, Wulandari and Erwinsyah [92] stated that the MC of the OPT decreased as the trunk height increased.

Due to the oil palm's monocotyledonous nature, its density values vary significantly across its trunk. Density, or specific gravity, is regarded as the most critical characteristic when utilizing OPT. It is defined as the quantity of material that is contained in a unit volume. Generally, the density is the highest at the bottom end's peripheral region and lowest at the top end's central core region. The density values range from 222 to 404 kg per m^3 [90]. According to Srivaro et al. [90], the density of an OPT decreases linearly with trunk height and increases toward the trunk's center; the outer region of the trunk has a density that is more than twice that of the inner region. A variety of factors cause these variations. For instance, the density across the trunk is primarily determined by the number of vascular bundles per square unit, which decreases towards the center. However, there were variations in the density along the trunk height since the vascular bundles at the top of the palm are younger.

6.2. Chemical Properties of Oil Palm Trunk

The chemical composition of OPT varies longitudinally and radially along the stem including the barks. No discernible trend exists for holocellulose, lignin, and ash content in bottom, middle, and upper parts of the oil palm meristem. However, they were significantly different from those of its bark [93]. The alpha-cellulose content, ash content, pentosan, and holocellulose within the OPT were lowest compared to those of OPF and EFB. Extractives within OPT which were indicated by water solubility (either cold or hot) and alcohol- and alkali-soluble components were in between of those of EFB and OPF [94].

Further, Onuorah et al. [94] examined the chemical properties of OPT. They reported that OPT contains more lignin than those of EFB and OPF. Ahmad et al. [95] observed that the bottom portion contained a higher lignin concentration and was significantly different than middle and top parts. This is probably due to the increased number of fibrous vascular bundles in the peripheral region [92]. These vascular bundles contain thicker and older cells, contributing to OPT's lower end's higher lignin content. The range of lignin content is between 17.21% and 20.47%. The ash content, which is also consistent throughout the trunk, ranges between 2.18 and 2.86% [95].

Additionally, free sugars between 2–14% were observed throughout the stem. The inner regions have a higher proportion of free sugars whereas the outer zones have the lowest free sugar content. Sucrose, glucose, and fructose are the three major free sugars in the OPT, as determined by high-performance liquid chromatography (HPLC), the most used method for analyzing sugar types and concentrations. Sucrose, the primary sugar component, remains relatively constant throughout the stem's length. On the other hand, the glucose levels decrease as the stem height increases, whereas the fructose levels increase. However, a high variation of sugar with different content that were found in different oil palms were influenced by different methods of analysis, quantifications on different locations of the palms (inner or outer part, height), including different storage and microbial infestation [96].

6.3. Mechanical Properties of Oil Palm Trunk

The MOE which measures the material's stiffness or rigidity, varies between 740 and 7960 MPa in the OPT, with the highest value found in the lower peripheral region. The lowest MOE is found in the central region's core, near the trunk's top end. The MOR, which varies similarly to the MOE, is extremely low compared to conventional timber species.

Srivaro et al. [89] examined the mechanical properties of a 25-year-old OPT and compared them to those of other species such as bamboo and hardwood. The mechanical properties of the OPT reflect the density variation that was observed in both the radial and vertical directions of the trunk. The peripheral lower portion of the trunk has the highest bending strength, while the central core of the top portion of the trunk has the lowest. OPT has had the lowest bending strength, therefore OPT is more likely to be recommended for non-structural purposes.

The variation in the MOR follows the same pattern as the variation in the MOE or bending strength. The value of each strength was lower compared to bamboo, they ranged from 53 to 275 and 7 to 58 MPa (MOR), and 5 to 22 and 0.5 to 7.0 GPa (MOE). OPT has had a lower density compared to bamboo. Wulandari and Erwinsyah [92] investigated the density and specific gravity of Dura x Pisifera species of oil palm. They reported that the average values at the peripheral, central, and inner were 0.73, 0.54, 0.48 g/cm^3 and 0.53, 0.26, 0.19, respectively.

6.4. The Anatomy of Oil Palm Trunk

The oil palm plant is not a woody plant. The oil palm is a monocotyledonous species; it lacks cambium, secondary growth, growth rings, ray cells, sapwood, heartwood, branches, and knots. Oil palm has an anatomical structure that is composed primarily of fibers, tracheids, vessels parenchyma, and ray parenchyma cells. OPT differs chemically from hardwood/softwood species, with differences in cellulose, hemicellulose, and lignin content [89].

The increase in the stem diameter occurs due to overall cell division and enlargement in the parenchymatous ground tissues and enlargement of the vascular bundle fibers. There are three primary sections. The cortex is first, followed by the peripheral zone, and finally by the central zone in the OPT [96,97]. Oil palm stands are typically replanted after 25 years. At replanting age, the OPT reaches a height of 7 to 13 m and a diameter of 45 to 65 cm, measured 1.5 m above ground level. When mature, the trunk tapers toward the crowns, which typically produce about 41 fronds. The anatomical characteristics of an oil palm trunk cross-section are described as follows:

1. Cortex

The cortex is a narrow zone measuring approximately 1.5–3.5 cm in width, covering most of the trunk. It is primarily composed of ground tissue parenchyma with numerous longitudinal fibrous strands and vascular bundles of small and irregular shape [90]. In order to better explain the two main parts within OPT, in Figure 5 exhibits the schematic separation between the two for optimization further.

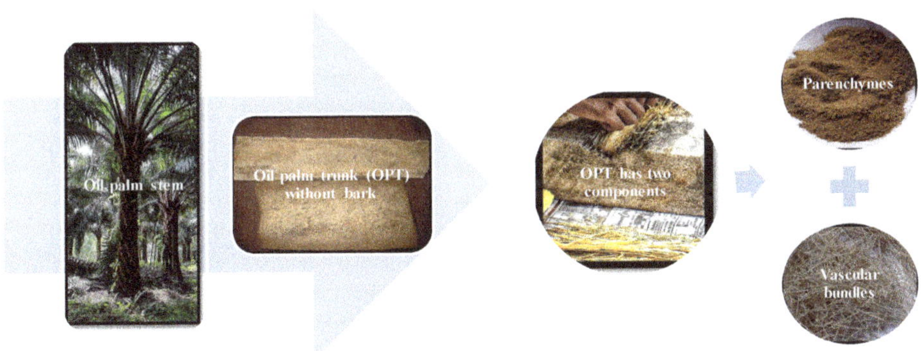

Figure 5. Two main parts of an oil palm trunk (OPT). For optimization, between the parenchyma and vascular bundles have been separated.

2. Periphery

The periphery is a region with thin parenchymal layers and dense vascular bundles that provides the palm trunk's primary mechanical support. The peripheral region is usually densely packed with radially extended fibrous sheaths, providing the palm with mechanical strength. This region accounts for approximately 20% of the total cross-section's area. The secondary walls of the fibers are multilayered and increase in length from the periphery to the pith. As the basal portions of the stem are older, they typically have more developed secondary walls than the top portions. Between the xylem and fiber strands are phloem cells in a single strand. According to Choowang [97], the areal number density of vascular bundles in the outer zone ranged from 68 to 115, much more compared to in the middle which had only 44 to 58 vascular bundles population/cm^2.

3. Central

The central zone accounts for approximately 80% of the total area and comprises of slightly larger and widely dispersed vascular bundles that are embedded in the parenchymatous ground tissues. Bundles become larger and more dispersed as they approach the trunk's core. An estimation of the vascular bundle density in central region's is approximately from 23 to 40 vascular bundles population/cm^2 [97].

4. Vascular bundles

The vascular bundles are thick, fibrous, and have a low hygroscopicity. The vascular bundle, on the other hand, is significantly less tightly packed toward the center zone, which contains more storage tissue. Fibers, xylem, vessels, protoxylem, sieve tubes, axial parenchyma, stegmata or spherical-shaped silica bodies form, and companion cells are all found inside the vascular bundle [91,98].

5. Parenchymatous tissue

The parenchymatous ground tissue is predominantly made up of soft, spongy, and very hygroscopic of thin-walled spherical cells. Food is stored as carbohydrates in the live parenchyma cells, mostly in the form of sugars and starch. The ground parenchymatous tissue is made up of thin-walled spherical parenchyma cells that are dense and thicker in the core than in the periphery. The diameter and length of the oil palm trunk increases as a result of cell division and elongation in the parenchymatous ground tissues as well as vascular fiber [91].

Picea abies (Norway spruce), a softwood species, has a different anatomy than the oil palm plant. Since it is a softwood, it contains earlywood (EW) and latewood (LW), which are readily visually visible. The LW is denser than the EW, and when viewed under a microscope, the cells of dense latewood have very thick walls and tiny cell cavities, whereas those of EW have thin walls and large cell cavities [99]. Macroscopically, the spruce wood consists of sapwood and heartwood whereas microscopy works revealed that wood was also arranged by tracheids which are typically ordered in a regular pattern and are nearly perfectly rectangular in cross-section [100]. Further, in spruce wood, abnormality in growth such as formation of hazel wood could be occurred [101]. In OPT, there is no differences among the bottom, middle or upper parts as shown in the scanning electron microscopy (SEM) image of Figure 6.

Therefore, whole parts utilization of OPT is beneficial compared to wood. However, further treatment should be taken, for instance in separating between the vascular bundles and the parenchyma. In the next sub chapter, the properties of OPT for raw material of panel products are discussed.

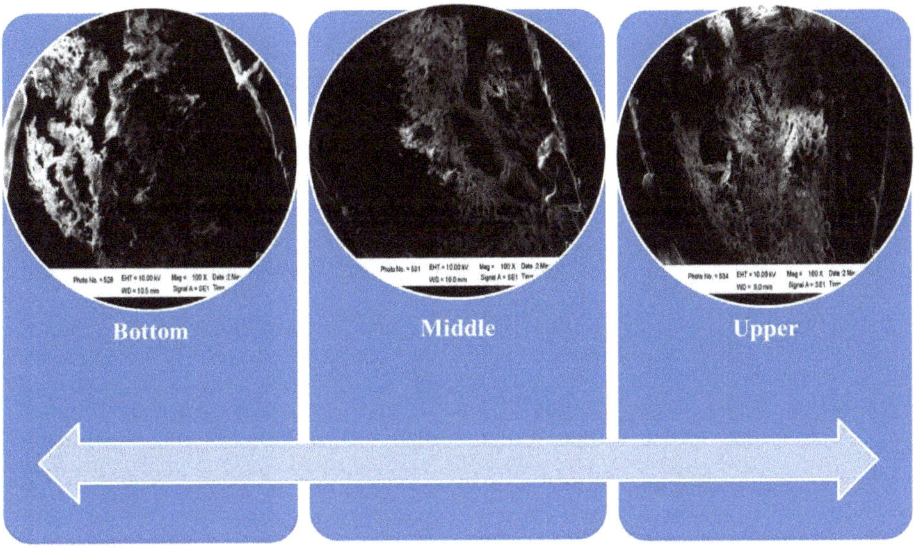

Figure 6. Vascular bundles of OPT showed no differences among the bottom, middle, and upper parts (100× magnifications).

7. Composite Panel Products Made of Oil-Palm Trunk and Its Properties

OPT has several disadvantages: low strength, low durability, poor dimensional stability, and poor machining [99]. Due to the high-density gradient along the stem's radial and longitudinal axes, OPT is less desirable for use as lumber. As a monocotyledon, oil palm differs significantly structurally from dicotyledonous woods. In comparison to wood, oil palm is much more porous, cellular, and anisotropic. As a result, liquids such as water and low molecular weight compounds can be rapidly absorbed and transported into the cell wall and lumen. OPT has been developed for composite panel products such as traditional panels (plywood and laminated veneer lumber), particleboard, and cement board.

7.1. LVL and Plywood

The first development for a traditional composite panel from the OPT was carried out by Sulaiman et al. [40]. The purpose of this study was to determine the physical and mechanical properties of LVL that was manufactured from OPT bonded by UF, PF, melamine urea formaldehyde (MUF), and phenol resorcinol formaldehyde (PRF) adhesives. Their shear strength constructed with PRF adhesive is greater than that of panels that were bonded with alternative adhesives. The shear strength of oil palm LVL, on the other hand, was lower than that of rubberwood LVL. There are few studies on OPT plywood constructed with thermoplastic adhesives. Recently, Hashim et al. [102] reported that LVL that was manufactured from OPT performed better when adhered with a cold setting adhesive, specifically PVAc and emulsion polymeric isocyanate (EPI). PVAc and EPI, with and without toluene, were superior to rubberwood LVL in TS and WA.

To improve the poor properties of traditional composite panels that are made from OPT, the raw materials have been modified. Numerous modifications were made, including treatment with low molecular weight phenol-formaldehyde (LmwPF) resin, impregnation of OPT with UF, hot air, and microwave drying, hybridization with another material, modification of the adhesive with nanoparticles, and compression of binderless veneer panels using response surface methodology (Table 3).

Table 3. Several modifications of OPT for LVL and plywood panel.

Product	Adhesive/Binder	Studies	Refs.
LVL	UF, PF, MF, and PRF	Variation of adhesive	[40]
Plywood	UF	UF impregnated OPT for core plywood	[36]
Plywood	PF	Treatment with LmwPF resin: effect of pressing pressure, the effect of resin content, and effect of hot-pressing time	[42,45,47]
Veneer	-	Effects of hot air and microwave drying	[46]
Composite-Plywood	UF and PF	Hybrid plywood from oil palm biomass	[57]
Composite-Plywood	MUF	Sandwich panel with OPT core overlaid with rubberwood veneer	[54]
Composite-Plywood	UF	Impregnated OPT with UF resin	[36]
Composite-Plywood	PF	Addition of OPA to hybrid plywood	[57]
Composite-Plywood	-	Binderless compressed veneer panel using response surface methodology	[26]

The pre-treatment of OPS veneers with LmwPF resin enhanced the surface characteristics, density, termite, and decay resistance. Loh et al. [38,50] investigated the wettability, surface roughness, and protection of OPS veneer from subterranean termites (*Coptotermes curvignathus*) and white-rot fungi (*Pycnoporous sanguineus*). For more than 30 s, the phenolic-treated veneer surface was capable of preventing liquid penetration. Veneer that was treated with LmwPF resin increased the durability of OPS plywood by 38% and 62% against termites and white-rot fungi, respectively. However, outer-layer OPS plywood is more resistant to termites and fungal decay than inner-layer OPS plywood.

Another study successfully developed the veneer treatment with LmwPF resin by varying the resin treatment, pressing time, solid resin content, and the resin concentration on formaldehyde emission, physical, and mechanical properties of plywood [42,45,47]. The plywood that was treated with LmwPF resin outperforms commercial OPS plywood in terms of physical and mechanical properties. The results indicated that veneers that were treated with a 30% solution of LmwPF provided the optimal treatment combination for plywood's physical and mechanical properties. The pressing time affected the pre-preg OPS plywood's mechanical properties and bonding performance. When LmwPF resins with a solid content of 32% or greater were used, the bonding quality met the minimum requirement for interior and exterior applications that are specified in Standard EN. The LmwPF resin can be used to treat OPT veneer with a solid content of up to 32% that meets the F standard for wood products as defined by Japanese Agriculture Standard (JAS).

Another successful treatment method to enhance the plywood's quality was used on oil palm veneer (OPV). Lekachaiworakul et al. [46] examined the effects of hot air and microwave drying on oil palm veneer's kinetics and mechanical properties of OPV. Additionally, drying with hot air and a high-power microwave revealed increased tensile stress and shear properties. However, the mechanical properties of hot air at 70–90 °C when combined with a microwave power level of 2000 watts were superior to those of hot air at 70–90 °C when combined with a microwave power level of 3000 watts.

Hybridization has been shown to affect the bending strength in variable studies. By crossbreeding of OPEFB and OPT, plywood's bending strength, screw withdrawal strength, and shear strength are increased [103]. Srivaro et al. [54] investigated the stiffness and bending strength of an OPT core sandwich panel that was overlaid with a rubberwood veneer when bent in the center. An increased OPT core density improved the stiffness and strength of the beam. Face fractures and core shear failures were observed, with the latter occurring more frequently when the OPT core density was low, the veneer face was relatively thick, and the span length was short.

Furthermore, H'ng et al. [36] examined the effects of UF impregnation on OPT's physical and mechanical properties by varying the UF solid content impregnation pressure and time. The three-layered OPT board that was impregnated with a higher resin solid content and a shorter impregnation time eventually reduced the degree of WA, resulting in changes in thickness and width. Additionally, impregnation treatments on the core layer increased the MOR and shear strength of the board but decreased the MOE of the board.

Previously, researchers manufactured plywood using formaldehyde-based resins adhesives. The immediate impact of the life cycle assessment (LCA) was due to the use of

UF and PF during the plywood manufacturing process [103]. As a result, formaldehyde causes health problems such as skin, eye, nose, and throat irritation, and excessive exposure may cause certain types of cancer [104]. Low formaldehyde emission of UF has been developed by our group [105]. However, application of this resin was not using OPT. Our group applied PF resin for bonding OPT and OPEFB to create a hybrid plywood with oil palm ash (OPA) nanoparticles and PF resin as a binder. The presence of OPA nanoparticles significantly altered the physical, mechanical, and thermal properties of the plywood panels. Thermal stability testing revealed that PF-filled OPA nanoparticles contributed to the plywood panels' thermal stability. Thus, the results of this study demonstrated that OPA nanoparticles significantly improved the properties of hybrid plywood.

Another study is being conducted to determine how to reduce formaldehyde emissions. Saari et al. [26] used a response surface methodology to produce compressed veneer panels from OPT without synthetic adhesive. Morphological analysis revealed that the shape of the vascular or parenchymatous tissue was altered as it was cranked, collapsed, or compressed together during the manufacturing process. Specific manufacturing parameters, such as pressing temperature, pressing time, and a portion of the oil palm trunk veneer, were demonstrated to bind the oil palm trunk veneers without using synthetic adhesives with acceptable mechanical and physical properties.

7.2. Particleboard

Table 4 summarizes the various studies and modifications that are made to the OPT panel. Hashim et al. [59] successfully developed the world's first particleboard that was made from OPT. Hashim et al. [35,37] investigated the properties of panels that were made entirely of OPT without the use of binders, focusing on the effect of particle geometry and press temperature. The particle's size and shape affect the sample's properties. According to the findings of this study, increasing the pressing temperature may be a viable method for improving the properties of binderless particleboard. Oil palm biomass contains significant amounts of holocellulose, lignin, starch, and sugar, that are necessary for self-bonding adhesion.

Table 4. Several studies and modifications oil palm trunk (OPT) for particleboard panel.

Product	Adhesive/Binder	Studies	Ref.
Binderless Particleboard	-	Effect of particle geometry on binderless particleboard	[37]
Binderless Particleboard	-	Influence of press temperature on the properties of binderless particleboard	[35]
Binderless Particleboard	-	Determine the chemical component suitability for binderless particleboard	[59]
Binderless Particleboard	-	Effect of oil palm age on properties of binderless particleboard	[49]
Particleboard	PHA	Effect addition of PHA	[34]
Particleboard	PHA	Influence of steam treatment and addition of PHA.	[31]
Particleboard	-	Optimization of press temperature and time for binderless particleboard	[30]
Particleboard	UF	Effect of treated with hot water and sodium hydroxide (NaOH) on the properties of particleboard	[32]
Binderless Particleboard	-	Steam treated on binderless particleboard	[26]
Particleboard	UF	Effects of two-step post heat-treatment on the particleboard properties.	[29]
Binderless Particleboard	ADP	Addition of ammonium dihydrogen phosphate (ADP)	[28]
Binderless Particleboard	-	Addition *Acacia mangium* particle	[44]
Particleboard	PVA	Addition citric acid and calcium carbonate	[25]

Additionally, Lamaming et al. [49] conducted a successful study which demonstrated superior mechanical and physical properties for binderless panels that were made from young oil palm trunk particles. Additionally, sugar was found to improve the boards' overall properties. The age of the oil palm affected the mechanical and physical properties of the boards that were manufactured, with sugars playing a significant role in the boards' self-bonding.

There are several drawbacks to using a binderless form of OPT. One of them is a lack of dimension stability [59]. Baskaran et al. [34] improved the properties of binderless materials by adding PHA using freeze-dried and pure samples. The addition of freeze-dried and pure PHA to the panels improved the overall properties of the samples. Additionally, compared to panels that were made without additives or controls, the panels that were made from steam-treated oil palm particles with varying amounts of PHA had improved physical and mechanical properties [31]. PHA appears to have some potential for enhancing the fundamental properties of experimental oil palm particleboard panels without adhesives. Baskaran et al. [30] optimized the press temperature and time for binderless particleboard that was manufactured from oil palm trunk biomass using a rotatable central composite design of response surface methodology. The optimization process was completed with the aid of the response surface method and software. Compared to the Japanese Industrial Standard (JIS), the optimal solutions that were proposed in this study were satisfactory.

Other techniques for enhancing the properties of binderless particleboard include being treated with hot water/sodium hydroxide (NaOH), steamed, or heated [26,28,29,32]. The MOE and MOR of samples that were prepared from raw material that was treated with hot water were greater than those of control panels that were prepared from NaOH-treated particles [32]. Saari et al. [26] reported that steaming improved the mechanical and physical properties of the boards. Steam exposure appears to be an inexpensive way to enhance the overall properties of such panels. The two-step post-heat treatment in OPT particles can enhance the dimensional stability and biological durability of the particleboard. Regrettably, the treatment significantly reduced the mechanical strength of the particleboard [29]. According to Komariah et al. [28], particleboard containing 10% ADP exhibited the highest MOR (8.9 MPa), IB (0.53 MPa), and thickness swelling (TS) values (5.9%). Additionally, particleboards containing 10% ADP demonstrated excellent resistance to water.

According to Yusof et al. [25] adding PVA, citric acid, and calcium carbonate to the composition of particleboards may also affect their fire resistance. Additionally, the addition of wood particles may improve the properties of binderless particleboard from OPT. The study was successfully developed by the addition of *Acacia mangium* wood which increased the fractal dimension of the particleboard, resulting in a decrease in hydrophilicity. The reduced percentage TS and WA properties of the binderless particleboard corroborate these findings [39,40].

7.3. Cement Board

OPT could be used as a natural fiber-based filler to manufacture gypsum composites with improved fire-retardant properties. The initial and final setting times of the OPT gypsum composite were significantly longer than the rice husk (RH) one. Compared to the RH gypsum composite, the OPT gypsum composite demonstrated superior mechanical properties [27]. Additionally, Phutthimethakul et al. [56] investigated the use of flue-gas desulfurization (FGD) gypsum, construction, and demolition waste (CDW), and OPT as part of the raw materials for the manufacture of concrete bricks. The compressive strength of the concrete brick specimens containing 5.5% gypsum without CDW and OPT was 45.18 MPa. In contrast, concrete brick specimens containing 5.5% FGD gypsum, 75% CDW, and 1% OPT had a compressive strength of 26.84 MPa.

8. Future Perspective

The future prospects of the utilization of OPT can be highlighted through the success of experimental works of engineered products such as the substitution of wood panel products, such as plywood, LVL, particleboard, fiberboard, and cement board as discussed above. In other words, OPT is potential to be a raw material for composite products which are usually made from wood. Challenges arose during the conversion of the OPT into veneer and lumber because of the monocotyledon characteristics which are rich in parenchyma and the irregular dimension of the fibers, thus after seasoning defects can be found remarkably. Board-based particles or fibers can be easily produced since transforming the OPT into both forms is now simply by applying a mechanical clearing system on the site plantation, where both types of dimensions are available afterwards. However, the determination of adhesive used still needed attention, whether its function as only a glue or at the same time performed as bonding agent, preservative chemical, and impregnation material. Even though both particleboard and fiberboard can be manufactured without adhesive, called binderless, care on hot-pressing stages should be carried out carefully. Wet processes or very high moisture within the particles was able to activate hydrogen bonding, thus the particles or fibers were interconnected each other. All this will be different to cement board for which the predominant component is cement or mineral. Cement board needs measurement of hydration temperature, which indicates the suitability of OPT with cement. A benefit would arise if both Indonesia and Malaysia, as the most oil palm producing countries, produced OSB, a structural panel product, since both countries have had no data on it.

9. Conclusions

Oil palm plantations have multiplied in Southeast Asia, particularly in Indonesia and Malaysia. Palm oil is used for the manufacture of multi-purpose products, ranging from edible to medicinal supplement and oleo-chemicals for household care products to biofuel, making it one of the most economically beneficial plants. After 25–30 years, the palms are felled and replaced due to declining oil production. OPT contributes significantly to waste because the trunks are then recycled or burned on the plantation. This increases insect and fungi populations, which poses difficulties for the new palm generation or contributes to the air pollution that is caused by the fire. If no effort is made to utilize oil palm biomass, it will become a problem for the environment. In fact, OPT is a sustainable non-wood source that is available in abundance at replanting operations, and would have seemed an alternative for raw materials for some of the products that are traditionally made from wood. Unfortunately, OPT is monocotyledon and consists of parenchyma and vascular bundles, which are the main obstacle in utilizing OPT as a solid. In contrast with wood, OPT has had higher moisture, dimensional instability, less durability, and poor machining. The oil palm industry must be prepared to capitalize on this situation and maximize the utilization of available OPT to convert waste to wealth, eliminate disposal issues, and generate value-added products by either raw material treatment or modification using resin. For the first, seasoning of OPT veneer using microwave, making hybrid plywood, and sandwiching or overlaying using rubberwood were examples for improving the dimensional stability of the final products. For the latter, impregnation using low molecular weight resin, treatment with several resins, and the addition of chemicals for binding activation were efforts for controlling the hygroscopicity and enhancing durability. Therefore, OPT would be utilized as alternative wood products, may be used in place of wood to manufacture wood-based panels because so far, many studies OPT could be engineered into the substitution of traditional wood products such as plywood, laminated veneer lumber (LVL), particleboard, and cement board or as the predominant matrix for end products of lumber, furniture, and lightweight construction. Unfortunately, the production of OSB in the two countries has not been carried out much. Therefore, this is the opportunity to develop the OSB industry that is made of vascular bundles in Indonesia and Malaysia.

Author Contributions: Conceptualization, A.N., R. and N.M.; methodology, A.N. and N.M.; software, A.N.; validation, R., J.S. and P.B.; formal analysis, A.N.; investigation, A.N., N.M. and, R.; resources, R. and N.M.; data curation, A.N., N.M. and P.B.; writing—original draft preparation, A.N.; writing—review and editing, A.N., J.S., N.M. and P.B.; visualization, A.N.; supervision, P.B.; project administration, R.; funding acquisition, A.N., R. and N.M. All authors have read and agreed to the published version of the manuscript.

Funding: This research and the APC were funded by Ministry of Education and Culture Republic of Indonesia under the World Class University Program of Universitas Sumatera Utara Year of 2020 grant number 1879/UN5.1.R/SK/PPM/2020.

Institutional Review Board Statement: Not applicable.

Informed Consent Statement: Not applicable.

Conflicts of Interest: The authors declare no conflict of interest.

References

1. Henson, I.E. A Brief History of the Oil Palm. In *Palm Oil*; AOCS Press: Urbana, IL, USA, 2012; pp. 1–29. [CrossRef]
2. Hassim, N.A.M.; Dian, N.L.H.M. Usage of palm oil, palm kernel oil and their fractions as confectionery fats. *J. Oil Palm Res.* **2017**, *29*, 301–310. [CrossRef]
3. Mohd, N.K.; Wen-Huei, L.; Abu Hassan, N.A.; Shoot-Kian, Y. Potential application of palm oil products as electrical insulating medium in oil-immersed transformers. *Environ. Prog. Sustain. Energy* **2021**, *40*, e13728. [CrossRef]
4. Shigetomi, Y.; Ishimura, Y.; Yamamoto, Y. Trends in global dependency on the Indonesian palm oil and resultant environmental impacts. *Sci. Rep.* **2020**, *10*, 20624. [CrossRef] [PubMed]
5. Singh, I.; Nair, R.S.; Gan, S.; Cheong, V.; Morris, A. An evaluation of crude palm oil (CPO) and tocotrienol rich fraction (TRF) of palm oil as percutaneous permeation enhancers using full-thickness human skin. *Pharm. Dev. Technol.* **2019**, *24*, 448–454. [CrossRef]
6. Rabiu, A.; Elias, S.; Oyekola, O. Oleochemicals from Palm Oil for the Petroleum Industry, Palm Oil, Viduranga Waisundara. *Palm Oil.* 2018. Available online: https://www.intechopen.com/books/palm-oil/oleochemicals-from-palm-oil-for-the-petroleum-industry (accessed on 25 April 2021). [CrossRef]
7. Zahan, K.A.; Kano, M. Biodiesel production from palm o0il, its by-products, and mill effluent: A review. *Energies* **2018**, *11*, 2132. [CrossRef]
8. Vijay, V.; Pimm, S.L.; Jenkins, C.N.; Smith, S.J. The impacts of oil palm on recent deforestation and biodiversity loss. *PLoS ONE* **2016**, *11*, e0159668. [CrossRef]
9. Sulaiman, O.; Salim, N.; Nordin, N.A.; Hashim, R.; Ibrahim, M.; Sato, M. The potential of oil palm trunk biomass as an alternative source for compressed wood. *Bioresources* **2012**, *7*, 2688–2706. [CrossRef]
10. Dungani, R.; Jawaid, M.; Khalil, H.A.; Jasni, J.; Aprilia, S.; Hakeem, K.R.; Hartati, S.; Islam, M.N. A review on quality enhancement of oil palm trunk waste by resin impregnation: Future materials. *BioResources* **2013**, *8*, 3136–3156. [CrossRef]
11. Kurnia, J.C.; Jangam, S.V.; Akhtar, S.; Sasmito, A.P.; Mujumdar, A.S. Advances in biofuel production from oil palm and palm oil processing wastes: A review. *Biofuel Res. J.* **2016**, *3*, 332–346. [CrossRef]
12. Food and Agriculture Organization. Forestry Production and Trade. Available online: http://www.fao.org/faostat/en/#data/FO (accessed on 25 December 2020).
13. Maluin, F.N.; Hussein, M.Z.; Idris, A.S. An overview of the oil palm industry: Challenges and some emerging opportunities for nanotechnology development. *Agronomy* **2020**, *10*, 356. [CrossRef]
14. Rowell, R.M.; Han, J.S.; Rowell, J.S. Characterization and factors effecting fiber properties. *Nat. Polym. Agrofibers Compos.* **2000**, 115–135.
15. Jacobus, P.H.; Wyk, V. Biotechnology and utilization of biowaste as a resource for bioproduct development. *Trends Biotechnol.* **2001**, *19*, 172–177.
16. Reddy, N.; Yang, Y. Biofiber from agricultural by-products for industrial applications. *Trends Biotechnol.* **2005**, *23*, 22–27. [CrossRef]
17. Samiran, N.A.; Jaafar, M.; Nazri, M.; Tung, C.C.; Ng, J.H. A review of palm oil biomass as a feedstock for syngas fuel technology. *J. Teknol.* **2015**, *72*, 13–18. [CrossRef]
18. Tow, W.-K.; Goh, A.P.-T.; Sundralingam, U.; Palanisamy, U.D.; Sivasothy, Y. Flavonoid composition and pharmacological properties of *Elaeis guineensis* Jacq. leaf extracts: A systematic review. *Pharmaceuticals* **2021**, *14*, 961. [CrossRef]
19. Rusli, N.D.; Ghani, A.A.A.; Mat, K.; Yusof, M.T.; Zamri-Saad, M.; Hassim, H.A. The potential of pretreated oil palm frond in enhancing rumen degradability and growth performance: A review. *Adv. Anim. Vet. Sci.* **2021**, *9*, 811.
20. Lim, A.; Chew, J.J.; Ngu, L.H.; Ismadji, S.; Khaerudini, D.S.; Sunarso, J. Synthesis, characterization, adsorption isotherm, and kinetic study of oil palm trunk-derived activated carbon for tannin removal from aqueous solution. *ACS Omega* **2020**, *5*, 28673–28683. [CrossRef]
21. Lim, S.C.; Gan, K.S. Characteristics and utilisation of oil palm stem. *Timber Technol. Bull.* **2005**, *35*, 1–7.

22. Yuliansyah, A.T.; Hirajima, T.; Rochmadi. Development of Indonesian palm oil industry and utilization of solid waste. *J. MMIJ* **2009**, *125*, 583–589. [CrossRef]
23. Hassan, K.; Kasim, J.; Mokhtar, A.; Aziz, A.A. Characterisation of oil palm trunks and their holocellulose fibres for the manufacture of industrial commodities. *Oil Palm Bull.* **2011**, *63*, 11–23.
24. Abdul Khalil, H.P.S.; Amouzgar, P.; Jawaid, M.; Hassan, A.; Ahmad, F.; Hadiyana, A.; Dungani, R. New approach to oil palm trunk core lumber material properties enhancement via resin impregnation. *J. Biobased Mater. Bioenergy* **2012**, *6*, 299–308. [CrossRef]
25. Yusof, M.; Lamaming, J.; Hashim, R.; Yhaya, M.F.; Sulaiman, O.; Selamat, M.E. Flame retardancy of particleboards made from oil palm trunk-poly(vinyl) alcohol with citric acid and calcium carbonate as additives. *Constr. Build. Mater.* **2020**, *263*, 120906. [CrossRef]
26. Saari, N.; Lamaming, J.; Hashim, R.; Sulaiman, O.; Sato, M.; Arai, T.; Nadhari, W.N. Optimization of binderless compressed veneer panel manufacturing process from oil palm trunk using response surface methodology. *J. Clean. Prod.* **2020**, *265*, 121757. [CrossRef]
27. Selamat, M.E.; Hashim, R.; Sulaiman, O.; Kassim, M.H.M.; Saharudin, N.I.; Taiwo, O.F.A. Comparative study of oil palm trunk and rice husk as fillers in gypsum composite for building material. *Constr. Build. Mater.* **2019**, *197*, 526–532. [CrossRef]
28. Komariah, R.N.; Miyamoto, T.; Tanaka, S.; Prasetiyo, K.W.; Syamani, F.A.; Subyakto Umemura, K. High-performance binderless particleboard from the inner part of oil palm trunk by addition of ammonium dihydrogen phosphate. *Ind. Crops Prod.* **2019**, *141*, 111761. [CrossRef]
29. Lee, S.H.; Ashaari, Z.; Ang, A.F.; Abdul Halip, J.; Lum, W.C.; Dahali, R.; Halis, R. Effects of two-step post heat-treatment in palm oil on the properties of oil palm trunk particleboard. *Ind. Crops Prod.* **2018**, *116*, 249–258. [CrossRef]
30. Baskaran, M.; Hashim, R.; Sulaiman, O.; Hiziroglu, S.; Sato, M.; Sugimoto, T. Optimization of press temperature and time for binderless particleboard manufactured from oil palm trunk biomass at different thickness levels. *Mater. Today Commun.* **2015**, *3*, 87–95. [CrossRef]
31. Baskaran, M.; Hashim, R.; Sudesh, K.; Sulaiman, O.; Hiziroglu, S.; Arai, T.; Kosugi, A. Influence of steam treatment on the properties of particleboard made from oil palm trunk with addition of polyhydroxyalkanoates. *Ind. Crops Prod.* **2013**, *51*, 334–341. [CrossRef]
32. Jumhuri, N.; Hashim, R.; Sulaiman, O.; Wan Nadhari, W.N.A.; Salleh, K.M.; Khalid, I.; Razali, M.Z. Effect of treated particles on the properties of particleboard made from oil palm trunk. *Mater. Des.* **2014**, *64*, 769–774. [CrossRef]
33. Nordin, N.A.; Sulaiman, O.; Hashim, R.; Salim, N.; Sato, M.; Hiziroglu, S. Properties of laminated panels made from compressed oil palm trunk. *Compos. Part B Eng.* **2013**, *52*, 100–105. [CrossRef]
34. Baskaran, M.; Hashim, R.; Said, N.; Raffi, S.M.; Balakrishnan, K.; Sudesh, K.; Sato, M. Properties of binderless particleboard from oil palm trunk with addition of polyhydroxyalkanoates. *Compos. Part B Eng.* **2012**, *43*, 1109–1116. [CrossRef]
35. Hashim, R.; Said, N.; Lamaming, J.; Baskaran, M.; Sulaiman, O.; Sato, M.; Sugimoto, T. Influence of press temperature on the properties of binderless particleboard made from oil palm trunk. *Mater. Des.* **2011**, *32*, 2520–2525. [CrossRef]
36. H'ng, P.S.; Chai, L.Y.; Chin, K.L.; Tay, P.W.; Eng, H.K.; Wong, S.Y.; Chai, E.W. Urea formaldehyde impregnated oil palm trunk as the core layer for three-layered board. *Mater. Des.* **2013**, *50*, 457–462. [CrossRef]
37. Hashim, R.; Saari, N.; Sulaiman, O.; Sugimoto, T.; Hiziroglu, S.; Sato, M.; Tanaka, R. Effect of particle geometry on the properties of binderless particleboard manufactured from oil palm trunk. *Mater. Des.* **2010**, *31*, 4251–4257. [CrossRef]
38. Loh, Y.F.; Paridah, T.M.; Hoong, Y.B.; Bakar, E.S.; Anis, M.; Hamdan, H. Resistance of phenolic-treated oil palm stem plywood against subterranean termites and white rot decay. *Int. Biodeterior. Biodegrad.* **2011**, *65*, 14–17. [CrossRef]
39. Danish, M.; Noor Aidawati Wan Nadhari, W.; Ahmad, T.; Hashim, R.; Sulaiman, O.; Ahmad, M.; Mohd Salleh, K. Surface measurement of binderless bio-composite particleboard through contact angle and fractal surfaces. *Measurement* **2019**, *140*, 365–372. [CrossRef]
40. Sulaiman, O.; Salim, N.; Hashim, R.; Yusof, L.H.M.; Razak, W.; Yunus, N.Y.M.; Azmy, M.H. Evaluation on the suitability of some adhesives for laminated veneer lumber from oil palm trunks. *Mater. Des.* **2009**, *30*, 3572–3580. [CrossRef]
41. Abdul Khalil, H.P.S.; Nurul Fazita, M.R.; Bhat, A.H.; Jawaid, M.; Nik Fuad, N.A. Development and material properties of new hybrid plywood from oil palm biomass. *Mater. Des.* **2010**, *31*, 417–424. [CrossRef]
42. Hoong, Y.B.; Paridah, M.T. Development a new method for pilot scale production of high grade oil palm plywood: Effect of hot-pressing time. *Mater. Des.* **2013**, *45*, 142–147. [CrossRef]
43. Mokhtar, A.; Hassan, K.; Aziz, A.A.; May, C.Y. Oil Palm Biomass for Various Wood-based Products. In *Palm Oil*; AOCS Press: Urbana, IL, USA, 2012; pp. 625–652. [CrossRef]
44. Nadhari, W.N.A.W.; Hashim, R.; Hiziroglu, S.; Sulaiman, O.; Boon, J.G.; Salleh, K.M.; Sugimoto, T. Measurement of some properties of binderless composites manufactured from oil palm trunks and Acacia mangium. *Measurement* **2014**, *50*, 250–254. [CrossRef]
45. Hoong, Y.B.; Loh, Y.F.; Nor Hafizah, A.W.; Paridah, M.T.; Jalaluddin, H. Development of a new pilot scale production of high grade oil palm plywood: Effect of pressing pressure. *Mater. Des.* **2012**, *36*, 215–219. [CrossRef]
46. Lekachaiworakul, P.; Dangvilailux, P.; Rattanamechaiskul, C. Effects of hot air and microwave drying on kinetic rate and mechanical property of oil palm veneer. *Energy Procedia* **2017**, *138*, 1093–1098. [CrossRef]

47. Hoong, Y.B.; Loh, Y.F.; Chuah, L.A.; Juliwar, I.; Pizzi, A.; Tahir Paridah, M.; Jalaluddin, H. Development a new method for pilot scale production of high grade oil palm plywood: Effect of resin content on the mechanical properties, bonding quality and formaldehyde emission of palm plywood. *Mater. Des.* **2013**, *52*, 828–834. [CrossRef]
48. Abdul Khalil, H.P.S.; Dungani, R.; Hossain, M.S.; Suraya, N.L.M.; Aprilia, S.; Astimar, A.A.; Davoudpour, Y. Mechanical properties of oil palm biocomposites enhanced with micro to nanobiofillers. In *Biocomposites*; Woodhead Publishing: Sawston, UK, 2015; pp. 401–435. [CrossRef]
49. Lamaming, J.; Hashim, R.; Sulaiman, O.; Sugimoto, T.; Sato, M.; Hiziroglu, S. Measurement of some properties of binderless particleboards made from young and old oil palm trunks. *Measurement* **2014**, *47*, 813–819. [CrossRef]
50. Saari, N.; Hashim, R.; Sulaiman, O.; Hiziroglu, S.; Sato, M.; Sugimoto, T. Properties of steam treated binderless particleboard made from oil palm trunks. *Compos. Part B Eng.* **2014**, *56*, 344–349. [CrossRef]
51. Nasir, M.; Khali, D.P.; Jawaid, M.; Tahir, P.M.; Siakeng, R.; Asim, M.; Khan, T.A. Recent development in binderless fiberboard fabrication from agricultural residues: A review. *Constr. Build. Mater.* **2019**, *211*, 502–516. [CrossRef]
52. Hashim, R.; Nadhari, W.N.A.W.; Sulaiman, O.; Kawamura, F.; Hiziroglu, S.; Sato, M.; Tanaka, R. Characterization of raw materials and manufactured binderless particleboard from oil palm biomass. *Mater. Des.* **2011**, *32*, 246–254. [CrossRef]
53. Siti Suhaily, S.; Abdul Khalil, H.P.S.; Asniza, M.; Nurul Fazita, M.R.; Mohamed, A.R.; Dungani, R.; Syakir, M.I. Design of green laminated composites from agricultural biomass. In *Lignocellulosic Fibre and Biomass-Based Composite Materials*; Woodhead Publishing: Sawston, UK, 2017; pp. 291–311. [CrossRef]
54. Srivaro, S.; Matan, N.; Lam, F. Stiffness and strength of oil palm wood core sandwich panel under center point bending. *Mater. Des.* **2015**, *84*, 154–162. [CrossRef]
55. Loh, Y.F.; Paridah, M.T.; Hoong, Y.B.; Yoong, A.C.C. Effects of treatment with low molecular weight phenol formaldehyde resin on the surface characteristics of oil palm (*Elaeis quineensis*) stem veneer. *Mater. Des.* **2011**, *32*, 2277–2283. [CrossRef]
56. Phutthimethakul, L.; Kumpueng, P.; Supakata, N. Use of flue gas desulfurization gypsum, construction and demolition waste, and oil palm waste trunks to produce concrete bricks. *Crystals* **2020**, *10*, 709. [CrossRef]
57. Nuryawan, A.; Abdullah, C.K.; Hazwan, C.M.; Olaiya, N.G.; Yahya, E.B.; Risnasari, I.; Masruchin, N.; Baharudin, M.S.; Khalid, H.; Abdul Khalil, H.P.S. Enhancement of oil palm waste nanoparticles on the properties and characterization of hybrid plywood biocomposites. *Polymers* **2020**, *12*, 1007. [CrossRef] [PubMed]
58. Megashah, L.N.; Ariffin, H.; Zakaria, M.R.; Hassan, M.A. Properties of cellulose extract from different types of oil palm biomass. In *IOP Conference Series: Materials Science and Engineering*; IOP Publishing: Selangor, Malaysia, 2018; p. 368. [CrossRef]
59. Yiin, C.L.; Yusup, S.; Quitain, A.T.; Uemura, Y.; Sasaki, M.; Kida, T. Thermogravimetric analysis and kinetic modeling of low-transition-temperature mixtures pretreated oil palm empty fruit bunch for possible maximum yield of pyrolysis oil. *Bioresour. Technol.* **2018**, *255*, 189–197. [CrossRef] [PubMed]
60. Rupani, P.F.; Alkarkhi, A.F.M.; Shahadat, M.; Embrandiri, A.; El-Mesery, H.S.; Wang, H.; Shao, W. Bio-optimization of chemical parameters and earthworm biomass for efficient vermicomposting of different palm oil mill waste mixtures. *Int. J. Environ. Res. Public Health* **2019**, *16*, 2092. [CrossRef] [PubMed]
61. Tahir, P.M.; Owolabi, F.A.T.; Shawkataly, A.K.H.P.; Alkarkhi, A.F.M.; Nwakaego, E.G.; Kamilu, O.K.F.; Chima, I.C.; Rizal, S.; Ghazali, A. Pulp and paper potentials of alkaline peroxide pre-treated of oil palm waste and industrial application. In *Palm Oil*; Waisundara, V.Y., Ed.; IntechOpen: London, UK, 2019; pp. 75–90. [CrossRef]
62. Lee, C.L.; H'ng, P.S.; Paridah, M.T.; Chin, K.L.; Rashid, U.; Maminski, M.; Go, W.Z.; Nazrin, R.A.R.; Rosli, S.N.A.; Khoo, P.S. Production of bioadsorbent from phosphoric acid pretreated palm kernel shell and coconut shell by two-stage continuous physical activation via N2 and air. *R. Soc. Open Sci.* **2018**, *5*, 180775. [CrossRef]
63. Rafatullah, M.; Ahmad, T.; Ghazali, A.; Sulaiman, O.; Danish, M.; Hashim, R. Oil palm biomass as a precursor of activated carbons: A review. *Crit. Rev. Environ. Sci. Technol.* **2013**, *43*, 1117–1161. [CrossRef]
64. Brito, P.L.; de Azevedo Ferreira, C.M.; Silva, A.F.F.; de Araújo Pantoja, L.; Nelson, D.L.; dos Santos, A.S. Hydrolysis, detoxification and alcoholic fermentation of hemicellulose fraction from palm press fiber. *Waste Biomass Valorization* **2018**, *9*, 957–968. [CrossRef]
65. Chan, C.M.; Chin, S.X.; Chook, S.W.; Chia, C.H.; Zakaria, S. Combined mechanical-chemical pre-treatment of oil palm empty fruit bunch (EFB) fibers for adsorption of methylene blue (MB) in aqueous solution. *Malays. J. Anal. Sci.* **2018**, *22*, 1007–1013. [CrossRef]
66. Hamzah, N.; Tokimatsu, K.; Yoshikawa, K. Solid fuel from oil palm biomass residues and municipal solid waste by hydrothermal treatment for electrical power generation in Malaysia: A review. *Sustainability* **2019**, *11*, 1060. [CrossRef]
67. Tajuddin, A.M.; Harun, S.; Sajab, M.S.; Zubairi, S.I.; Jahim, J.M.; Markom, M.; Nor, M.T.M.; Abdullah, M.A.; Hashim, N. Influence of deep eutectic solvent (DES) pre-treatment on various chemical composition of empty fruit bunch (EFB). *Int. J. Eng. Technol.* **2019**, *8*, 266–274. [CrossRef]
68. Rizal, N.F.A.A.; Ibrahim, M.F.; Zakaria, M.R.; Abd-Aziz, S.; Yee, P.L.; Hassan, M.A. Pre-treatment of oil palm biomass for fermentable sugars production. *Molecules* **2018**, *23*, 1381. [CrossRef] [PubMed]
69. Kaniapan, S.; Hassan, S.; Ya, H.; Patma Nesan, K.; Azeem, M. The utilisation of palm oil and oil palm residues and the related challenges as a sustainable alternative in biofuel, bioenergy, and transportation sector: A review. *Sustainability* **2021**, *13*, 3110. [CrossRef]
70. Onoja, E.; Chandren, S.; Razak, F.I.A.; Mahat, N.A.; Wahab, R.A. Oil palm (*Elaeis guineensis*) biomass in Malaysia: The present and future prospects. *Waste Biomass Valorization* **2019**, *10*, 2099–2117. [CrossRef]

71. Latif, A.A.; Harun, S.; Sajab, M.S.; Markom, M.; Jahim, J.M. Ammonia- based pre-treatment for ligno-cellulosic biomass conversion—An overview. *J. Eng. Sci. Technol.* **2018**, *13*, 1595–1620.
72. Loehr, R. *Agricultural Waste Management: Problems, Processes, and Approaches*; Elsevier: Ithaca, NY, USA, 2012.
73. Siddiqui, Y.; Surendran, A.; Paterson, R.R.M.; Ali, A.; Ahmad, K. Current strategies and perspectives in detection and control of basal stem rot of oil palm. *Saudi J. Biol. Sci.* **2021**, *28*, 2840–2849. [CrossRef]
74. Paterson, R.R.M. *Ganoderma boninense* disease of oil palm to significantly reduce production after 2050 in Sumatra if projected climate change occurs. *Microorganisms* **2019**, *7*, 24. [CrossRef]
75. Paterson, R.R.M. Oil palm survival under climate change in Malaysia with future basal stem rot assessments. *For. Pathol.* **2020**, *50*, e12641. [CrossRef]
76. Haryadi, D.; Hendra, H.; Sidhu, M.S.; Panjaitan, T.; Chong, K.P. The first report on basal stem rot disease causal pathogen in Asian Agri Group, North Sumatra, Indonesia. *Trans. Sci. Technol.* **2019**, *6*, 141–149.
77. Fernanda, R.; Siddiqui, Y.; Ganapathy, D.; Ahmad, K.; Surendran, A. Suppression of *Ganoderma boninense* using benzoic acid: Impact on cellular ultrastructure and anatomical changes in oil palm wood. *Forests* **2021**, *12*, 1231. [CrossRef]
78. Sahebi, M.; Hanafi, M.M.; Wong, M.-Y.; Idris, A.; Azizi, P.; Jahromi, M.F.; Shokryazdan, P.; Abiri, R.; Mohidin, H. Towards immunity of oil palm against Ganoderma fungus infection. *Acta Physiol. Plant* **2015**, *37*, 195. [CrossRef]
79. Cooper, R.M.; Flood, J.; Rees, R. Ganoderma boninense in oil palm plantations: Current thinking on epidemiology, resistance and pathology. *Planter* **2011**, *87*, 515–526.
80. Othman, N.Q.; Paravamsivam, P.; Tan, J.S.; Lee, Y.P.; Kwan, Y.; Alwee, S.S.R.S. Validation of differential gene expression of transcriptome assembly via Nanostring®technologies analysis platform. *J. Oil Palm Res.* **2018**, *30*, 36–46. [CrossRef]
81. Hushiarian, R.; Yusof, N.A.; Dutse, S.W. Detection and control of Ganoderma boninense: Strategies and perspectives. *SpringerPlus* **2013**, *2*, 555. [CrossRef] [PubMed]
82. Ng, W.P.Q.; Lam, H.L.; Ng, F.Y.; Kamal, M.; Lim, J.H.E. Waste-to-wealth: Green potential from palm biomass in Malaysia. *J. Clean. Prod.* **2012**, *34*, 57–65. [CrossRef]
83. Arno, F.; Katja, F.-K. The use of oil palm trunks for wood products. *Mater. Res. Proc.* **2019**, *11*, 69–80. [CrossRef]
84. Ibrahim, Z.; Alias, A.H.; Ramli, R.; Wahab, N.A.; Ahmad, M.; Osman, S.; Sheng, E.L. Effects of refining parameters on the properties of oil palm frond (OPF) fiber for Medium Density Fibreboard (MDF). *J. Adv. Res. Fluid Mech. Therm. Sci.* **2021**, *87*, 64–77. [CrossRef]
85. Murphy, D.J.; Goggin, K.; Paterson, R.R.M. Oil palm in the 2020s and beyond: Challenges and solutions. *CABI Agric. Biosci.* **2021**, *2*, 39. [CrossRef]
86. Food and Agriculture Organization (FAO). Global Forest Products Facts and Figures (2018–2019). 2018. Available online: http://www.fao.org/forestry/statistics/80938/en/ (accessed on 11 January 2021).
87. Thoemen, H.; Irle, M.; Sernek, M. *Wood-Based Panels: An Introduction for Specialists*; Brunel University Press: London, UK, 2010.
88. Bekhta, P.; Sedliačik, J. Environmentally-friendly high-density polyethylene-bonded plywood panels. *Polymers* **2019**, *11*, 1166. [CrossRef]
89. Srivaro, S.; Rattanarat, J.; Noothong, P. Comparison of the anatomical characteristics and physical and mechanical properties of oil palm and bamboo trunks. *J. Wood Sci.* **2018**, *64*, 186–192. [CrossRef]
90. Srivaro, S.; Matan, N.; Lam, F. Property gradients in oil palm trunk (*Elaeis guineensis*). *J. Wood Sci.* **2018**, *64*, 709–719. [CrossRef]
91. Lamaming, J.; Hashim, R.; Leh, C.P.; Sulaiman, O.; Sugimoto, T.; Nasir, M. Isolation and characterization of cellulose nanocrystals from parenchyma and vascular bundle of oil palm trunk (*Elaeis guineensis*). *Carbohydr. Polym.* **2015**, *134*, 534–540. [CrossRef] [PubMed]
92. Wulandari, A.; Erwinsyah, E. Distribution of vascular bundles and physical properties analysis of variety D x P oil palm trunk based on various zones and trunk heights. *J. Penelit. Kelapa Sawit* **2020**, *28*, 1–14. [CrossRef]
93. Wistara, N.J.; Rohmatullah, M.A.; Febrianto, F.; Pari, G.; Lee, S.-H.; Kim, N.-H. Effect of bark content and densification temperature on the properties of oil palm trunk-based pellets. *J. Korean Wood Sci. Technol.* **2017**, *45*, 671–681. [CrossRef]
94. Onuorah, E.O.; Nwabanne, J.T.; Nnabuife, E.L.C. Pulp and paper making potentials of *Elaeis guineensis* (oil palm) grown in South East, Nigeria. *World J. Eng.* **2015**, *12*, 1–12. [CrossRef]
95. Ahmad, N.; Kasim, J.; Yunus, N.Y.M.; Zaki, J.A.; Munirah, A. Chapter 73 Physical and Chemical Properties of Different Portions of Oil Palm Trunk. In *Regional Conference on Science, Technology and Social Sciences (RCSTSS 2016)*; Ahmad, N., Kasim, J., Mohd Yunus, N.Y., Ahmad Zaki, J., Munirah, A., Eds.; Springer: Singapore, 2018; pp. 759–766. [CrossRef]
96. Dirkes, R.; Neubauer, P.R.; Rabenhorst, J. Pressed sap from oil palm (*Elaeis guineensis*) trunks: A revolutionary growth medium for the biotechnological industry? *Biofuels Bioprod. Biorefining* **2021**, *15*, 931–944. [CrossRef]
97. Choowang, R. Correlation of the areal number density of vascular bundles with the mechanical properties of oil palm wood (*Elaeis guineensis* Jacq.). *Wood Res. J.* **2018**, *9*, 4–7. [CrossRef]
98. Rosli, F.; Ghazali, C.M.R.; Abdullah, M.M.A.B.; Hussin, K. A review: Characteristics of oil palm trunk (OPT) and quality improvement of palm trunk plywood by resin impregnation. *BioResources* **2016**, *11*, 5565–5580.
99. Myśkow, E.; Błaś, M.; Sobik, M.; Godek, M.; Owczarek, P. The efect of pollutant fog deposition on the wood anatomy of subalpine Norway spruce. *Eur. J. For. Res.* **2019**, *138*, 187–201. [CrossRef]
100. Ganguly, S.; Balzano, A.; Petri˘c, M.; Kržišnik, D.; Tripathi, S.; Žigon, J.; Merela, M. Effects of different energy intensities of microwave treatment on heartwood and sapwood microstructures in Norway spruce. *Forests* **2021**, *12*, 598. [CrossRef]

101. Račko, V.; Kačík, F.; Mišíková, O.; Hlaváč, P.; Čunderlík, I.; Ďurkovič, J. The onset of hazel wood formation in Norway spruce (Picea abies [L.] Karst.) stems. *Ann. For. Sci.* **2018**, *75*, 82. [CrossRef]
102. Hashim, R.; Sarmin, S.N.; Sulaiman, O.; Yusof, L.H.M. Effects of cold setting adhesives on properties of laminated veneer lumber from oil palm trunks in comparison with rubberwood. *Eur. J. Wood Wood Prod.* **2011**, *69*, 53–61. [CrossRef]
103. Ahmad Shamim, M.; Subramaniam, V.; Mohammad, H.; Mokhtar, A.; Ismail, B. Life cycle assessment for oil palm based plywood: A gate-to-gate case study. *Am. J. Environ. Sci.* **2014**, *10*, 86–93. [CrossRef]
104. United States Environmental Protection Agency (USEPA). Facts about Formaldehyde. 2015. Available online: http://www.epa.gov/formaldehyde/facts-about-formaldehyde#howcan (accessed on 11 January 2021).
105. Nuryawan, A.; Rahmawaty; Tambun, K.D.; Risnasari, I.; Masruchin, N. Hydrolysis of particleboard bonded with urea-formaldehyde resin for recycling. *Heliyon* **2020**, *6*, e03936. [CrossRef] [PubMed]

polymers

MDPI

Article

Effects of Wood Particles from Deadwood on the Properties and Formaldehyde Emission of Particleboards

Pavlo Bekhta [1,2,*], Ruslan Kozak [1], Vladimír Gryc [2], Václav Sebera [2] and Jan Tippner [2]

1 Department of Wood-Based Composites, Cellulose and Paper, Ukrainian National Forestry University, 79057 Lviv, Ukraine
2 Department of Wood Science and Technology, Faculty of Forestry and Wood Technology, Mendel University in Brno, Zemědělská 3, 613 00 Brno, Czech Republic
* Correspondence: bekhta@nltu.edu.ua

Abstract: The volume of deadwood increases annually because of changes in environmental, climatic, and hydrological conditions. On the other hand, during the last decade, manufacturers of wood-based boards have been facing an acute problem of a shortage of conventional raw materials. The purpose of this study was to evaluate the possibility of using wood particles from deadwood in the production of particleboards. Three-layer particleboards with different content of deadwood particles (0%, 25%, 50%, 75%, 100%) were produced. Conventional urea-formaldehyde (UF) resin was used for gluing the particles. The physical and mechanical properties of the boards, as well as the formaldehyde content in the boards, were determined. In addition, the effect of adding melamine-urea-formaldehyde (MUF) resin to UF adhesive on the properties of the boards was investigated. Replacing conventional sound wood particles with deadwood particles leads to deterioration of the physical and mechanical properties of the boards. The boards from deadwood particles absorb more water and swell more. The bending strength (MOR), modulus of elasticity in bending (MOE), and internal bonding (IB) values for boards with 100% deadwood particles are reduced by 26.5%, 23.1%, and 72.4%, respectively, compared to reference boards from sound wood particles. Despite this, a significant advantage is that boards made from 100% deadwood particles are characterized by 34.5% less formaldehyde content than reference boards made from conventional sound wood. Moreover, adding 3% of MUF resin to UF adhesive increases MOR, MOE, and IB by 44.1%, 43.3%, and 294.4%, respectively.

Keywords: particleboards; deadwood; wood particles; formaldehyde emission; urea-formaldehyde adhesive; bending strength; internal bond strength; modulus of elasticity; thickness swelling

Citation: Bekhta, P.; Kozak, R.; Gryc, V.; Sebera, V.; Tippner, J. Effects of Wood Particles from Deadwood on the Properties and Formaldehyde Emission of Particleboards. *Polymers* **2022**, *14*, 3535. https://doi.org/10.3390/polym14173535

Academic Editor: Gianluca Tondi

Received: 15 August 2022
Accepted: 26 August 2022
Published: 28 August 2022

1. Introduction

An intensive consumption of wood has led to a shortage of wood of industrial value in many countries of the world. This encourages wood processing firms to search for additional wood resources suitable for industrial use. Deadwood (dead fallen and standing trees, as well as felled by a windstorm etc.) (Figure 1) can be such an unused resources of raw materials. Nowadays, the challenge of tree drying in the forests is one of the acute ones for the whole of Europe. An increase in the average annual temperature and a decrease in the average annual precipitation contribute to the rapid reproduction of fungal diseases, various bacteria and pests and the growth of stocks of deadwood. Millions of cubic meters of infested trees (beetle-killed trees are very common) are now standing in Europe forests (Figure 1). With such rapid climate changes observed in recent years, the area of such deadwood can be expected to increase. Currently, the major European forest product companies harvest green wood. However, it is possible that in the near future, the majority of available wood will be infested wood. In unmanaged Central European forests, deadwood usually comprises up to 25% of the entire volume of wood in the forest [1]. Over the last 25 years, the amount of deadwood has increased in all European regions

(Figure 2) [2,3]. This could be explained by more frequent disturbances such as storms, insects' outbreaks and forest fires caused also by changing climatic conditions. It seems that temperature and moisture are the driving factors. The average volume of deadwood in 2015 is above 11 m^3/ha, equal to above 7% of the average volume of the growing stock density of European forests [2,3]. From Figure 2, it can be seen that in countries with mountain forests, the amount of deadwood is 2–3 times higher than in lowlands.

Figure 1. Dead trees in Šumava National Park, Czech Republic (Ing. Tomáš Koutecký, Ph.D).

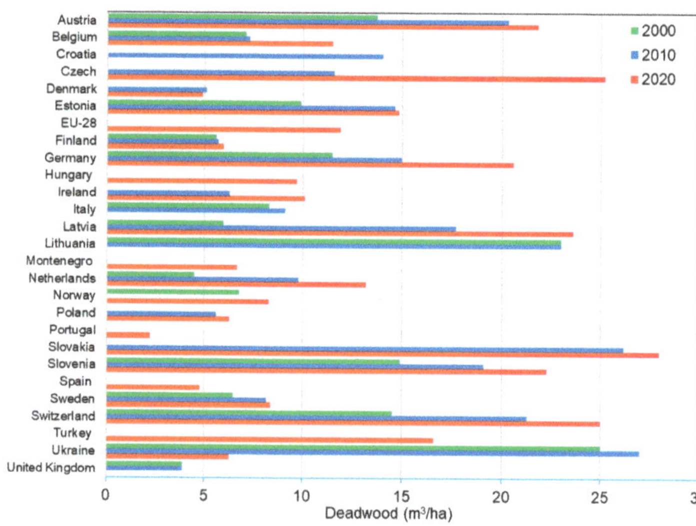

Figure 2. Average volumes of deadwood in European countries, 2000–2020 [2,3].

We have to remember that deadwood is a crucial component of forest ecosystems and plays an essential role in sustaining biodiversity as well as in processes such as soil formation and nutrient cycling [4]. However, we cannot also forget that wood is an important renewable commercial commodity, irreplaceable in numerous applications necessary for society's development [2]. Therefore, considering that deadwood represents a significant economic resource, economic uses of this resource need to be carefully considered [5].

Over the years, research efforts have focused on the study of how well deadwood is suited for the manufacture of a range of products (power poles, pallets, round timbers, lumber, laminated wood, panels, pulp and paper, fuel etc.) and to better understand how such wood will affect different products. Despite that, the use of deadwood is currently limited. This is often because certain challenges exist through all phases of the production of wood products from deadwood, including harvesting, transportation, log storage, processing, and end-product marketing [6].

Woo et al. [7] found that infested wood compared to sound wood is characterized by substantial loss of moisture, lower density, significantly lower concentrations of extractives, lower lignin and hemicellulose contents and better permeability. Another authors [8] noted that standing dead trees lost a 50% more of moisture content compared to fallen trees or trees on the ground. Such excessive dryness of deadwood creates technical problems for its use [5]. For example, the reduction in the moisture content of the wood makes it prone to checking and warping. This negatively affects the value of lumber and chip quality for pulping [7,9], as well as causes problems with the quality of flakes (or strands) used in composite boards and with maintenance of cutting tools [10]. However, on the other hand, as deadwood dries, it becomes lighter, which reduces transportation costs, and requires less drying time, which can save production costs [5].

Increased permeability of deadwood indicates possible irregular absorption or over-absorption of finishes and glues [11]. Troxell et al. [12] observed an increase of up to three times in chemical uptake when deadwood is considered.

Wood density decreases significantly over the course of decomposition [7,13–15]. For example, while wood density of living Norway spruce trees is about 0.43 g/cm^3 [13], the average density of the most decayed Norway spruce deadwood is only 0.138 g/cm^3 [14]. Other authors mentioned similar results. At one year after death, the density of beetle-killed southern pine was 60% of green timber [15].

According to the literature, the influence of different types of pests on the strength of wood is unclear. Some authors observed a reduction of 30–40% in toughness, of 11% in stiffness or modulus of elasticity, and of 19% in breaking strength or modulus of rupture in southern pine beetle-killed timber [16]. Walters [17] pointed out that southern pines had shown a reduction of 12% in bending strength (MOR) and 13% in modulus of elasticity in bending (MOE) after one year since death. On the contrary, several authors showed that infested wood has quite similar properties (including strength characteristics) to those of green timber [18] or has no effect on wood mechanical properties [19].

In the deadwood, because of the action of biologically active organisms, in addition to changes in physical and mechanical parameters, there are also changes in the chemical composition. Seifert [20] found losses of 7% cellulose and 3–4% hemicellulose in blue-stained timber, which could be related to changes in permeability and/or toughness. Thus, with a deep degree of mycological destruction, the physical and mechanical properties of wood deteriorate to such an extent that it becomes unsuitable for use as a construction material. However, the morphological structure and chemical composition allow its use as an active filler for the manufacture of wood composite materials.

Several authors [21] concluded that dead pines after an outbreak is a suitable feedstock for the production of lignocellulose micro-/nanofibrils. They found that the tree with advanced decay, which has no value for lumber, produced lignocellulose nanofibrils similar to those from the live tree. Studies also found that the infested trees have great potential for wood-plastic composites [22,23] and cement-bonded particleboard [24].

Various researchers have studied the possibility of using infested wood to make veneers and plywood. The most serious problems that were observed when processing dead wood into veneer were reduced veneer yield and reduction in full-sheet recovery. For example, Walters and Weldon [25] found a 9% less veneer volume, fewer full sheets and a higher percentage of random-width veneer, whereas Snellgrove and Ernst [26] found a 30% reduction in volume recovery and a higher percentage of random-width veneer for wood after three years since tree death. On the contrary, several authors found that there was no significant difference in veneer recovery between green and dead timber, especially when the affected trees are used immediately after the attack [25]. To improve veneer recovery from beetle-killed logs, some authors recommend proper conditioning of the logs [27]. They also showed that beetle-wood veneer can be dried faster, with a reduction in drying time by about 35% and a 27% increase in productivity from veneer drying.

Some authors note that the costs are similar in making particleboards from beetle-killed timber and from green trees, and significant equipment modifications for the production of such boards (using deadwood) is not required [28]. Moreover, studies showed an improved quality of particleboard when adding blue-stained timber into the furnish [29]. However, on the other hand, the use of beetle-killed wood leads to an increase in the amount of small fine fraction produced, the need for extra screening capacity and additional extra maintenance for cutting knives, the decrease in slenderness, and the tendency of the flakes to become folded [28,30].

A positive aspect regarding the use of deadwood in the production of particleboards is its significantly lower cost and limited application currently, compared to the traditional sound wood, which makes such a raw material attractive from an economic point of view. A negative aspect is the reduction of physical and mechanical properties of wood composites made from deadwood. It would be possible to compensate the loss in mechanical properties of composites from deadwood by using more reactive adhesive compositions than those traditionally used. Urea-formaldehyde (UF) resins are the most widely used adhesives in the manufacture of particleboards. In practice, melamine-urea-formaldehyde (MUF) resins are often added to UF adhesives to improve its adhesive strength and water resistance properties [31]. However, the modification of UF adhesives with MUF resin for the production of particleboards from deadwood requires additional research. Moreover, the UF adhesives have a major drawback, connected to the hazardous emission of volatile organic compounds (VOCs) and free formaldehyde from the finished particleboards [32]. As a result, new formaldehyde emission restrictions have been set for wood-based composites in Europe, the United States, and Japan. From the other hand, it was found that the emission of volatile organic compounds from wood decreases with wood storage tremendously. The emission of VOC from pine wood decreased by 50% on storage for 14 days [33]. Furthermore, Schäfer and Roffael [34] found that with increasing storage time, the spruce and pine particles emit less formaldehyde than non-stored wood. Based on this, we can assume that the addition of deadwood to the traditional wood in the production of particleboards can reduce the formaldehyde emission of the boards.

As follows from the literature resources, the general strategy would be to use deadwood as soon after death as possible, because the longer it's dead, the more it deteriorates, and the fewer are the options for its utilization [35]. Increasing amounts of deadwood and related literature provide insights into the feasibility of converting deadwood into composite wood panel products including particleboard. However, the use of deadwood for particleboards will not be possible without a comprehensive knowledge of the physical and chemical characteristics of this wood and the impact that these characteristics would have on a manufactured board's properties [6].

Thus, the objective of the present study was to evaluate the possibility of using wood particles from deadwood in the production of particleboards and to find out how the amount of deadwood particles affect the physical and mechanical properties, as well as the formaldehyde content of the boards.

2. Materials and Methods

2.1. Materials

Factory-produced wood particles from deadwood and traditional (sound) wood comprised of coniferous (75%) and deciduous (25%) species (originated from the Ukrainian Carpathians, Ivano-Frankivsk region) were obtained from the local particleboard plant. The deadwood was stored in the raw material warehouse for approximately four months prior to processing. Pine (*Pinus sylvestris* L.) and beech (*Fagus sylvatica* L.) woods mostly prevail among conifers and deciduous species, respectively. The moisture content of the particles, determined by the drying-weighing method, was approximately 3%. The fractional composition of the particles from deadwood and sound wood for the outer and core layers of the boards is presented in Table 1.

Table 1. Fraction analysis (by % weight) of wood particles from deadwood and sound wood.

Outer Layers				Core Layer			
Screen Hole Size (mm)	Content (%)		Difference (±)	Screen Hole Size (mm)	Content (%)		Difference (±)
	Deadwood	Sound Wood			Deadwood	Sound Wood	
1.25	4.42	8.8	−4.38	5.0	9.90	12.0	−2.1
1.0	9.05	1.2	+7.85	3.15	20.69	25.6	−4.91
0.8	12.19	9.4	+2.79	2.0	30.57	31.4	−0.83
0.63	15.24	12.2	+3.04	1.25	24.47	10.6	+13.87
0.4	27.44	26.4	+1.04	0.63	11.78	8.4	+3.38
0.2	19.14	17.6	+1.54	0.32	1.69	1.4	+0.29
Dust	12.52	14.5	−1.98	Dust	0.90	0.6	+0.3
Total	100	100	-	Total	100	100	-

UF and MUF resins were used in the experiments. UF adhesive consisted UF resin grade A (density 1.28 g/cm^3, solid content 66%, Ford cup (4 mm, 20 °C) viscosity 98 s, pH = 7.8, gel time 50 s) (producer LLC "Karpatsmoly", Kalush, Ukraine), paraffin emulsion, urea, and ammonium sulfate. A 33% aqueous solution of ammonium sulfate [$(NH_4)_2SO_4$] was used as a hardener and was mixed with the UF resin before spraying onto the wood particles. A 43% aqueous solution of urea [$CO(NH_2)_2$] and paraffin emulsion were mixed with UF resin.

MUF resin (density 1.29 g/cm^3, solid content 64.3%, viscosity 224 (Brookfield)/41 KF, pH = 9.32, gel time 83 s), due to its high reactivity towards wood surface and UF resin molecules, was used as an additional component to UF adhesive to improve water resistance and mechanical properties of particleboards. To find out how the addition of MUF resin to UF adhesive affects the properties of particleboards manufactured from deadwood, 1% and 3% of MUF were added to UF adhesive.

2.2. Manufacture of Particleboards

Three-layered particleboards of 290 × 290 mm dimensions and a thickness of 16 mm with a target density of 650 kg/m^3 were made. The boards contained particles from deadwood and sound wood. Particles from deadwood were added to the outer and core layers of the boards in the amount of 25%, 50%, 75% and 100%. The mass share of the outer layers was 33%, and the core layer was 67%. The amounts of UF resin, urea, hardener, and paraffin emulsion that were required for the mixing process were different for the core layer and the outer layers. This is due to the temperature difference between the surface and the core caused by heat transfer from the surface to the core of the board. In addition, the different amount of resin and additives used is related to the difference in the surface area of the particles used in the core and outer layers of the board. The amount of solid UF resin was 14 wt.% and 9 wt.% based on the weight of oven-dried wood particles for the outer and middle layers, respectively. During resin mixing, 2.3% and 0.5% of urea solution and 0.2% and 0.6% of ammonium sulfate were added, based on the weight of dry particles, to UF resin for the outer and core layers, respectively. A 0.8% of paraffin emulsion based on the weight of dry particles was also included in the resin mixture. The MUF resin was added to the UF adhesive used for the core layer. Wood particles were mixed with adhesive by hand. After mixing, the resinated particles were evenly distributed by hand in a 290 mm × 290 mm rectangular wooden mold. Pre-pressing of the formed mat was carried out manually in the wooden box. Next, the mat (Figure 3a) was subjected to hot pressing in an automatically controlled hydraulic laboratory press "xoM o" (LLC "ODEK" Ukraine, Ukraine) (Figure 3b) at the pressure of 2.5 MPa, temperature of 190 °C and the pressing time of 22.5 s/mm. During the last 30 s of the pressing cycle, the pressure was continuously reduced to 0 MPa. The experimental design for this study is summarized in Table 2.

(**a**) (**b**)

Figure 3. The formed particle mat (**a**) and hydraulic laboratory press (**b**).

Table 2. Manufacturing parameters of particleboards.

Board Type	Content (%) of Wood Particles from		Amount of MUF Resin (%)
	Sound Wood	Deadwood	
A	100	0	0
B	75	25	0
C	50	50	0
D	25	75	0
E	0	100	0
F	0	100	1
G	0	100	3

2.3. Particleboards Testing

After pressing, the boards were stored in air until reaching room temperature. Then, before evaluating their properties, the boards were conditioned for one week in a conditioning room, where the relative humidity of 65 ± 5% and 20 °C were maintained. The moisture content of the boards was within 6%. Three boards were made for each type of particleboard in the experimental design (Table 2), i.e., 21 boards. The conditioned boards were cut into required testing size according to relevant standards. Three samples of each board were tested according to European standards for density (EN 323) [36], bending strength (EN 310) [37], modulus of elasticity in bending (EN 310) [37], internal bond (IB) strength (EN 319) [38], thickness swelling (TS) (EN 317) [39] and water absorption (WA). On the other hand, for each batch, one board was randomly selected for analysis of formaldehyde content (FC) based on EN ISO 12460-5 (perforator method) [40].

The effects of wood particles content from deadwood and the amount of MUF resin on the properties of the laboratory-made boards was evaluated using an analysis of variance (ANOVA) at a significance level of 0.05. Duncan's range tests were performed to determine significant differences between means.

3. Results

In this study, a great difference in the amount of fine fractions of particles between deadwood and sound wood was not observed (Table 1). However, other authors indicate that logs dried to an average 50% moisture content produced nearly double the fines relative to green logs [41]. Fractional analysis of the wood particles (Table 1) used in this study showed that more fine particles prevailed in the particles obtained from deadwood. This can be considered as a factor that can significantly affect the properties of the boards. It is generally known that the quality of wood particles is the key factor in limiting production of quality particleboard as the geometry of particles affects the board's physical properties and internal bond strength characteristics [28,30].

3.1. Physical Properties of Boards

Table 3 presents mean values of density, WA and TS after 2 and 24 h immersion in the water for boards manufactured with adding different amount of deadwood particles and various amount of MUF resin into the UF adhesive. Deviations of the average values of the densities of the boards from the target density of 650 kg/m^3 are caused by the effect of material loss during the formation of the carpet, the uneven laying of wood particles over the area of the carpet, as well as the manual formation itself. However, these deviations of 1.2–5.3% were only marginally significant and did not affect significantly the results of the board's property values.

Table 3. Physical properties of particleboards.

Board Type	Density (kg/m^3)	Water Absorption 2 h (%)	Water Absorption 24 h (%)	Thickness Swelling 2 h (%)	Thickness Swelling 24 h (%)
Effects of wood particles content					
A	630.5 ± 32.1 ab	29.01 ± 8.61 a	90.24 ± 10.05 a	11.03 ± 3.55 a	42.01 ± 6.98 a
B	657.7 ± 28.6 d [1]	32.59 ± 10.19 ab	90.33 ± 7.68 a	12.49 ± 4.36 ab	45.08 ± 7.20 ab
C	639.3 ± 27.4 bc	37.57 ± 13.31 bc	97.85 ± 8.20 b	16.14 ± 5.30 c	48.55 ± 8.36 bc
D	615.7 ± 70.5 a	35.68 ± 6.27 c	92.18 ± 10.88 a	13.76 ± 4.27 b	41.36 ± 13.00 a
E	649.3 ± 28.6 cd	38.96 ± 9.00 c	99.76 ± 13.31 b	16.17 ± 4.20 c	50.82 ± 7.32 c
Effects of amount of MUF resin					
E	649.3 ± 28.6 ab	38.96 ± 9.00 b	99.76 ± 13.31 b	16.17 ± 4.20 c	50.82 ± 7.32 c
F	656.9 ± 32.7 b	36.39 ± 11.59 b	96.07 ± 9.10 b	12.31 ± 3.34 b	38.63 ± 3.41 b
G	638.4 ±32.8 a	25.29 ± 7.78 a	81.63 ± 7.86 a	7.73 ± 2.25 a	29.00 ± 3.15 a

[1] Averages followed by the same letter at the column are statistically equal by the Duncan test at 95% probability.

It was found that wood particles content from deadwood and the amount of MUF resin added into the UF adhesive have a significant effect on the TS and WA of the boards after soaking in water for 2 and 24 h. It can be stated that the influence of wood particles content from deadwood was less pronounced than the influence of the amount of MUF resin. Replacing sound wood particles with deadwood particles results in increased WA after 2 and 24 h of soaking in water. The lowest values of WA 2 h (29.01%) and WA 24 h (90.24%) were observed for boards made from sound wood particles. The highest WA after 2 h (38.96%) and 24 h (99.76%) of soaking in water was observed in the boards made with 100% deadwood particles. In addition, as can be seen (Figure 4a), more than a third of all the water absorbed by the samples is absorbed by the samples during the first 2 h. In addition, within the first 2 h, samples from sound wood particles (type A) absorb 32.1% of water, and samples from deadwood particles (type E) absorb 7% more water (39.1%).

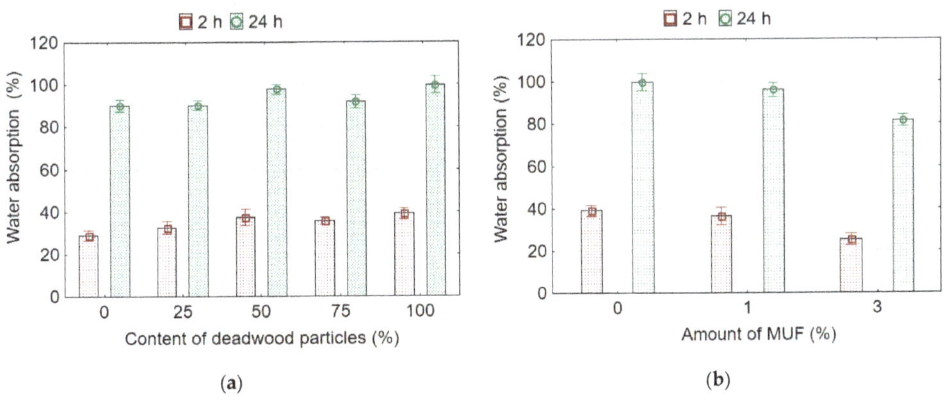

Figure 4. Water absorption of boards samples depending on content of deadwood particles (**a**) and amount of MUF resin in adhesive (**b**).

A similar trend is observed for TS. The presence of deadwood particles in the boards negatively affects the indicators of its swelling in water (Figure 5a). The samples with the addition of deadwood particles swell more than samples made from conventional sound wood particles. The lowest TS 2 h value 11.03% was found for the boards made of sound wood particles, and the highest value 16.17% for the boards made of deadwood particles. The lowest TS 24 h values were found for the boards with 75% deadwood particles (41.36%) and conventional sound wood particles (42.01%), and the highest 48.55% and 50.82%, respectively, in boards with 50% and 100% deadwood particles.

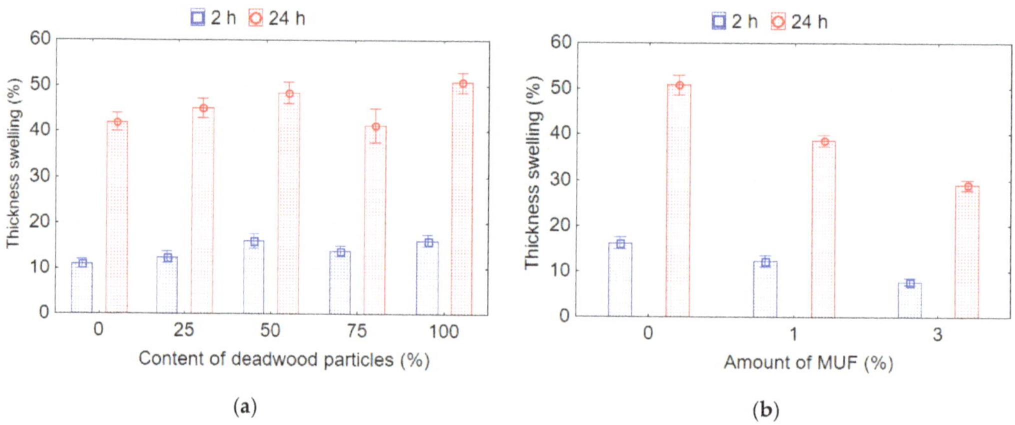

Figure 5. Thickness swelling of boards samples depending on content of deadwood particles (**a**) and amount of MUF resin in adhesive (**b**).

The boards with 25% (type B) and 75% (type D) of deadwood particles in terms of WA 2 h, WA 24 h, TS 2 h and TS 24 h do not differ significantly from each other and reference boards (except WA 2 h) made from conventional sound wood particles (type A). In contrast, the values of WA 2 h and TS 2 h for the boards D differ significantly from those values for the reference boards (type A). Likewise, the WA 2 h, WA 24 h, TS 2 h and TS 24 h of the boards with 50% (C) and 100% (E) deadwood particles do not differ from each other but differ significantly from the reference boards made from conventional sound wood particles (type A).

The addition of MUF resin into UF adhesive reduces the WA and TS of board samples made from deadwood particles. With an increase in the amount of MUF in the adhesive up to 3%, the values of WA 2 h, WA 24 h, TS 2 h, and TS 24 h compared to the adhesive without the addition of MUF resin decrease by 35.1%, 18.2%, 52.2%, and 42.9%, respectively.

Similar results were obtained by other authors who indicate that the oriented strand boards (OSB) derived from 100% mountain pine beetle-killed wood (standing dead for 2 or for 20 years) had greatly reduced water-resistance properties and dimensional stability [5]. The increase in WA and TS of boards containing deadwood particles is caused by the fact that the deadwood particles, due to destructive changes in its structure, have a large number of cracks that are formed during chipping. Water penetrates through the cracks in the board structure, destroys the UF adhesive joints, causes swelling of not only particles in outer layers, but also particles in the core layer, and fills additional voids that are formed because of destructive processes. The obtained results and their explanations are in good agreement with the results of other authors. For example, several authors found as much as 14% of the density losses due to decay for white pine with deep checks and no twigs, which is a sign of the loss of wood substance and the increase of its porosity [30]. In contrast, other authors mentioned that at one year after death, the density of beetle-killed southern pine was 60% of green timber [15]. In addition, increased permeability of wood particles from deadwood allows for greater water penetration [42]. These authors [42] suggest that the mechanism for increased permeability is probably the opening up of ray parenchyma cells by blue-stain fungi, and the microcracking that could be observed on some lumber samples.

The adding of water-resistant MUF resin into UF adhesive compositions allows reducing the negative impact of water on the values WA and TS of the boards containing deadwood particles. Adhesive bonds are destroyed less, which makes it difficult for water to penetrate the core of the board and causes a decrease in the swelling of the particles in the core layer and the formation of additional voids.

3.2. Mechanical Properties of Boards

Table 4 presents mean values of mechanical properties for boards manufactured with adding different amount of deadwood particles and various amount of MUF resin into the UF adhesive. It was found that the content of deadwood particles and the amount of MUF resin added into UF adhesive significantly affect the mechanical properties of the boards, including MOR, MOE, and IB. In addition, it was observed that the addition of MUF resin into the UF adhesive has a significantly stronger effect on the mechanical properties than the content of deadwood particles. However, all produced boards did not meet the respective European standard EN-312 requirements [43] (MOR > 11.5 MPa, IB > 0.24 MPa) for applications in dry conditions. To some extent, the low density (≈650 kg/m^3) of the produced boards can explain this. A graphic representation of the effect of deadwood particle content and amount of MUF resin on the mechanical properties of particleboards is shown in Figures 6–8.

Table 4. Mechanical properties of particleboards.

Board Type	MOR (MPa)	MOE (MPa)	IB (MPa)
	Effects of wood particles content		
A	9.75 ± 0.97 b [1]	1978.21 ± 257.63 c	0.19 ± 0.02 e
B	9.33 ± 1.48 b	1928.68 ± 157.98 bc	0.14 ± 0.03 d
C	9.17 ± 1.00 b	1838.84 ± 247.09 bc	0.12 ± 0.03 c
D	8.72 ± 0.96 b	1747.08 ± 189.10 b	0.08 ± 0.02 b
E	7.16 ± 1.49 a	1521.48 ± 341.64 a	0.05 ± 0.01 a
	Effects of amount of MUF resin		
E	7.16 ± 1.49 a	1521.48 ± 341.64 a	0.05 ± 0.01 a
F	9.25 ± 0.95 b	1329.41 ± 213.61 a	0.12 ± 0.02 b
G	10.32 ± 1.24 c	2180.54 ± 328.35 b	0.21 ± 0.03 c

[1] Averages followed by the same letter at the column are statistically equal by the Duncan test at 95% probability.

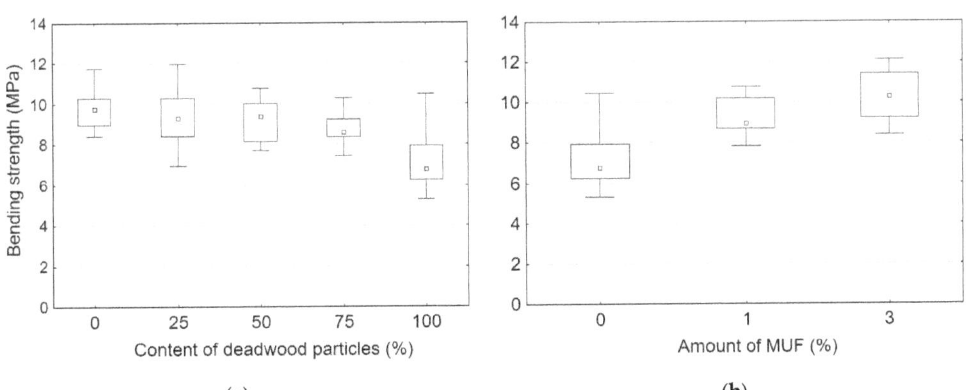

Figure 6. Bending strength of boards samples depending on content of deadwood particles (**a**) and amount of MUF resin in adhesive (**b**).

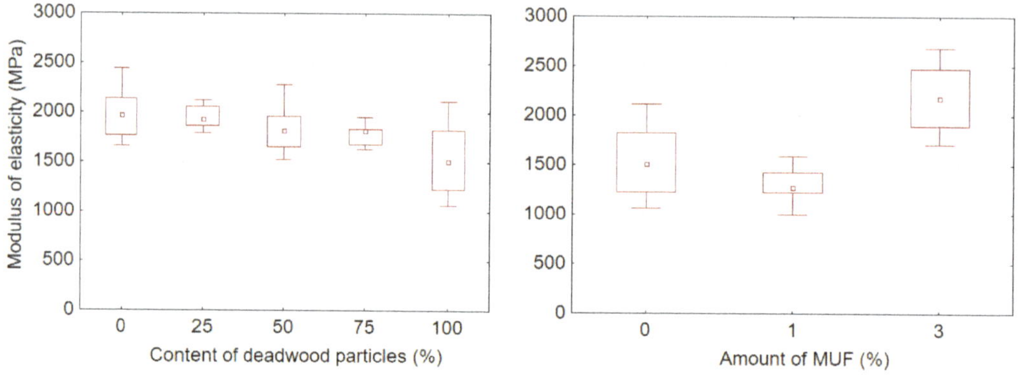

(a) (b)

Figure 7. Modulus of elasticity of boards samples depending on content of deadwood particles (**a**) and amount of MUF resin in adhesive (**b**).

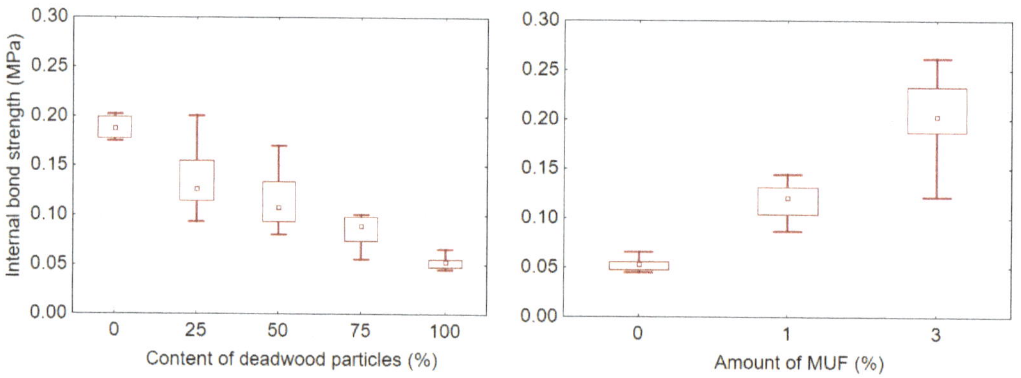

(a) (b)

Figure 8. Internal bond strength of board samples depending on the content of deadwood particles (**a**) and amount of MUF resin in adhesive (**b**).

It was found that an increase in the deadwood particle content causes a decrease in MOR (Figure 6a), MOE (Figure 7a) and IB (Figure 8a), while an increase in the amount MUF resin, on the contrary, leads to an improvement in these properties. The highest MOR, MOE, and IB values for samples with sound wood particles were 9.75 MPa, 1978.21 MPa, and 0.19 MPa, respectively. The lowest MOR, MOE, and IB values for samples with 100% deadwood particle content were 7.16 MPa, 1521.48 MPa, and 0.05 MPa, respectively. Thus, compared to reference boards made of sound wood particles, the values of MOR, MOE and IB for boards with 100% of deadwood particles are reduced by 26.5%, 23.1% and 72.4%, respectively. The addition of deadwood particles has the strongest effect on the quality of bonding (IB), reducing it by almost four times. The opposite results were obtained by other authors [12,29,30]. Some authors [30] did not find significant differences between particleboard produced from beetle-killed wood and those produced from green wood. Moreover, the composite board made from beetle-killed showed good internal bond test values, acceptable values for MOR and MOE, and a slight increase in thickness swelling and water absorption [12,30]. Adjusting the particle mixture to include 25% material from beetle-killed wood increased both MOR and MOE compared to the 100% green-wood mixture. The water-soak test results were better as well [29].

In general, it was observed that the addition of MUF resin into UF adhesive enables to improve the properties of boards made from 100% of deadwood particles (Figures 6b, 7b and 8b). Already the amount of 1% of MUF resin in the adhesive mix, the MOR of samples made from deadwood particles was higher than that of samples made from sound wood particles. The addition of 3% MUF resin increased MOR, MOE and IB by 44.1%, 43.3%, and 294.4%, respectively.

It can be stated that such changes in MOR for boards from deadwood particles are caused by a greater content of smaller chip fractions (Table 1) in outer layer than for boards from sound wood particles. Due to the large surface area of fine particles, the percentage of its coverage with adhesive is smaller than that of particles of larger fractions. This reduces the total chip bonding area and, as a result, reduces the bending strength of the boards. It is generally known that acceptable wood-based panels require quality particles and the smallest amount of fines, because fines consume excess amounts of resin binder and contribute little to mechanical properties [5].

Significant loss of IB strength (Figure 8a) occurs due to destructive changes in the structure of deadwood, an increased content of small particles fractions in the core layer (Table 1), as well as a large number of cracks that are formed during chipping. The existing adhesive contacts are not enough for strong bonding of particles. One of the reasons for low IB strength is that the wood particles from dead trees were difficult to glue because of apparent surface quality damage due to chipping dry wood [28]. In addition, the presence of a larger number of smaller particles from deadwood (Table 1) also impairs the mechanical properties of the boards. After all, it is well known that as the amount of fines increases, board property values decrease [44]. Moreover, the amount of resin required increases, thereby increasing product manufacturing costs [44]. Study showed that at least 30% more adhesive would be needed to produce commercially acceptable OSB panel products from 100% mountain pine beetle wood; however, such increase of adhesive content is uneconomical [5].

3.3. Formaldehyde Release of Boards

Markedly, the boards (type E) made from deadwood particles like as the control board, reached the E1 emission class (≤8.0 mg/100 g) but characterized by a much lower FC than the reference boards (type A) made from conventional sound wood particles. In the boards made from deadwood particles (type E), the formaldehyde content is lower by 34.5% compared to the reference samples (type A) made from conventional sound wood particles (Figure 9).

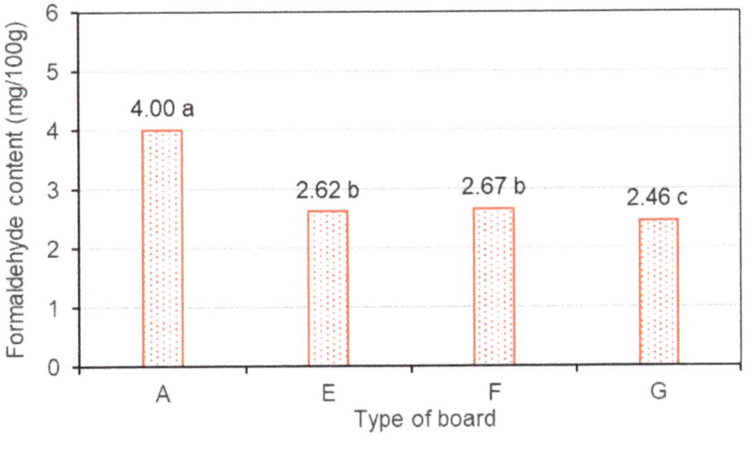

Figure 9. Effect of the amount of MUF resin in adhesive on the formaldehyde content of boards made with wood particles from dead trees. (Averages with the same letter are statistically equal.).

Addition of 1% or 3% MUF resin to the UF adhesive did not significantly affect the reduction of formaldehyde content in the boards (F and G). At the maximum added amount 3% of MUF resin, the formaldehyde content in the boards made from deadwood particles decreased by 38.5% compared to the reference boards made from sound wood. Others [45] found that the use of MUF resin increases the formaldehyde emission of the boards from thermo-mechanical and chemo–thermomechanical pulping.

Therefore, we can assume that excessive dryness and low content of extractive substances, as well as long storage time of deadwood, are the factors that led to a decrease in formaldehyde release. It is known that the emission levels of formaldehyde depend on numerous factors such as wood species, moisture content, content of extractives, outside temperature, and storing time [32,34]. The removal of extractives decreases the formaldehyde emitted from the wood. The results reveal that extracted chips release significantly lower amounts of formaldehyde compared to unextracted chips [34]. The air-dried wood produces low emissions of formaldehyde [46].

4. Conclusions

This preliminary study confirmed the possibility of manufacturing particleboards using wood particles from deadwood. However, the findings suggest that replacing conventional sound wood particles with deadwood particles leads to deterioration of the physical and mechanical properties of the boards with using UF adhesive. The particleboards from deadwood particles absorb more water and swell more. The MOR, MOE, and IB values for boards with 100% deadwood particles are lower by 26.5%, 23.1%, and 72.4%, respectively, compared to reference boards from sound wood particles. However, the modification of UF adhesive with MUF resin significantly improves the physical and mechanical properties of the boards. Adding 3% of MUF resin to UF adhesive increases MOR, MOE, and IB by 44.1%, 43.3%, and 294.4%, respectively whereas decreases WA 24 h and TS 24 h by 18.2% and 42.9%, respectively. A significant advantage is that boards made from 100% deadwood particles are characterized by 34.5% less formaldehyde content than reference boards made from conventional sound wood.

Further studies on the morphological structure and chemical composition of the deadwood are required, taking into account the age and time since death of trees. These data will help to find out the gluing mechanism and to choose the appropriate adhesive and mode parameters for pressing boards using such wood.

Author Contributions: Conceptualization, P.B.; methodology, P.B., R.K., V.G., V.S. and J.T.; investigation, P.B. and R.K.; writing—original draft preparation, P.B.; writing—review and editing, P.B., V.G., V.S., J.T. and R.K. All authors have read and agreed to the published version of the manuscript.

Funding: This research received no external funding.

Institutional Review Board Statement: Not applicable.

Informed Consent Statement: Not applicable.

Data Availability Statement: Not applicable.

Acknowledgments: Pavlo Bekhta acknowledges the Mendel University in Brno, Czech Republic for support of his research under the project "MENDELU international development II" No. CZ.02.2.69/0.0/0.0/18_053/0016930. The authors express their sincere thanks to Ing. Tomáš Koutecký, for providing photos of the dead forest in the Šumava National Park, Czech Republic.

Conflicts of Interest: The authors declare no conflict of interest.

References

1. Bobiec, A.; Gutowski, J.M.; Laudenslayer, W.F.; Pawlaczyk, P.; Zub, K. *The Afterlife of a Tree*; WWF Poland: Warszawa, Poland, 2005; 253p.
2. *Forest Europe, UNECE and FAO 2011: State of Europe's Forests 2011. Status and Trends in Sustainable Forest Management in Europe*; Ministerial Conference on the Protection of Forests in Europe: Bonn, Germany, 2011; ISBN 978-82-92980-05-7.
3. *Forest Europe, 2020: State of Europe's Forests 2020*; Ministerial Conference on the Protection of Forests in Europe: Bonn, Germany, 2020.

4. Vandekerkhove, K.; De Keersmaeker, L.; Menke, N.; Meyer, P.; Verschelde, P. When nature takes over from man: Dead wood accumulation in previously managed oak and beech woodlands in North-western and Central Europe. *For. Ecol. Manag.* **2009**, *258*, 425–435. [CrossRef]
5. Byrne, A.; Stonestreet, C.; Peter, B. *Current Knowledge of Characteristics and Utilization of Post-Mountain Pine Beetle Wood in Solid Wood Products*; Mountain Pine Beetle Initiative, Working Paper 2005–8; Forintek Canada Corporation: Vancouver, BC, Canada; Natural Resources Canada, Canadian Forest Service, Pacific Forestry Centre: Victoria, BC, Canada, 2005.
6. Hartley, I.D.; Pasca, S. *Evaluation and Review of Potential Impacts of Mountain Pine Beetle Infestation to Composite Board Production and Related Manufacturing Activities in British Columbia*; Mountain Pine Beetle Initiative, Working Paper 2006-12; University of Northern British Columbia: Prince George, BC, Canada; Natural Resources Canada, Canadian Forest Service, Pacific Forestry Centre: Victoria, BC, Canada, 2006.
7. Woo, K.L.; Watson, P.; Mansfield, S.D. The effects of mountain pine beetle attack on lodgepole pine wood morphology and chemistry: Implications for wood and fiber quality. *Wood Fiber Sci.* **2005**, *37*, 112–126.
8. Barron, E.H. Deterioration of southern pine beetle-killed trees. *For. Prod. J.* **1971**, *21*, 57–59.
9. Chow, S.; Obermajer, A. Moisture and blue stain distribution in mountain pine beetle infested lodgepole pine trees and industrial implications. *Wood Sci. Technol.* **2007**, *41*, 3–16. [CrossRef]
10. Nielson, R.W.; Wright, D.M. *Utilization of Beetle-Killed Lodgepole Pine*; Forintek Western Laboratory, Special Publication: Vancouver, BC, Canada, 1984.
11. Levi, M.P. *Southern Pine Beetle Handbook: A Guide for Using Beetle-Killed Southern Pine Based on Tree Appearance*; Agriculture Handbook 572; U.S. Department of Agriculture: Washington, DC, USA, 1981; 19p.
12. Troxell, H.E.; Tang, J.L.; Sampson, G.R.; Worth, H.E. *Suitability of Beetle-Killed Pine in Colorado's Front Range for Wood and Fiber Products*; USDA, Forest Service Resource Bulletin RM 2. Rocky Mountain Forest and Range Experiment Station: Fort Collins, CO, USA, 1980.
13. Morelli, S.; Paletto, A.; Tosi, V. Deadwood in forest stands: Assessment of wood basic density in some tree species, Trentino, Italy. *Forest* **2007**, *4*, 395–406. [CrossRef]
14. Merganičová, K.; Merganič, J. Coarse woody debris carbon stocks in natural spruce forests of Babia hora. *J. For. Sci.* **2010**, *56*, 397–405. [CrossRef]
15. Walters, E.; Weldon, D. *Weight Loss in Southern Pine Beetle-Killed Timber*; Texas Forest Service: Lufkin, TX, USA, 1982; Volume 258.
16. McLain, T.E.; Ifju, G. Strength properties of bluestained wood from beetle-killed southern pine timber. In *How the Environment Affects Lumber Design—Assessments and Recommendations: Proceedings of a Workshop Sponsored by: Society of Wood Science and Technology, USDA Forest Service, Forest Products Laboratory, Mississippi Forest Products Utilization Laboratory, May 28–30, 1980*; Forest Products Laboratory: Madison, WI, USA, 1982; pp. 55–67.
17. Walters, E. *Bending Strength Loss for SPB-Killed Timber*; Texas Forest Service: Lufkin, TX, USA, 1982; Volume 260.
18. Lemaster, R.L.; Troxell, H.E.; Sampson, G.R. Wood utilization potential of beetle-killed lodgepole pine for solid wood products. *For. Prod. J.* **1983**, *33*, 64–68.
19. Lum, C. *Characterising the Mechanical Properties of Wood Containing Beetle-Transmitted Bluestain*; Report to Forest Innovation Investment; [W-1984]; Forintek Canada, Western Division: Vancouver, BC, Canada, 2003; 17p.
20. Seifert, K. Changes of the chemical wood components by the blue rot *Pullularia pullulans* (de bary) berkhout (=*Aureobasidium pullulans* (de Bary) arnaud). *Holz Als Roh-Und Werkst* **1964**, *22*, 405–409. [CrossRef]
21. Hoeger, I.; Gleisner, R.; Negron, J.; Rojas, O.J.; Zhu, J.Y. Mountain Pine Beetle-Killed Lodgepole Pine for the Production of Submicron Lignocellulose Fibrils. *For. Sci.* **2014**, *60*, 502–511. [CrossRef]
22. Kim, J.-W.; Matuana, L.M.; McCullough, D.G. Ash trees infested by emerald ash borers as raw material for wood-based composites. *For. Prod. J.* **2005**, *55*, 89–92.
23. Lam, F.; Chang, F.C. Feasibility of using mountain pine beetle-attacked wood to produce wood-plastic composites: Preliminary work. *Wood Fiber Sci.* **2010**, *42*, 107–116.
24. Chang, F.C.; Lam, F. Use of mountain pine beetle killed wood to produce cement-bonded particleboard. *Wood Fiber Sci.* **2009**, *41*, 291–299.
25. Walters, E.; Weldon, D. *Veneer Recovery from Green and Beetle-Killed Timber in East Texas*; Texas Forest Service: Lufkin, TX, USA, 1982; Volume 257.
26. Snellgrove, T.A.; Ernst, S. Veneer recovery from live and dead lodgepole pine. *For. Prod. J.* **1983**, *33*, 21–26.
27. Wang, B.; Dai, C. *Maximizing Value Recovery from Mountain Beetle-Killed Pine for Veneer Products*; Working Paper 2005-9; Natural Resources Canada, Canadian Forest Service, Pacific Forestry Centre: Victoria, BC, Canada, 2005.
28. Maloney, T.M. Comparative economics of manufacturing composition boards from dead timber. *For. Prod. J.* **1981**, *31*, 28–36.
29. Kelly, M.W.; Barefoot, J.E.; Swint, W.H.; Levi, M.P. Properties of particle and hardboard made from healthy and beetle-killed southern pine. *For. Prod. J.* **1982**, *32*, 33–39.
30. Maloney, T.M.; Talbott, J.W.; Strickler, M.D.; Lentz Martin, T. Composition board from standing dead white pine and dead lodgepole pine. In *The Dead Softwood Timber Resource: Proceedings of Symposium Head in May 22–24, 1978. Spokane, WA*; Washington State University: Pullman, WA, USA, 1978; pp. 19–51.
31. Pizzi, A. *Advanced Wood Adhesives Technology*; Marcel Dekker, Inc.: New York, NY, USA, 1994; 289p.

32. Salem, M.Z.M.; Böhm, M. Understanding of formaldehyde emissions from solid wood: An overview. *BioResources* **2013**, *8*, 4775–4790. [CrossRef]
33. Dix, B.; Roffael, E.; Schneider, T. *Abgabe von Flüchtigen Verbindungen (Volatile Organic Compounds, VOC) von Strands, Hergestellt aus Kernund Splintholz der Kiefer*; WKI-Kurzbericht 6/2004, WKI Short Report 6/2004; Fraunhofer Institute for Wood Research (WKI): Braunschweig, Germany, 2004.
34. Schäfer, M.; Roffael, E. On the formaldehyde release of wood. *Holz Als Roh-Und Werkst.* **2000**, *58*, 259–264. [CrossRef]
35. Dobie, J. An overview of dead timber potential in Canada. In *Symposium the Dead softwood Timber Resource*; Engineering Extension Service Washington State University: Pullman, WA, USA, 1978.
36. *EN 323*; Wood-Based Panels—Determination of Density. European Committee for Standardization: Brussels, Belgium, 1993.
37. *EN 310*; Wood-Based Panels—Determination of Modulus of Elasticity in Bending and of Bending Strength. European Committee for Standardization: Brussels, Belgium, 1993.
38. *EN 319*; Particleboards and Fibreboards—Determination of Tensile Strength Perpendicular to the Plane of the Board. European Committee for Standardization: Brussels, Belgium, 1993.
39. *EN 317*; Particleboards and Fibreboards. Determination of Swelling in Thickness after Immersion in Water. European Committee for Standardization: Brussels, Belgium, 1993.
40. *EN ISO 12460-5*; Wood-Based Panels-Determination of Formaldehyde Release—Part 5. Extraction Method (Called the Perforator Method). European Committee for Standardization: Brussels, Belgium, 2015.
41. Knudson, R.M.; Chen, L. *Effect of Aspen LOG Moisture Content on Stranding, Strand Quality and Properties of OSB*; Contract No. 2001–2322; Forintek Canada Corp., Western Division: Vancouver, BC, Canada, 2001.
42. McFarling, S.; Byrne, A. *Characterizing the Dimensional Stability, Checking, and Permeability of Wood Containing Beetle-Transmitted Bluestain*; Report to Forest Innovation Investment; [W-1985]; Forintek Canada, Western Division: Vancouver, BC, Canada, 2003; 13p.
43. *EN 312*; Particleboards—Specifications. European Committee for Standardization: Brussels, Belgium, 2010.
44. Feng, M.W.; Knudson, R.M. Effect of log rehydration on quality of OSB strands manufactured from beetle-killed lodgepole pine. *For. Prod. J.* **2007**, *57*, 35–42.
45. Roffael, E.; Dix, B.; Schneider, T. Influence of pulping process on the emission of formaldehyde and volatile organic acids from pulps and medium density fiberboards (MDF). *Holz Als Roh–Und Werkst.* **2007**, *65*, 145–148. [CrossRef]
46. Young, S. Formaldehyde emission from solid wood—Will it become an issue? Timber Test Laboratories. Unpublished data. 2004.

polymers

MDPI

Article

Multi-Scale Evaluation of the Effect of Thermal Modification on Chemical Components, Dimensional Stability, and Anti-Mildew Properties of Moso Bamboo

Xiao Xiao [1,2,†], Xingyu Liang [1,2,†], Haozhe Peng [1,2], Kaili Wang [1,2], Xiaorong Liu [1,2,*] and Yanjun Li [1,2,*]

1 Jiangsu Co-Innovation Center of Efficient Processing and Utilization of Forest Resources, Nanjing Forestry University, Nanjing 210037, China
2 Bamboo Engineering and Technology Research Center, State Forestry and Grassland, Nanjing 210037, China
* Correspondence: lxr2020@njfu.edu.cn (X.L.); lalyj@njfu.edu.cn (Y.L.)
† These authors contributed equally to this work.

Abstract: By promoting greenhouse gas sequestration, bamboo and bamboo-based products can improve carbon storage, and thus help decrease greenhouses gas emission through replacing traditional products like concrete, steel, and alloy. Thermal modification is a useful way to effectively enhance the dimensional stability and mold-resistance property of bamboo and bamboo-based products compared with chemical treatment. This work investigates the change in anti-mildew properties, micro-structure, and chemical composition of bamboo after heat treatment. Saturated steam heat treatment was applied for this project. SEM results showed that the structural damage of parenchyma cells resulted in the separation of thin-walled cells and vascular bundles. Thus, the original regular structure of bamboo, characterized by plump and intact cells, changed markedly. After thermal modification, bamboo samples exhibited improved dimensional stability and anti-fungal properties due to the decrement of hemicellulose and cellulose. The hardness and MOE of the modified bamboo were 0.75 and 20.6 GPa, respectively.

Keywords: bamboo; thermal modification; bamboo cell wall; anti-mildew property

Citation: Xiao, X.; Liang, X.; Peng, H.; Wang, K.; Liu, X.; Li, Y. Multi-Scale Evaluation of the Effect of Thermal Modification on Chemical Components, Dimensional Stability, and Anti-Mildew Properties of Moso Bamboo. *Polymers* **2022**, *14*, 4677. https://doi.org/10.3390/polym14214677

Academic Editors: Satoshi Komasa and Pavlo Bekhta

Received: 11 October 2022
Accepted: 1 November 2022
Published: 2 November 2022

1. Introduction

There are 78 genera divided into 1500 species of bamboo all over the world, and the area of bamboo forests is about 20 million square hectares. By promoting greenhouse gas sequestration, bamboo and bamboo-based products can improve carbon storage, and thus help decrease greenhouses gas emission through replacing traditional products like concrete, steel, and alloy [1–4]. In the past decades, bamboo has caused wide concern in the construction, building, decoration, and other fields due to its advantages such as excellent mechanical properties, easy harvesting, low density, and good safety [5]. When bamboo and bamboo-based products are applied in decoration and construction projects, they can be easily affected by fungi [6], water [7], and UV [8]. Therefore, it is of great important to find a modification method that can effectively improve the dimensional stability, physical properties, and mold-resistance property of bamboo, and thus expand the application field of bamboo and bamboo-based products and extend its service life [9–11].

Heat treatment is a useful way of effectively enhancing the dimensional stability and mold-resistance of bamboo-based products compared with that of chemical treatment [12]. In the thermal modification process, the bamboo becomes more dimensionally stable and less hygroscopic due to the decomposition of hemicelluloses, crystallinization of cellulose, remification of lignin, and removal of extractives. These changes can help bamboo [5] reduce moisture content re-absorption, and thus improve weathering and durability, and increase the dimensional stability [13]. More specifically, thermal modification can effectively improve the hygroscopicity, swelling, and shrinkage when the treatment temperature is over 150 °C. At 180 °C or higher, the anti-fungal property can be positively enhanced due to the decomposition of hemicellulose and cellulose [5].

Modern spectroscopic techniques, such as the Wet chemistry method, X-ray diffractometer, Fourier transform infrared spectroscopy, and anti-fungal test, are effectively methods for analyzing the relationship between thermal modification and the change in chemistry components [14]. Thus, the thermal modification mechanism can be deeply investigated through these surface spectroscopic techniques. Previous works focused on the changes in the surface chemical properties and the micro-morphology of woody resources after high-temperature heat treatment [15–18], ignoring the change in anti-mildew properties, dimensional stability, and micro-mechanical property [6]. Changes in micro-morphology, chemical composition, crystallinity index, anti-mildew properties, and dimensional stability of bamboo after heat treatment are still unclear. In addition, bamboo cell walls are composed of hemicellulose, cellulose, and lignin, and the relationship between chemical composition change and micro-mechanics have rarely been investigated [19]. Nanoindentation (NI) can directly reveal the response between the thermal modification parameters and micro-mechanical properties of bamboo cell walls [20]. Thus, it is meaningful to investigate the change in nano-mechanics of bamboo at the cell-wall level [21].

This work investigates the change in anti-mildew properties, micro-structure, and chemical components of bamboo after heat treatment. High temperature steam was applied for this project. We revealed the thermal modification mechanism of bamboo through the Wet chemistry method, X-ray diffractometer (XRD), Fourier transform infrared spectroscopy (FTIR), physical/mechanical properties test, and anti-fungal test.

2. Materials and Methods

2.1. Sample Preparation

Six-year-old moso bamboo was used in this work. Bamboo was harvested in JiangXi Provience, China. The arc-shaped bamboo sheets were 1050 mm × 11 mm × 120 mm, and the surfaces of the bamboo sheets were smooth and no defects. The bamboo sheets were directly transformed by heat treatment at different thermal modification treatment temperatures and the same durations (6 min, 8 min, and 10 min). Saturated steam was used as a thermal modification medium. Pressure tank (RDW-1.5-D, Rongda Boiler Container Co., Nanjing Ltd., Nanjing, China) was provided by Hangzhou Rongda Boiler Container Co., Ltd., Hangzhou, China.

2.2. Scanning Electron Microscopy

Bamboo specimens with average dimensions of 5 mm (length) × 5 (width) × 1 mm (thickness) were cut from different bamboo samples. The cross-section surface of different bamboo specimens were polished with a knife and coated with gold for observation by scanning electron microscope (Quanta 200, FEI, Tokyo, Japan).

2.3. XRD

The bamboo powders used in the Wet chemistry method were also used here. During the XRD test, the crystallinity degree of different bamboo specimens was determined by X-ray diffractometer (Ultima IV, Tokyo, Japan). The bamboo powders were exposed to X-ray radiation. The 2-theta and scan rate were set from 5° to 45° and 2 min^{-1}, respectively. Segal's method was used to calculate the crystallinity index [22].

2.4. FTIR

Fourier transform infrared (FTIR) spectroscopy was used to investigate the chemical reaction between bamboo and thermal modification parameters, using the VERTEX 80V Spectrum (Bruker, Berlin, Germany) at room temperature. Untreated and treated bamboo samples were milled to powders of 100 mesh and then pressed together with KBr powders into transparent film. The resolution of the device was 2 cm^{-1} with 32 accumulations. The range of the wavelength was recorded from 400 to 4000 cm^{-1}.

2.5. EMC, ASE, and Mass Loss Test

The different bamboo samples were cut into small blocks with average dimensions of 20 × 20 × t mm (length × width × thickness). Then, all bamboo specimens were placed in an incubator (HWS-250, Jinhong Co., Ltd., Nanjing, China) under relative humidity of 65% and a temperature of 20 °C for 14 days. Lastly, the EMC, ASE, and mass loss of untreated and treated bamboo samples were tested according to the Chinese National Standard GB/T 15780–1995 "Testing methods for physical and mechanical properties of bamboo" [20].

2.6. Wet Chemistry Method

Before the chemical components analysis, the different bamboo samples were ground into powder and passed through a sieve with an average size of 40–80. Approximately 200–300 mg of the bamboo samples were placed in a tube, and 3 mL of 72% H_2SO_4 was added to submerge the bamboo specimens. The detailed experimental process can be found in National Renewable Energy Laboratory (NREL).

2.7. Anti-Mildew Property Test

The anti-mildew property of the different bamboo specimens was analyzed according to the National standard GB/T 18261-2000, "The method for control of wood mold and cyanobacteria by mildew inhibitor", with Aspergillus niger as representative mold. In the national standard, one month is a complete measurement period.

3. Results and Discussion

3.1. Micro-Morphology Analysis

SEM was used to detect the micro-morphology of bamboo specimens. As seen from Figure 1, the treated bamboo samples exhibited a deformation of the thin-walled cell after thermal modification. The shape of parenchyma cells changed from "round and smooth" to "flat and distort" [23]. These changes indicated that the treated bamboo's thin-walled cell became fragile and loose in comparison to the control [24]. This is because of the high-temperature steam inserted into bamboo inner tissue, which lead to the degradation of starch and hemicellulose [7–10]. The Wet chemistry method, XRD, FTIR, and physical tests are needed to further explore the effects of thermal modification on functional groups of bamboo samples.

3.2. Chemical Components and Mass Loss Ratio Analysis

In the Wet chemistry method test, the relative content of hemicellulose, cellulose, and lignin of the control were 21.5%, 41.5%, and 22.3%, respectively. Modified bamboo samples had a low hemicellulose and cellulose content and high lignin content, indicating that thermal modification can accelerate the decomposition of hemicellulose, which is conducive to the decomposition of polysaccharide and starch, allowing enhancement of mechanical and mold resistance. Especially compared to the decreasing tendency of hemicellulose and cellulose content, the lignin content exhibited an increasing tendency. According to previous research [11–13], the main composition of hemicellulose is xylan, which easily decomposes and dehydrates due to its branched and amorphous structure. Therefore, hemicellulose is easier to degrade than cellulose and lignin. At 160 °C or higher, the lignin content increases quickly with the increment of treatment temperature, which can be attributed to the lignin condensation reaction [14]. For detail, the polysaccharides in hemicellulose have low thermal stability under high temperatures due to their branched structure and amorphous structure, making it easier to decompose hemicellulose than other chemical components in bamboo [15]. The results of mass loss ratio are shown in Figure 2B. As shown in Figure 2B, with the increasing treatment temperature and duration, the mass loss increased. This is due to the decomposition of carbohydrate polymers in bamboo samples. Additionally, the decomposition of extractives can also contribute to this conclusion [16–18]. Figure 3 shows the starch content of different bamboo samples. The starch content of bamboo samples decreased from 3.25% to 2.90%. This is due to the high temperature and high pressure provided by saturated steam.

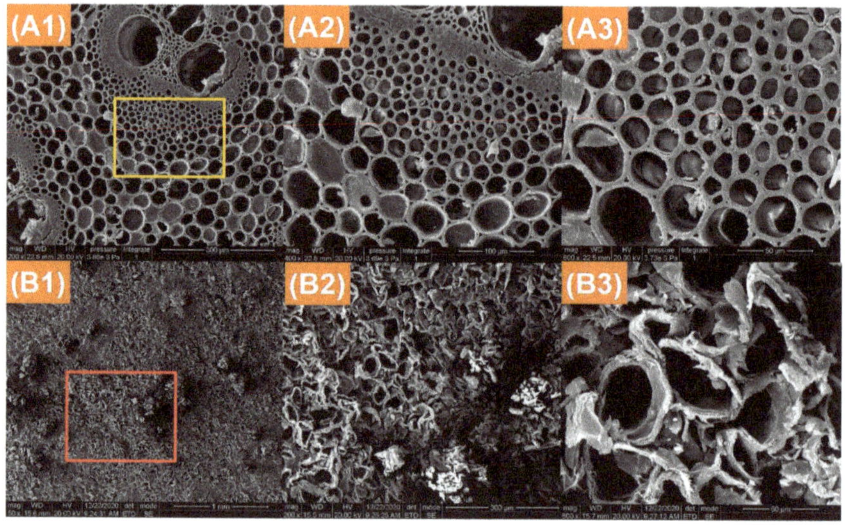

Figure 1. (**A1–A3**) Images of untreated bamboo in the cross-section and (**B1–B3**) thermal modified bamboo treated under 180 °C/10 min.

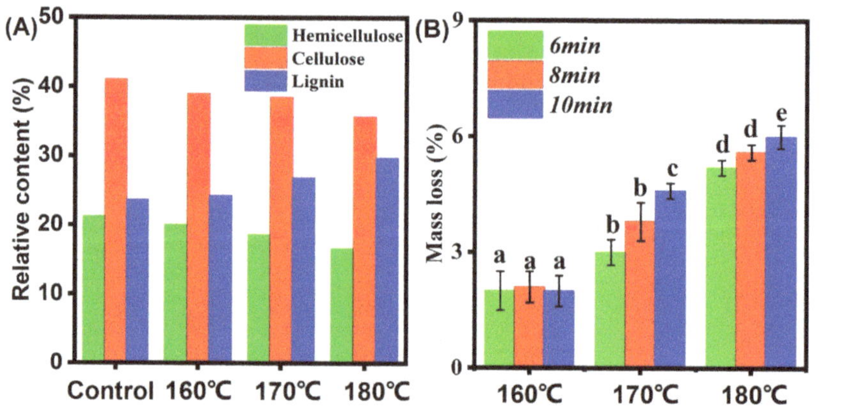

Figure 2. (**A**) Change in three major chemical compositions; (**B**) Mass loss. Different small (a–e) represent the significant difference between heat treatment groups. The error bar in the picture represents the standard deviation.

3.3. XRD and FTIR Analysis

Figure 4A shows the XRD patterns and CrI of different bamboo samples. In the high temperature-treated bamboo samples, the CrI increased from 40.5% to 59.5%. It is well known that biomass cellulose contains quasi crystalline regions, this being attributed to rearrangement or reorientation of cellulose molecules inside these regions. In addition, the degradation of cellulose in the amorphous region may happen at high temperatures, which results in more crystallization. Therefore, the crystallinity index of treated bamboo specimens increased. In addition, the decomposition of the para-crystalline part of cellulose can also make a positive contribution to the increment of cellulose crystallinity index [19–22].

As shown in Figure 4C,D, the FTIR curves of untreated and treated bamboo specimens from 500 cm^{-1} to 4500 cm^{-1} were presented. The spectra of different bamboo samples presented the typical peaks of bamboo: 1731 cm^{-1} (C=O strength vibrations of hemicellulose), 1230 cm^{-1} (C-O strength vibrations peaks), 1590 cm^{-1} (stretching of carboxylic acid), and 1425 cm^{-1} (strengthening of acetyl acid). It can be seen from the (C) that there are no significant differences between the FTIR spectra of untreated and treated bamboo specimens. The relative intensities of peak at 1370 cm^{-1} do not change too much due to the stability of C-H, illustrating that C-H can remain unchanged during the thermal modification. The relative intensity of peaks at 1230 cm^{-1} and 1730 cm^{-1} show decreasing tendency in comparison to that of the control, which was due to the decomposition of the cellulose and hemicellulose in bamboo samples. During the saturated steam heat treatment process, the relative intensity of peaks at 1630 cm^{-1} and 1590 cm^{-1} increased with the increasing treatment temperature. This can be attributed to the increase in lignin content. The condensation reaction of lignin positively contributes to the increase in relative lignin content and can enhance the dimensional stability of the bamboo specimen. Lastly, the relative intensity of peak at 898 cm^{-1} obviously decreased, possibly resulted from acid environment provided by the decomposition of hemicellulose.

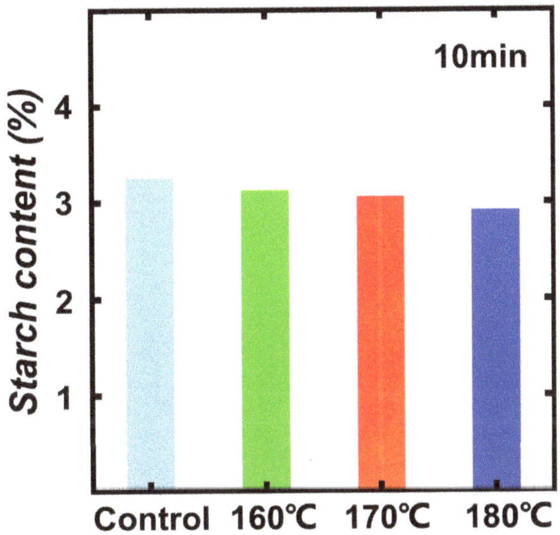

Figure 3. The starch content of different bamboo samples.

3.4. EMC and ASE Analysis

Figure 5A,B shows the EMC and ASE of the saturated steam treated bamboo samples. The EMC and ASE of the control were 13.5% and 8.4%, respectively. The EMC and ASE of treated bamboo specimens presented similar change stages. In Figure 5A, the EMC decreased with the increment of thermal modification parameters. In detail, we can find that the bamboo samples exhibited the lowest value (7.45%) of EMC under 180 °C and 10 min. These results suggest the degradation of extractives, ash, and starch in bamboo inner tissue and thus increases hygroscopicity of treated bamboo samples. This is due to the decrement of OH groups in hemicellulose [23–27].

3.5. Bamboo Cell Wall Mechanics

Modulus of elasticity and hardness are two measurement indexes to evaluate cell wall mechanics in Nanoindentation tests. The tested bamboo sample pictures nanoindentation curves were inserted in Figure 6A,B. Figure 6A,B shows the MOE and H of the saturated steam treated bamboo samples. The MOE and H of the control were 15.5 GPa and 0.59 GPa, respectively. The elastic modulus and hardness of the saturated steam-treated bamboo samples presented a similar increasing tendency. In Figure 6A, the MOE and H increased with the increment of thermal modification parameters. For example, the bamboo specimens (180 °C/10 min) exhibited highest MOE (20.6 GPa) compared with that of the control. At the same time, the hardness of the bamboo exhibited an increasing tendency, from 0.59 GPa to 0.75 GPa. The lignin condensation can endow the increment of bamboo cell wall mechanics. Additionally, the condensation polymerization and decreased EMC can also positively affect the bamboo cell wall mechanics [2,28–33].

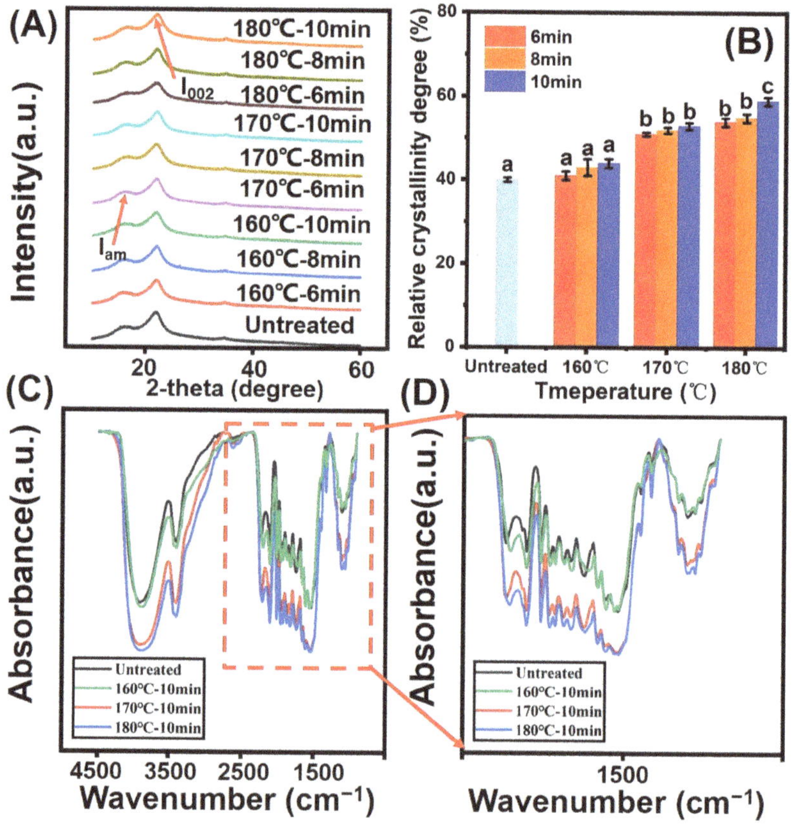

Figure 4. (**A**) XRD patterns of different specimens with the 2-theta angle ranges of 5–60; (**B**) Crystallinity index; (**C**,**D**) FTIR and enlarged FTIR curves. Different small letters shows significant differences between groups.

3.6. Mold-Resistance Property Analysis

Figure 7A,B show the anti-mildew property and correspondeding pictures. In order to evaluate the mold-resistance property of the bamboo samples, the bamboo samples were incubated by *Aspergillus niger* for 30 days. It was observed that the infection ratio of the control was 0% in the first test day. After 8 days' infection, the *Aspergillus niger* appeared in the bamboo surface, illustrating that the bamboo samples were incubated in a very short time. As shown in Figure 7A, after only 8 days' incubation, the infection ratio of the control was 100%. However, the saturated steam-treated bamboo samples (180 °C and 10 min) exhibited better anti-mildew properties in comparison to the control. This is because starch and polysaccharide decreased in bamboo specimens after thermal modification. Additionally, the enhanced relative content of lignin in bamboo samples can inhibit the adhesion between *Aspergillus niger* and the bamboo surface, which can improve the anti-mildew property of bamboo samples [34–40].

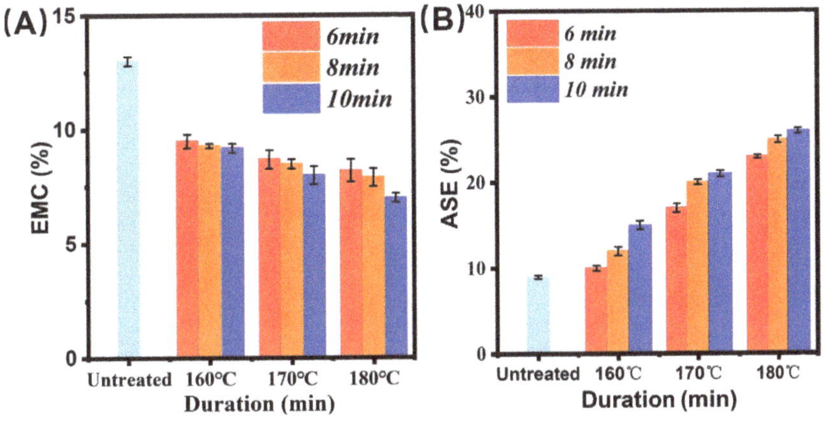

Figure 5. Physical property characterization. (**A**) EMC; (**B**) ASE.

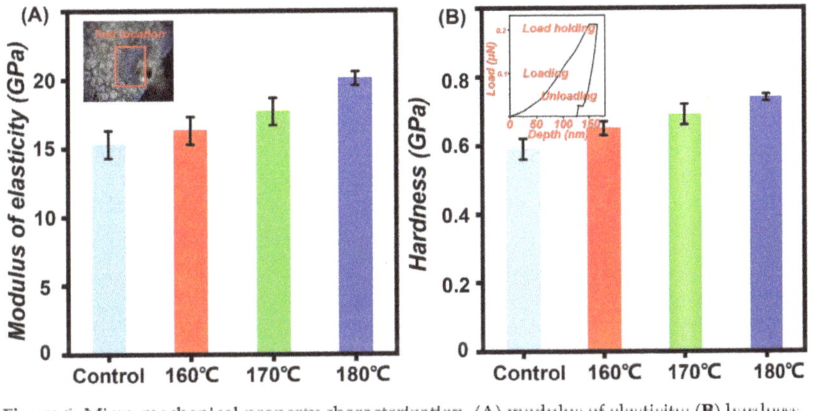

Figure 6. Micro-mechanical property characterization. (**A**) modulus of elasticity; (**B**) hardness.

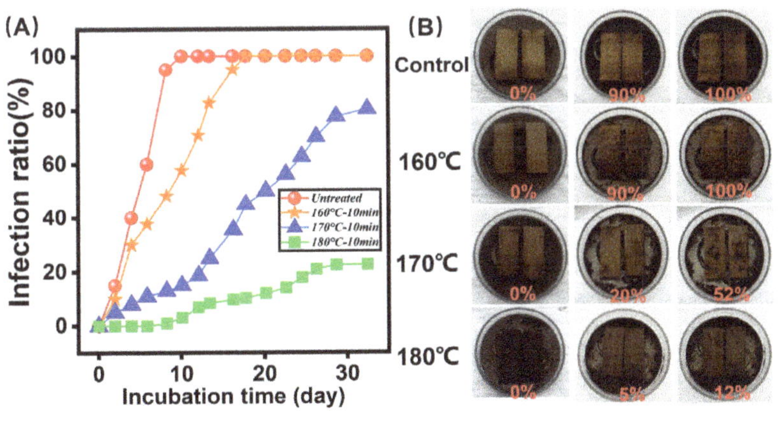

Figure 7. Micro-mechanical property characterization. (**A**) Infection ratio in one month; (**B**) Picture of anti-mold test in one month.

4. Conclusions

By promoting greenhouse gas sequestration, bamboo and bamboo-based products can improve carbon storage, and thus help decrease greenhouses gas emission through replacing traditional products like concrete, steel, and alloy. In summary, a green and cost-effective method was proposed for bamboo thermal modification. The mold-resistance property of bamboo was enhanced due to the degradation of starch and hemicellulose. The MOE of the modified bamboo increased from 15.5 GPa to 20.6 GPa. At the same time, the hardness of the bamboo exhibited an increasing tendency, from 0.59 GPa to 0.75 GPa. Because of the degradation of hemicellulose, the dimensional stability of the modified bamboo was enhanced and the mass loss test confirmed the degradation of ash, chemical composition, and extractives in bamboo tissue. We believe that this work can help people focus on bamboo resources and deeply understand the thermal modification mechanisms.

Author Contributions: X.X.: Writing & Conceptualization; X.L. (Xingyu Liang): Polished the bamboo samples for SEM analysis; H.P.: Nanoindentation test; X.L. (Xiaorong Liu): FTIR and XRD analysis; K.W.: methodology; Y.L.: Conceptualization. All authors have read and agreed to the published version of the manuscript.

Funding: The authors acknowledge the funding support from the National Natural Science Foundation of China (Nos. 31971740), Jiangxi Forestry Bureau Forestry Science and Technology Innovation Special Project (No.202134); Nanping Science and Technology Plan Project (N2020Z001), Zhejiang Provincial Key R&D Program Project (2019C02037).

Conflicts of Interest: The authors declare no conflict of interest.

References

1. Yuan, T.; Wang, X.; Liu, X.; Lou, Z.; Mao, S.; Li, Y. Bamboo flattening technology ebables efficient and value-added utilization of bamboo in the manufacture of furniture and engineered composites. *Compos. Part B Eng.* **2022**, *242*, 110097. [CrossRef]
2. Yuan, T.; Wang, X.; Lou, Z.; Zhang, T.; Han, X.; Wang, Z.; Hao, X.; Li, Y. Comparison of the fabrication process and macro and micro properties of two types of crack-free, flatten bamboo board. *Constr. Build. Mater.* **2022**, *317*, 125949. [CrossRef]
3. Huang, Y.; Yu, Y.; Zhang, C.; Wang, X.; Yang, Z.; Yang, Y. Investigation of the relationship between surface colour, contact angle and chemical properties of heat-treated bamboo. *Wood Mater. Sci. Eng.* **2020**, *27*, 1–9. [CrossRef]
4. Yuan, Z.; Wu, X.; Wang, X.; Zhang, X.; Yuan, T.; Liu, X.; Li, Y. Effects of One-Step Hot Oil Treatment on the Physical, Mechanical, and Surface Properties of Bamboo Scrimber. *Molecules* **2020**, *25*, 4488. [CrossRef] [PubMed]
5. Viglasova, E.; Galambos, M.; Divis, D.; Dankova, Z.; Dano, M.; Krivosudsky, L.; Lengauer, C.L.; Matik, M.; Briancin, J.; Soja, G. Engineered biochar as a tool for nitrogen pollutants removal: Preparation, characterization and sorption study. *Desalin. Water Treat.* **2020**, *191*, 318–331. [CrossRef]
6. Dano, M.; Viglasova, E.; Stamberg, K.; Galambos, M.; Galanda, D. Pertechnetate/Perrhenate Surface Complexation on Bamboo Engineered Biochar. *Materials* **2021**, *14*, 486. [CrossRef]
7. Kuai, B.; Tong, J.; Zhang, Y.; Zhan, T.; Lu, J.; Cai, L. Analysis of micro-morphology, mechanical properties, and dimensional stability of densified faber fir infused with paraffin. *Holzforschung* **2022**, *76*, 451–462. [CrossRef]
8. Kuai, B.; Wang, Z.; Gao, J.; Tong, J.; Zhan, T.; Zhang, Y.; Lu, J.; Cai, L. Development of densified wood with high strength and excellent dimensional stability by impregnating delignified poplar by sodium silicate. *Constr. Build. Mater.* **2022**, *344*, 128282. [CrossRef]
9. Tong, J.; Wang, X.; Kuai, B.; Gao, J.; Zhang, Y.; Huang, Z.; Cai, L. Development of transparent composites using wheat straw fibers for light-transmitting building applications. *Ind. Crop. Prod.* **2021**, *170*, 113685. [CrossRef]
10. Liang, R.; Zhu, Y.-H.; Wen, L.; Zhao, W.-W.; Kuai, B.-B.; Zhang, Y.-L.; Cai, L.-P. Exploration of effect of delignification on the mesopore structure in poplar cell wall by nitrogen absorption method. *Cellulose* **2020**, *27*, 1921–1932. [CrossRef]
11. Tjeerdsma, B.F.; Boonstra, M.; Pizzi, A.; Tekely, P.; Militz, H. Characterisation of thermally modified wood: Molecular reasons for wood performance improvement. *Holz Als Roh-Werkst.* **1998**, *56*, 149–153. [CrossRef]
12. Boonstra, M.J.; Tjeerdsma, B. Chemical analysis of heat treated softwoods. *Holz Als Roh-Werkst.* **2006**, *64*, 204–211. [CrossRef]
13. Bekhta, P.; Niemz, P. Effect of High Temperature on the Change in Color, Dimensional Stability and Mechanical Properties of Spruce Wood. *Holzforschung* **2003**, *57*, 539–546. [CrossRef]
14. Tiwari, S.K.; Bystrzejewski, M.; de Adhikari, A.; Huczko, A.; Wang, N. Methods for the conversion of biomass waste into value-added carbon nanomaterials: Recent progress and applications. *Prog. Energy Combust. Sci.* **2022**, *92*, 101023. [CrossRef]
15. Iswanto, A.H.; Madyaratri, E.W.; Hutabarat, N.S.; Zunaedi, E.R.; Darwis, A.; Hidayat, W.; Susilowati, A.; Adi, D.S.; Lubis, M.A.R.; Sucipto, T.; et al. Chemical, Physical, and Mechanical Properties of Belangke Bamboo (*Gigantochloa pruriens*) and Its Application as a Reinforcing Material in Particleboard Manufacturing. *Polymers* **2022**, *14*, 3111. [CrossRef]

16. Li, Y.; Yin, L.; Huang, C.; Meng, Y.; Fu, F.; Wang, S.; Wu, Q. Quasi-static and dynamic nanoindentation to determine the influence of thermal treatment on the mechanical properties of bamboo cell walls. *Holzforschung* **2015**, *69*, 909–914. [CrossRef]
17. Meng, Y. Optimization of cellulose nanofibrils carbon aerogel fabrication using response surface methodology. *Eur. Polym. J.* **2015**, *73*, 12. [CrossRef]
18. Li, Y.; Huang, C.; Wang, L.; Wang, S.; Wang, X. The effects of thermal treatment on the nanomechanical behavior of bamboo (*Phyllostachys pubescens* Mazel ex H. de Lehaie) cell walls observed by nanoindentation, XRD, and wet chemistry. *Holzforschung* **2017**, *71*, 7. [CrossRef]
19. Yang, X.; Liu, H.; Chai, Y.; Sun, Z.; Fei, B.; Jiang, Z. Indentation Coefficient and Indentation Behavior of Bamboo. *Wood Fiber Sci.* **2020**, *52*, 346–355. [CrossRef]
20. Su, M.; Zhang, R.; Li, J.; Jin, X.; Zhang, X.; Qin, D. Tailoring growth of MOF199 on hierarchical surface of bamboo and its antibacterial property. *Cellulose* **2021**, *28*, 11713–11727. [CrossRef]
21. Piao, X.; Zhao, Z.; Guo, H.; Wang, Z.; Jin, C. Improved properties of bamboo by thermal treatment with wood wax oil. *Colloids Surf. A Physicochem. Eng. Asp.* **2022**, *643*, 128807. [CrossRef]
22. Zhang, Y.; Yu, W.; Kim, N.; Qi, Y. Mechanical Performance and Dimensional Stability of Bamboo Fiber-Based Composite. *Polymers* **2021**, *13*, 1732. [CrossRef] [PubMed]
23. Ju, Z.; Zhan, T.; Brosse, N.; Wei, Y.; Zhang, H.; Cui, J.; Lu, X. Interfacial properties of windmill palm (*Trachycarpus fortunei*) fiber reinforced laminated veneer lumber (LVL) composites under high voltage electrostatic field (HVEF). *Ind. Crop. Prod.* **2022**, *180*, 114795. [CrossRef]
24. He, Q.; Zhan, T.; Ju, Z.; Zhang, H.; Hong, L.; Brosse, N.; Lu, X. Influence of high voltage electrostatic field (HVEF) on bonding characteristics of Masson (*Pinus massoniana* Lamb.) veneer composites. *Eur. J. Wood Wood Prod.* **2019**, *77*, 105–114. [CrossRef]
25. He, Q.; Zhan, T.; Zhang, H.; Ju, Z.; Hong, L.; Brosse, N.; Lu, X. Facile preparation of high anti-fungal performance wood by high voltage electrostatic field (HVEF). *J. Clean. Prod.* **2020**, *260*, 120947. [CrossRef]
26. Ju, Z.; Zhan, T.; Zhang, H.; He, Q.; Hong, L.; Yuan, M.; Cui, J.; Cheng, L.; Lu, X. Strong, Durable, and Aging-Resistant Bamboo Composites Fabricated by Silver In Situ Impregnation. *ACS Sustain. Chem. Eng.* **2020**, *8*, 16647–16658. [CrossRef]
27. Ju, Z.; He, Q.; Zhan, T.; Zhang, H.; Hong, L.; Li, S.; Chen, L.; Lu, X. Silver electrochemical treatment of bamboo and its effect on decay fungi. *Holzforschung* **2021**, *75*, 288–301. [CrossRef]
28. Yuan, T.; Han, X.; Wu, Y.; Hu, S.; Wang, X.; Li, Y. A new approach for fabricating crack-free, flattened bamboo board and the study of its macro-/micro-properties. *Eur. J. Wood Wood Prod.* **2021**, *79*, 1531–1540. [CrossRef]
29. Yuan, T.; Xiao, X.; Zhang, T.; Yuan, Z.; Wang, X.; Li, Y. Preparation of crack-free, non-notched, flattened bamboo board and its physical and mechanical properties. *Ind. Crop. Prod.* **2021**, *174*, 114218. [CrossRef]
30. Yuan, T.; Wang, Z.; Han, X.; Yuan, Z.; Wang, X.; Li, Y. Multi-scale evaluation of the effect of saturated steam on the micromechanical properties of Moso bamboo. *Holzforschung* **2021**, *75*, 1052–1060. [CrossRef]
31. Xing, D.; Li, J.; Wang, X.; Wang, S. In situ measurement of heat-treated wood cell wall at elevated temperature by nanoindentation. *Ind. Crop. Prod.* **2016**, *87*, 142–149. [CrossRef]
32. Xing, D.; Wang, X.; Wang, S. Temperature-Dependent Creep Behavior and Quasi-Static Mechanical Properties of Heat-Treated Wood. *Forests* **2021**, *12*, 968. [CrossRef]
33. Qin, L.; Lin, L.; Fu, F.; Fan, M. Micromechanical properties of wood cell wall and interface compound middle lamella using quasi-static nanoindentation and dynamic modulus mapping. *J. Mater. Sci.* **2018**, *53*, 549–558. [CrossRef]
34. Yang, D.; Li, H.; Xiong, Z.; Lorenzo, R.; Corbi, I.; Corbi, O. Fiber alignment angles effect on the tensile performance of laminated bamboo lumber. *Eur. J. Wood Wood Prod.* **2022**, *80*, 829–840. [CrossRef]
35. Yang, D.; Li, H.; Wei, D.; Lorenzo, R.; Corbi, I.; Corbi, O.; Yuan, C.; Xiong, Z.; Hong, C.; Zhang, H. Length effect on bending properties and evaluation of shear modulus of parallel bamboo strand lumber. *Eur. J. Wood Wood Prod.* **2021**, *79*, 1507–1517. [CrossRef]
36. Zhou, K.; Li, H.; Hong, C.; Ashraf, M.; Sayed, U.; Lorenzo, R.; Corbi, I.; Corbi, O.; Yang, D.; Zuo, Y. Mechanical properties of large-scale parallel bamboo strand lumber under local compression. *Constr. Build. Mater.* **2021**, *271*, 121572. [CrossRef]
37. Lorenzo, R.; Mimendi, L.; Yang, D.; Li, H.; Mouka, T.; Dimitrakopoulos, E.G. Non-linear behaviour and failure mechanism of bamboo poles in bending. *Constr. Build. Mater.* **2021**, *305*, 124747. [CrossRef]
38. Yang, D.; Li, H.; Xiong, Z.; Mimendi, L.; Lorenzo, R.; Corbi, I.; Corbi, O.; Hong, C. Mechanical properties of laminated bamboo under off-axis compression. *Compos. Part Appl. Sci. Manuf.* **2020**, *138*, 106042. [CrossRef]
39. Long, Z.; Zhang, L.; Wu, Q.; Tan, Z.; Guo, P. Effect of temperature on color and chemical composition of poplar powder compacts during warm-press forming. *Eur. J. Wood Wood Prod.* **2021**, *79*, 1461–1468. [CrossRef]
40. Liu, X.; Xiao, X.; Zhang, T.; Li, Y.; Peng, H.; Dong, Y.; Wang, K.; Li, J. Construction of thorough cross-linked networks in soybean meal adhesive system by biomimetic boronic acid-anchored cellulose nanofibril for multifunctionality of high-performance, mildew resistance, anti-bacterial, and flame resistance. *Ind. Crop. Prod.* **2022**, *180*, 114791. [CrossRef]

polymers MDPI

Article

Fabrication and Characterization of EVA Resins as Adhesives in Plywood

Yu Zhang [1,2], Ye He [1], Jiayan Yu [1], Yuxin Lu [1], Xinhao Zhang [1] and Lu Fang [1,2,*]

1 College of Furnishings and Industrial Design, Nanjing Forestry University, Nanjing 210037, China
2 Co-Innovation Center of Efficient Processing and Utilization of Forest Resources, Nanjing Forestry University, Nanjing 210037, China
* Correspondence: fanglu@njfu.edu.cn

Abstract: The practical problem of free formaldehyde pollution in the plywood industry is that polyethylene films have been shown to be able to replace some urea–formaldehyde resins for wood adhesives. To broaden the variety of thermoplastic plywood, reduce the hot-press temperature, and save energy consumption, an ethylene–vinyl acetate (EVA) film was selected as a wood adhesive to manufacture a novel wood–plastic composite plywood via hot-press and secondary press processes. The effects of the hot-press and secondary press processes at different levels on the physical–mechanical properties of EVA plywood (tensile shear strength, 24 h water absorption, and immersion peel performance) were evaluated. The results showed that the properties of the resulting plywood using the EVA film as an adhesive could meet the type III plywood standard. The optimum hot-press time was 1 min/mm, the hot-press temperature was 110–120 °C, the hot-press pressure was 1 MPa, the dosage film was 163 g/m^2, the secondary press time was 5 min, the secondary press pressure was 0.5 MPa, and the secondary press temperature was 25 °C. EVA plywood can be used in indoor environments.

Keywords: wood–plastic plywood; hot-press; secondary press; thermoplastic resin; physical–mechanical properties; aldehyde-free

Citation: Zhang, Y.; He, Y.; Yu, J.; Lu, Y.; Zhang, X.; Fang, L. Fabrication and Characterization of EVA Resins as Adhesives in Plywood. *Polymers* **2023**, *15*, 1834. https://doi.org/10.3390/polym15081834

Academic Editor: Pavlo Bekhta

Received: 21 February 2023
Revised: 31 March 2023
Accepted: 3 April 2023
Published: 10 April 2023

1. Introduction

With the rapid development of the construction and furniture industries, the demand for wood has intensely increased. Wood-based panels could effectively improve the broad utilization rate of wood. Wood-based panels mainly include plywood, particle board, and fiberboard. Among them, plywood is widely used in the furniture, construction, packaging, car, and boat manufacturing industries owing to its excellent physical and mechanical properties. However, plywood production was mainly formaldehyde-based-material-type adhesives, of which urea–formaldehyde (UF) resin adhesives account for 80% of the total use [1]. UF resins are characterized by low cost, mature technology, and a great gluing effect, but they release free formaldehyde during their production and use. To reduce the formaldehyde emission from wood-based composites, several researchers have investigated the optimization process of aldehydes in adhesives through the reduction of the formaldehyde–urea molar ratio [2], the control of the reaction temperature and pH [3], and the addition of a formaldehyde trapping agent [4,5] of UF resins during synthesis. These can effectively reduce the emission of free formaldehyde from wood-based panels.

With the improved living conditions of people, most countries have issued more stringent environmental protection standards. The formaldehyde emission standard for composite wood products issued by the United States in 2017 stipulates that the formaldehyde emission of hardwood plywood manufactured with single or composite cores should be ≤0.05 ppm. According to the GB/T 39600-2021 classification of formaldehyde emissions from wood-based panels and their products implemented in China, the formaldehyde content has been newly classified, and the highest electric network frequency level limits the formaldehyde emission from exceeding 0.025 mg/m^3. Under the advocacy of energy conservation and emission reduction and green environmental protection policies, green and environment-friendly adhesives have received considerable attention [6–8]. Presently, isocyanate (MDI) adhesives [9], soybean protein adhesives [10,11], starch adhesives [11–13], and other biological adhesives [14] have been widely used in the production of wood-based panels, and related products have been marketed. In addition, inorganic adhesives, such as silicate, magnesium oxychloride, and phosphate, have received attention from scientific researchers owing to their mildew resistance, water resistance, and flame retardancy. Zheng et al. [15] prepared bamboo chips/magnesium oxychloride composites with higher mechanical properties and water resistance using magnesium oxychloride gel, bamboo chips, and 0.3% polycarboxylate superplasticizer, providing research guidance for new wall materials.

In recent years, thermoplastic resin films, such as polyethylene (PE), polypropylene (PP), and polyvinyl chloride (PVC), have been widely used for producing plywood owing to their excellent water resistance, flexibility, easy processing, and secondary melting characteristics. PE has the simplest structure and has been widely studied as a wood adhesive [16–19]. Fang et al. [17] systematically evaluated the adhesive properties of high-density polyethylene films. The results showed that the plywood manufactured with PE as an adhesive featured similar bonding strength and elastic modulus with the UF resin plywood containing similar resin contents. In addition, because its plasticity can endow the plywood with stronger resistance to bending damage, PE thermoplastic plywood exhibits a higher modulus of rupture. PVC is a thermoplastic resin characterized by high flame retardancy and chemical resistance and can be used for wood veneer bonding. Gao [20] obtained the optimal process conditions for the production of PVC–thermoplastic plywood via a response surface method. The result showed that the properties of the produced plywood can meet the type II plywood standard, and the optimum hot-press conditions were 170 °C and 1 min/mm. Compared with PE and PVC films, the PP film features higher heat resistance. The bonding strength of the PP thermoplastic plywood is 1.5 MPa after three treatment cycles (immersing in boiling water for 4 h, then drying at 63 °C for 20 h and immersing again in boiling water for 4 h), which can meet the requirements of ordinary plywood category I in the GB/T 9846-2015 standard [21]. However, owing to the high melting temperature of the PP film, the temperature of the PP thermoplastic plywood should not be lower than 180 °C. In the manufacturing process of thermoplastic resin plywood, the heat transfer rate and uniformity can directly affect the plywood performance. Li et al. [22] not only increased the mechanical interlocking between the plastic and wood but also significantly reduced the hot-press time of the plywood through the perforation of the PVC film and then bonding with the wood veneer. Ye et al. [23] used mechanical methods to punch holes in the wood veneer surface and then combined it with the PE film. The increase in holes on the veneer was highly conducive to the penetration of PE, increasing the number of dendritic glue nails, which can form a highly stable microstructure, and thus, the panel strength is improved. Conventional wood–plastic composites can be obtained by extrusion molding or injection molding [24,25]; however, thermoplastic plywood is usually prepared by hot-press and secondary press processes. Bekhta et al. [26] used a high-density polyethylene film as an adhesive to manufacture alder plywoods, investigated the effects of hot-press temperatures and hot-press times on the physical and mechanical properties of alder plywood panels, and compared these properties with UF resin and phenol–formaldehyde resin plywoods. In a study on a hot-press process factor of a wood–plastic composite plywood, Chang [27] selected the secondary press conditions as follows: a secondary press time of 5 min, a secondary press pressure of 1 MPa, and a secondary press temperature of 30 °C; Fang [28] used a PE film to manufacture a poplar plywood and selected a hot-press pressure and secondary press pressure of 1 MPa. Among them, the selection of the secondary press conditions has a certain impact on the plywood performance, but studies on the secondary press process are few at present.

The main bonding mechanism between wood veneer and thermoplastic film is mechanical interlock, no chemical reaction. When a specific thermoplastic film is selected as a plywood adhesive, a suitable hot-press temperature needs to be selected so that the thermoplastic resin can flow into the wood pores of the wood to form glue nails that give the plywood mechanical strength. Usually, the hot-press temperature should be higher than the melting point of the thermoplastic film. Fang et al. [17] chose a hot-press temperature of 160 °C for the preparation of PE plywood. At this time, the PE film has better mobility and forms a tight bond with the wood veneer. Compared with the PE film, the melting temperature of the PP film was higher. Song et al. [29] prepared PP plywood at a hot-press temperature of 165–195 °C based on the melting temperature of the PP film. At 185 °C, it was found to have higher tensile strength than those formed at 165 °C due to the lower viscosity and deeper penetration of PP at 185 °C. Poly-β-hydroxybutyrate film (PHBF) is a biodegradable thermoplastic that can also be used as an adhesive for aldehyde-free plywood. Chen et al. [30] prepared veneer–PHBF composite properties under a hot-press temperature at 170 °C. Results showed that the properties of the produced plywood can meet the type II plywood standard. The mechanical properties and water resistance of these thermoplastic plywoods are excellent, but the processing temperature is usually above 160 °C, which not only consumes more energy to process, but also causes surface discoloration of the plywood. Ethylene–vinyl acetate copolymer (EVA) is a thermoplastic resin with a relatively low melting temperature and remarkable flexibility. In this study, an ethylene–vinyl acetate copolymer (EVA) film was selected as an adhesive instead of the traditional formaldehyde adhesive to prepare the EVA wood–plastic plywood. The effects of the hot-press and secondary press processes at different levels on the physical–mechanical properties of the EVA plywood were evaluated. The purpose of this study was to provide a new type of thermoplastic plywood with excellent performance, simple preparation, and no formaldehyde environmental protection, and to provide a theoretical basis for production practice.

2. Materials and Methods

2.1. Materials

Poplar veneers were purchased from Minsheng Wood Industry Co., Ltd. (Shandong, China). The dimensions were 300 × 300 × 1.6 mm^3, and the moisture content was 6–8%. The EVA film purchased from Huakai Supply Chain Co., Ltd. (Shenzhen, China), had a thickness of 0.1 mm and a density of 0.91 g/cm^3.

2.2. Production of the EVA Wood–Plastic Plywood

A three-layer wood-based plywood was assembled by three poplar veneers and EVA films. The plywood was prepared via a combination of hot-press and secondary press methods (Figure 1). The hot-press pressure was controlled to 1 MPa, and the hot-press temperatures were 90, 100, 110, 120, 130, and 140 °C. The numbers of EVA films were 1 layer, 2 layers, and 3 layers (one layer of EVA film was equivalent to 81.5 g/m^2 of double-sided sizing) during the hot-press process. Then, the plywood was secondary-pressed at room temperature for 5 min under 1 MPa.

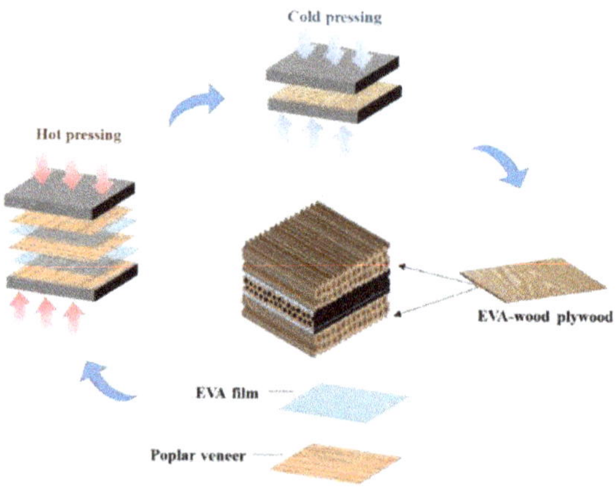

Figure 1. The process of plywood preparation.

The best secondary press time for a pretest was 5 min. Under the same hot-press conditions, the EVA wood–plastic plywood was prepared under secondary press temperatures of 25, 45, 65, and 85 °C and cold pressures of 0.5, 1, 1.5, and 2 MPa.

2.3. *Characterization*

2.3.1. Thermal Properties of the EVA Film

Melting temperature: A differential scanning calorimetry (DSC) analysis of the EVA film (5–10 mg) was conducted under a nitrogen atmosphere using DSC-250 (TA Instruments, New Castle, DE, USA). The EVA film was heated up from 0 to 180 °C at a rapid heating rate of 10 °C/min and preserved at 180 °C for 5 min to remove the thermal history of the sample. Then, the EVA film was cooled down from 180 to 0 °C at a rate of 10 °C/min and preserved at 0 °C for 5 min. Finally, it was heated up again from 0 to 180 °C at a rapid heating rate of 10 °C/min to obtain the melting point of the EVA film.

Heat stability: A thermogravimetric analysis of the EVA film was conducted using TGA-250 (TA Instrument) under a nitrogen atmosphere at a heating rate of 10 °C/min from 25 to 800 °C. Then, 5–10 mg of the EVA film was used for the test.

2.3.2. Physical–Mechanical Characterization

The physical–mechanical properties of the EVA plywood (tensile shear strength, wood failure ratio, 24 h WA, and immersion peel performance) were evaluated according to the Chinese National Standard (GB/T 17657-2013) [31]. Before the test was conducted, all specimens were conditioned at 20 °C and 65% relative humidity for 48 h.

(1) Mechanical strength: According to the requirements of GB/T 9846-2015 [32] "Ordinary Plywood" type II and III plywood standard, the tests for plywood were performed under the conditions of 63 ± 3 °C hot water immersion for 3 h and 20 ± 3 °C cold water immersion for 24 h. The result revealed that the plywood could not meet the type II bonding strength test.

(2) Twenty-four-hour water absorption: The size of three-layer plywood specimens with dimensions of 100 mm × 100 mm was weighed in 20 °C water before and after 24 h of soaking mass m_1 and m_2. The 24 h water absorption is calculated using the following equation:

$$WA(\%) = \frac{m_2 - m_1}{m_1} \times 100\% \quad (1)$$

(3) Immersion peel performance: The immersion peel performance of the EVA wood–plastic plywood was tested according to the Chinese National Standard GB/T 9846-

2015 [32], and the peeling delamination between the adhesive layers of the specimen was observed.

2.3.3. Scanning Electron Microscopy (SEM)

Two layers of plywood with parallel structures were prepared as SEM observation samples under all process conditions. The interface structure of the plywood was examined using a Quanta-200 ESEM (Hillsboro, OR, USA). The specimens were fixed onto the copper sheet with adhesive tape and sprayed with gold.

3. Results and Discussion

3.1. Characterization of the EVA Film

The thermal properties of the EVA film had a vital impact on the preparation and performance of the EVA wood–plastic plywood. The preparation process of the EVA thermoplastic plywood indicated that EVA films melted and softened at a higher temperature and pressure to penetrate the wood veneer, which endows the EVA wood–plastic plywood with mechanical strength. Figure 2a shows the melting curve of the EVA film and its peak temperature (melting temperature) of 84.8 °C. Based on this, hot-press temperatures of 90, 100, 110, 120, 130, and 140 °C were set.

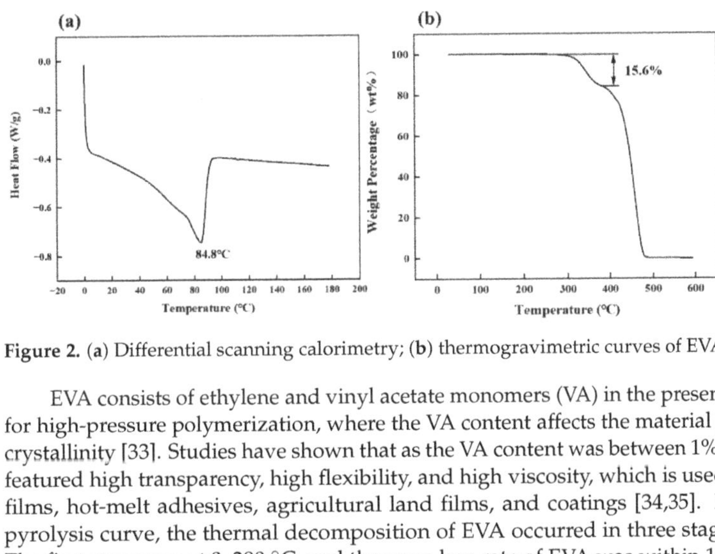

Figure 2. (**a**) Differential scanning calorimetry; (**b**) thermogravimetric curves of EVA film.

EVA consists of ethylene and vinyl acetate monomers (VA) in the presence of initiators for high-pressure polymerization, where the VA content affects the material properties and crystallinity [33]. Studies have shown that as the VA content was between 1% and 40%, EVA featured high transparency, high flexibility, and high viscosity, which is used in packaging films, hot-melt adhesives, agricultural land films, and coatings [34,35]. From the EVA pyrolysis curve, the thermal decomposition of EVA occurred in three stages (Figure 2b). The first stage was at 0–280 °C, and the mass loss rate of EVA was within 1%. The second stage occurred between 300 and 380 °C, at a temperature of 380 °C, and the weight loss rate of EVA was 15.6%; moreover, EVA ester bond breakage released an acetic acid and generated a mixture of CO_2 and CH_4 [36]. The VA content of the EVA film used in this study was 22.3% based on the weight loss rate of the acetic acid. In the final stage, PE began to decompose at 420–550 °C and ended as the temperature reached 600 °C, indicating that EVA has a wide processing temperature range.

A previous study noted that the thermal decomposition of most wood and other natural fibers is between 215 and 310 °C [37]. In order to avoid pyrolysis of plywood, the hot-press temperature should be chosen below 215 °C.

3.2. Effect of the Hot-Press Process on the Performance of the EVA Wood–Plastic Plywood

3.2.1. Effect of the Hot-Press Process on the Bonding Strength of the EVA Wood–Plastic Plywood

Hot-Press Temperature

In order for the thermoplastic resin to fully flow into and between the wood pores and form a glue nail structure, the hot-press temperature should be 15 to 35 °C higher than the melting temperature of the thermoplastic. Therefore, in this paper, the hot-pressing temperature for EVA wood–plastic plywood ranged from 90 to 140 °C.

The hot-press temperature showed a low effect on the dry tensile shear strength of the EVA wood–plastic plywood (Figure 3). With increasing hot-press temperatures from 90 to 110 °C, the dry strength of the EVA plywood increased from 1.09 to 1.2 MPa with a 11.7% variation, and the hot-press temperature continuously increased, with no significant change in dry strength. Because EVA had melted and entered the porous structure of the plywood, it formed a mechanical interlock structure, which endows the plywood with dry strength.

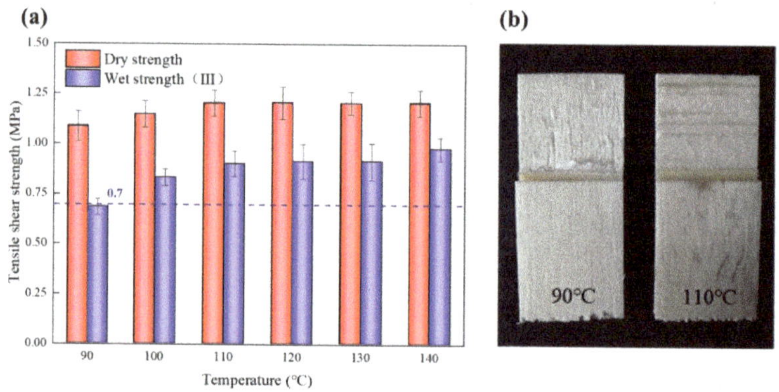

Figure 3. (**a**) Tensile shear strength of plywood at different hot-press temperatures; (**b**) wood failure rate of EVA–wood plywood (III).

The main bonding mechanism between wood veneer and EVA film was mechanical interlock, not chemically bonded [29,38–40], and its bonding interface has poor resistance to water molecule damage, resulting in the wet strength of the plywood being significantly lower than the dry strength. The hot-press temperature had an excellent influence on the stability of the glued structure (Figure 3a). At a very low hot-press temperature (90 °C), the stability of the bonding structure of the wood was poor, and the bonding strength could not meet the type III standard of plywood in GB/T 9846-2015. The wood failure ratio of the EVA wood–plastic plywood prepared under this condition was almost 0 (Figure 3b). There was a large gap between the bonding interfaces of plywood (Figure 4a,b). At hot-press temperatures between 110 and 120 °C, the bonding strength of type III plywood was 0.9 MPa, reaching a stable state, and its wood failure ratio significantly increased, indicating that the fluidity and permeability of EVA films were enhanced at higher hot-press temperatures. The mechanical interlock structure formed by the wood veneer was more stable, and the gap between the bonding interface was smaller (Figure 4c,d). Because the EVA melt viscosity was high, the continuous increase in the hot-press temperature on its permeability improvement was not significant (Figure 4e,f). With increasing hot-press temperatures from 110 to 140 °C, the improvement rate of wet strength of the plywood was <10%. Song et al. [29] prepared wood veneer/PP film composites using wood veneer and PP film. When the hot-pressing temperature was 20 °C higher than the melting temperature of the PP film, the tensile shear strength of the wood veneer/PP film composites was the best. This conclusion is consistent with the conclusions of this study.

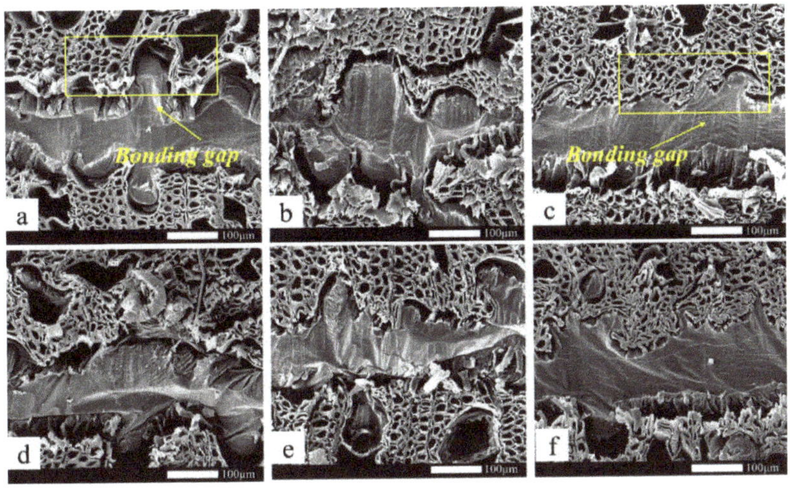

Figure 4. The interfacial surface of plywood with different hot-press temperatures: (**a**) 90 °C, (**b**) 100 °C, (**c**) 110 °C, (**d**) 120 °C, (**e**) 130 °C, (**f**) 140 °C.

Dosage of the EVA Film

EVA film was used as an adhesive in the EVA wood–plastic plywood, and its dosage had a significant effect on the number of mechanical nails and the thickness of the wood interface layer. The plywood with different film dosages was prepared under a hot-press temperature of 120 °C, hot-press time of 1 min/mm, and hot pressure of 1 MPa, and its bonding strength is shown in Figure 5. With increasing dosages from 81.5 to 244.5 g/m^2, the dry strength of the plywood increased from 1.19 to 1.24 MPa, with a 4% increase. Therefore, the EVA film that could endow the bonding strength of the panel was limited. With increasing EVA dosages, the number of mechanical nails increased, but the change in bonding strength of the wood was small. An EVA film dosage of 81.5 g/m^2 could meet the requirements of the bonding interface of the mechanical interlock structure.

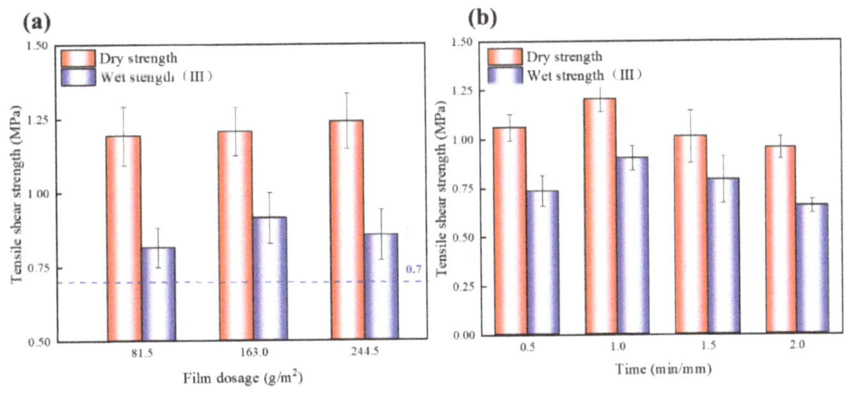

Figure 5. (**a**) Tensile shear strength of plywood at different film dosages; (**b**) tensile shear strength of plywood at different hot-press times.

The hydrophobicity of the EVA film blocked the entry of water molecules and resisted the damage of water molecules to the bonding interface. All samples exhibited strength higher than 0.7 MPa for wet tests, which met the requirement for type III–grade plywood of GB/T 9846-2015 standard (Figure 5a). With increasing EVA dosages from 81.5 to 163 g/m^2, the wet strength of the wood increased from 0.82 to 0.91 MPa. However, with an increasing film dosage of 244.5 g/m^2, the wood adhesive strength decreased by 7%; because of the high viscosity of EVA, its fully molten state was attained after a longer time, and the shorter hot-press time led to an incomplete molten state. Moreover, too thick a layer of glue would weaken the adhesion between veneers. Therefore, with increasing EVA film dosages, its bonding strength decreases. The effect of the dosage of the thermoplastic film used on the bonding properties of another thermoplastic plywood is similar. Fang et al. [41] laminated the silane-modified poplar veneer with PE film, and when the PE thin film increased from 1 to 4 layers, the adhesive layer was more likely to detach from the modified poplar veneer, causing a decrease in bonding strength.

Hot-Press Time

The EVA thermoplastic plywood was prepared with different hot-press times under the conditions of a hot-press temperature of 110 °C, hot-press pressure of 1 MPa, and EVA film dosage of 163 g/m^2. The hot-press time had a significant impact on the bonding strength of the panel (Figure 5b). The dry and wet strength of the EVA plywood first increased and then decreased with the hot-press time. With increasing hot-press times from 0.5 to 1 min/mm, the dry and wet strength of the plywood increased by 14% and 22%, respectively. This phenomenon is elucidated as follows: At the same hot-press temperature, the heat transfer from the surface layer of the slab to the core layer occurred at a certain amount of time, the short hot-press time resulted in a low slab temperature of the core layer, and incomplete molten EVA and the veneer could not form a mechanical interlock structure. With increasing hot-press times, the heat was completely transferred to the core layer, and the permeability of the EVA film in the wood pores became stronger, resulting in the thinning of the panel glued interface layer, which reduces the wood bonding strength. Therefore, with an increasing hot-press time of 2 min/mm, the dry and wet strength (type III) of the plywood decreased. The result showed that the hot-press time affects the penetration depth of the thermoplastic film. Overpenetration may occur if the hot-press time is too long, which will have a negative impact on the bonding strength of the thermoplastic plywood [22].

3.2.2. Effect of the Hot-Press Process on the Water Absorption of the EVA Wood–Plastic Plywood

The water absorption could evaluate the dimensional stability of the plywood, which has a vital impact on the long-term use of the plywood. With increasing EVA film dosages, hot-press temperature, and hot-press time, the water absorption of the plywood gradually decreased, and the water resistance increased. Among them, the water absorption of the plywood was mainly affected by the amount of EVA film. With increasing EVA film dosages from 81.5 to 244.5 g/m^2, the 24 h water absorption decreased by 23%, because water-repellent materials did not absorb water (Figure 6). The water absorption mainly occurred in the EVA wood–plastic plywood. As the EVA film was combined with the poplar veneer, one part of the EVA film fully penetrated the micropores of the wood veneer, and the other part covered the wood surface; therefore, the contact area between the wood and water molecules became smaller, which can reduce the absorption rate of water molecules in the wood. With increasing EVA film contents, the effect of blocked water molecules increased, and thus, the water absorption rate of the panel became lower. This conclusion was confirmed in the paper by another researcher [40], who found that plywood density also has an effect on water absorption: the water absorption of plywood panels decreased with increasing density.

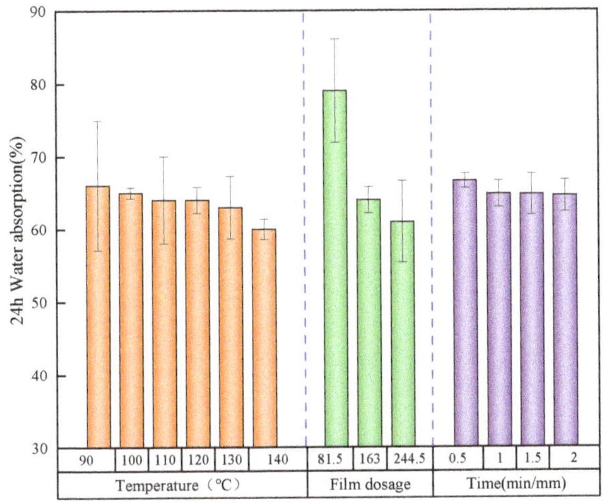

Figure 6. 24 h water absorption of plywood under different conditions.

The hot-press time and the hot-press temperature had a slight effect on the water absorption of the panel as the amount of the EVA film and wood components was constant. The increase in these conditions increased the penetration depth of the EVA film, and the interfacial compatibility of the board was also improved owing to the increase in the hydrophobicity, the volatilization of a small number of extracts in the wood, and the reduction of hydrophilic hydroxyl groups under a high temperature. However, the water absorption of the wood was still dominant; therefore, with increasing hot-press temperatures from 90 to 140 °C, the 24 h water absorption decreased by 6%, and with increasing hot-press times from 0.5 to 2 min/mm, the 24 h water absorption decreased by 3%.

3.2.3. Effect of the Hot-Press Process on the Immersion Peel Performance of the EVA Wood–Plastic Plywood

The immersion peel performance was a vital indicator for evaluating the water resistance and gluing properties of the plywood. The immersion peel test (type III) performed on all EVA plywood samples showed no evidence of delamination and degumming, which still met the lowest requirement of the GB/T 9846-2015 standard (the total length of each side of each specimen peeled from the same adhesive layer should not exceed 25 mm). Although the glue layer could resist the damage of water molecules, some water molecules in the glue layer could not destroy the board glue interface, which preserved its glue interface.

The type II impregnation peel test (soaking at 63 ± 3 °C for 3 h and then drying at 63 ± 3 °C for 3 h) produced stresses that caused the plywood to peel to varying degrees (Figure 7). Among them, the plywood manufactured at 90 °C could not meet the standard requirement of type II plywood. This phenomenon is elucidated as follows: At a lower hot-press temperature, the adhesive layer of the bonding strength of specimens was less than the wet expansion and dry shrinkage stress, and the panel layer produced peels. With increasing hot-press temperatures, the veneer and EVA film closely combined, which can partly resist stress. The panel with different EVA film dosages was peeled at the ends, and the peeling length was approximately the same, indicating that the amount of the EVA film has little impact on the immersion peel performance of the plywood (Figure 8). However, the EVA film was not completely melted at a low hot-press time, resulting in cracking and delamination on the panel. At hot-press times of 0.5 min/mm and 1 min/mm, the immersion peel strength of the panel was 22 and 14.4 mm (Figure 9). With increasing hot-press times from 1.5 to 2 min/mm, a little crack was observed on the panel.

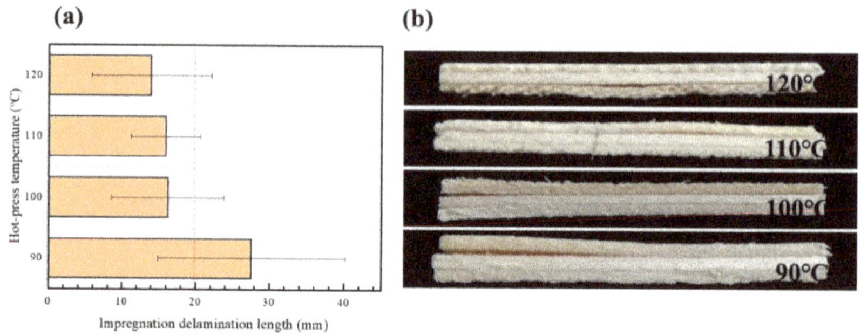

Figure 7. (**a**) Effect of hot-pressing temperature on impregnation peel performance; (**b**) impregnation delamination length under different hot-pressing temperature conditions.

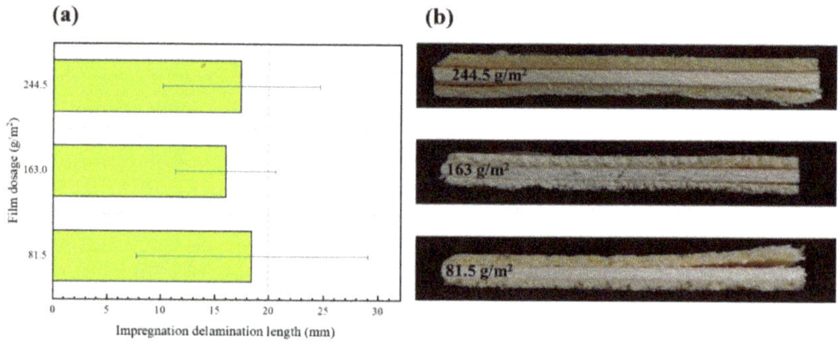

Figure 8. (**a**) Effect of EVA film dosage on impregnation peel performance; (**b**) impregnation delamination length under different film dosage conditions.

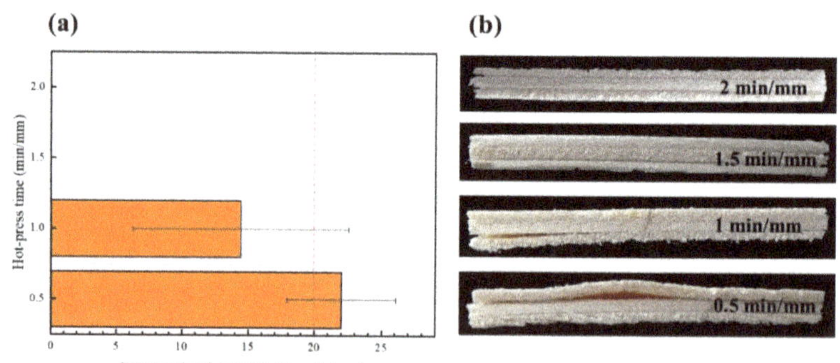

Figure 9. (**a**) Effect of hot-press time on impregnation peel performance; (**b**) impregnation delamination length under different hot-press times.

3.3. *Effect of the Secondary Press Process on the Performance of the EVA Wood–Plastic Plywood*

The temperature of the wood panel decreased after the EVA thermoplastic plywood was subjected to a hot-press process. The thermoplastic film shrinks during the formation of the adhesive interface, and the shrinkage residual stress affects the stability of the adhesive interface [42]. A suitable secondary press process could reduce the shrinkage stress of the panel bonding interface and ensure the mechanical strength of the plywood. Therefore, accurate control of the secondary press process is a key factor for maintaining the stability of the mechanical mesh structure of the interface layer and the mechanical properties of the panel. A literature search showed that most researchers have systematically studied the hot-press process of thermoplastic plywood [17,22,29,43], but there is no in-depth discussion on the secondary press process. Based on the hot-press process, the effect of the secondary press process on the tensile shear strength of EVA wood–plastic plywood was studied. In this study, the optimum conditions for preparing the EVA thermoplastic plywood were a hot-press temperature of 110 °C, hot-press pressure of 1 MPa, hot-press time of 1 min/mm, EVA film dosage of 163 g/m^2, and secondary press time of 5 min, varying with the secondary press temperature and secondary press pressure.

At low temperatures, the crystalline curing of the meshing structure at the glued interface of the sheet was complete. At a lower secondary press temperature, the dry and wet strength of the EVA thermoplastic plywood was relatively large, because the EVA film was completely crystallized at 25 °C, and the glued interface of the adhesive nail structure maintained a stable state (Figure 10a). With increasing cold press temperatures, the EVA film at a certain cold press time was not completely crystallized, but a part of the crystallization process was complete under no pressure conditions; moreover, the interface layer of the EVA film was damaged by shrinkage stress (Figure 11). At a cold press temperature of 85 °C, close to the melting point of EVA, the increased hot-press times without cold press could not eliminate the contraction stress of the glued interface of the EVA film.

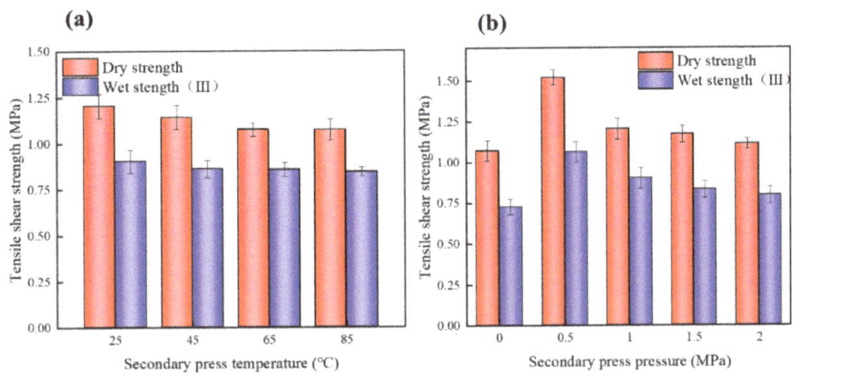

Figure 10. (**a**) Effect of secondary press temperature on tensile shear strength; (**b**) effect of secondary press pressure on tensile shear strength.

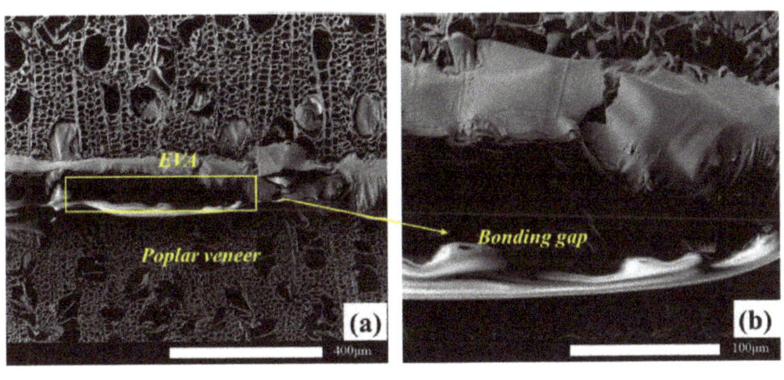

Figure 11. The bonding interface of nonsecondary press plywood. (**a**) ×150; (**b**) ×600.

A suitable secondary press pressure could resist the shrinkage stress of the panel bonding interface and ensure the mechanical strength of the plywood. Figure 10b shows the influence of the secondary press pressure on the bonding strength of the panel. The bonding strength of the secondary pressed panels was higher than that of nonsecondary pressed panels; particularly, a significant difference was observed between the bonding strengths of the secondary pressed panels (0.5 MPa) and nonsecondary pressed panels. With increasing secondary press pressures from 0.5 to 2 MPa, the dry and wet bonding strength of the panel decreased, because the greater the secondary press pressure, the faster the plate cooling rate, resulting in a highly intense shrinkage of the EVA film, and the interface layer gaps partly caused the glue nail failure; moreover, the panel from the hot-press to the secondary press processes without pressure caused the EVA film to rebound. With excessive secondary press pressure, the molten EVA film from the glue interface overflow formed a thin glue layer, which reduces the bonding strength of the panel.

4. Conclusions

EVA film can be used as an ideal wood adhesive to produce formaldehyde-free wood–plastic plywood. The hot-press and secondary press conditions have a significant impact on the physical and mechanical properties of EVA wood–plastic plywood. The hot-press temperature and hot-press time have an interactive effect on the EVA wood–plastic plywood tensile shear strength. With a hot-press temperature of 110–120 °C and a hot-press time of 1 min/mm, EVA melted fully and did not overpenetrate, and poplar veneer formed a good mechanical interlock structure. With the increased EVA film dosage, the number of bonded joints formed also increased. With a film dosage of 163 g/m^2, the plywood bonding strength was optimal. Moreover, to reduce the shrinkage stress of the panel interface layer, a secondary press temperature of ~25 °C and a secondary press pressure of 0.5 MPa should be selected. According to the GB/T 9846-2015 standard, the tensile shear strength of the plywood should meet the type III plywood standard. EVA wood–plastic plywood can be used as a green panel for an indoor environment. Compared with thermoplastic plywood, such as PE, PP, and PVC plywood, the hot-press temperature of EVA plywood is 40 to 60 °C lower than other thermoplastic plywoods, which is more energy saving and consumption reducing. However, it also led to EVA plywood samples showing evidence of delamination and degumming after immersion in hot water at 63 °C for 3 h. In order to improve the mechanical properties of EVA plywood, chemical modification or using EVA/PE blends as plywood adhesives can be used to improve the interfacial adhesion of plywood in a subsequent work.

Author Contributions: Writing—original draft, Y.Z.; writing—review and editing, L.F.; software, Y.H.; data curation, Y.L.; investigation, X.Z.; methodology, J.Y. All authors have read and agreed to the published version of the manuscript.

Funding: This research was funded by the Postdoctoral Science Preferential Funding of Zhejiang Province (ZJ2022139), Postgraduate Research and Practice Innovation Program of Jiangsu Province (Grant Number SJCX22_0315), College Students Innovation and Entrepreneurship Training Program Project of Nanjing Forestry University (Grant Number 2021NFUSPITP0425), and Industry University Cooperation Collaborative Education Project (Grant Number 202101148003).

Institutional Review Board Statement: Not applicable.

Informed Consent Statement: Not applicable.

Data Availability Statement: Not applicable.

Acknowledgments: The authors are grateful to the Advanced Analysis and Testing Center of Nanjing Forestry University.

Conflicts of Interest: The authors declare no conflict of interest.

References

1. Duan, X.X.; Shi, J.Y. Preparation and application of formaldehyde-free adesives from biomass feedstock. *China Wood-Based Panels* **2016**, *23*, 19–23.
2. Park, B.-D.; Lee, S.-M.; Roh, J.-K. Effects of formaldehyde/urea mole ratio and melamine content on the hydrolytic stability of cured urea-melamine-formaldehyde resin. *Eur. J. Wood Wood Prod.* **2009**, *67*, 121–123. [CrossRef]
3. Ferra, J.M.; Mena, P.C.; Martins, J.; Mendes, A.M.; Costa, M.R.N.; Magalhaes, F.D.; Carvalho, L.H. Optimization of the synthesis of urea-formaldehyde resins using response surface methodology. *J. Adhes. Sci. Technol.* **2010**, *24*, 1455–1472. [CrossRef]

4. He, Z.; Zhang, Y. Control of formaldehyde emission from wood-based panels by doping adsorbents: Optimization and application. *Heat Mass Transf.* **2013**, *49*, 879–886. [CrossRef]
5. Song, J.X.; Tian, H.; Lei, H.; Xu, G.X.; Wang, J.S.; Pu, L.; Chen, Q. Modification of urea-formaldehyde resin by Nano MnO_2. *J. For. Eng.* **2022**, *7*, 91–96. [CrossRef]
6. Chang, L.; Gou, W.J.; Chen, Y.P. Review of research and application of non-formaldehyde adhesives for producing wood-based panels. *China For. Prod. Ind.* **2014**, *41*, 3–6, 12. [CrossRef]
7. Gu, M.; Wu, J. Green product evaluation standard of furniture advocates green development. *China Qual. Certif.* **2020**, *3*, 59–61. [CrossRef]
8. Xiong, X.; Ma, Q.; Yuan, Y.; Wu, Z.; Zhang, M. Current situation and key manufacturing considerations of green furniture in China: A review. *J. Clean. Prod.* **2020**, *267*, 121957. [CrossRef]
9. Lubis, M.A.R.; Falah, F.; Harini, D.; Sudarmanto; Kharisma, A.; Tjahyono, B.; Fatriasari, W.; Subiyanto, B.; Suryanegara, L.; Iswanto, A.H. Enhancing the performance of natural rubber latex with polymeric isocyanate as secondary pressing and formaldehyde free adhesive for plywood. *J. Adhes.* **2023**, *99*, 58–73. [CrossRef]
10. Lei, Z.; Jiang, K.; Chen, Y.; Yi, M.; Feng, Q.; Tan, H.; Qi, J.; Xie, J.; Huang, X.; Jiang, Y.; et al. Study on the bonding performance and mildew resistance of soy protein-based adhesives enhanced by hydroxymethyl L-tyrosine cross-linker. *Int. J. Adhes. Adhes.* **2022**, *117*, 103167. [CrossRef]
11. Xiong, X.-Q.; Yuan, Y.-Y.; Niu, Y.-T.; Zhang, L.-T. Development of a cornstarch adhesive for laminated veneer lumber bonding for use in engineered wood flooring. *Int. J. Adhes. Adhes.* **2020**, *98*, 102534. [CrossRef]
12. Zhang, Y.; Guo, Z.; Chen, X.; Ma, Y.; Tan, H. Synthesis of grafting itaconic acid to starch-based wood adhesive for curing at room temperature. *J. Polym. Environ.* **2021**, *29*, 685–693. [CrossRef]
13. Xiong, X.Q.; Bao, Y.L.; Guo, W.J.; Fang, L.; Wu, Z.H. Preparation and application of high performance corn starch clue in straw decorative panel. *Wood Fiber Sci.* **2018**, *50*, 88–95.
14. Liu, Z.; Kou, F.J.; Duan, Y.C.; Wang, W.J.; Peng, L.; Li, J.Z.; Gao, Q. Preparation and Investigation of Distillers-dried Grains with Solubles-based Wood Adhesive. *J. For. Eng.* **2021**, *6*, 105–111.
15. Zheng, L.; Zuo, Y.F.; Li, P.; Wang, S.; Sheng, G.A.; Wu, Y.Q. Construction of homogeneous structure and chemical bonding in bamboo scrap/magnesium oxychloride composites by polycarboxylate superplasticizer. *J. Mater. Res. Technol.* **2021**, *12*, 2257–2266. [CrossRef]
16. Fang, L.; Zeng, J.; Zhang, X.H.; Wang, D. Effect of Veneer Initial Moisture Content on the Performance of Polyethylene Film Reinforced Decorative Veneer. *Forests* **2021**, *12*, 102. [CrossRef]
17. Fang, L.; Chang, L.; Guo, W.-j.; Ren, Y.-p.; Wang, Z. Preparation and characterization of wood-plastic plywood bonded with high density polyethylene film. *Eur. J. Wood Wood Prod.* **2013**, *71*, 739–746. [CrossRef]
18. Grillo, C.C.; Saron, C. Wood-plastic from pennisetum purpureum fibers and recycled low-density polyethylene. *J. Nat. Fibers* **2022**, *19*, 858–871. [CrossRef]
19. Sun, Y.X.; Liu, Y.; Wang, H.W.; Li, C.F.; Liu, M.L. Study on the Preparation Techology of Low Density Polyethylene Film Plywood. *China For. Prod. Ind.* **2020**, *57*, 24–26, 36. [CrossRef]
20. Gao, Y.L. Process Optimization of Plywood Made of PVC Film and Eucalyptus Veneer. Master's Thesis, Fujian Agriculture and Forestry University, Fuzhou, China, 2018.
21. Huang, Z.W. Study on the Preparation and Performance of Eucalyptus Veneer Plywood Glued by Thin Polyethylene Film. Master's Thesis, Fujian Agriculture and Forestry University, Fuzhou, China, 2017.
22. Li, Z.; Qi, X.; Gao, Y.; Zhou, Y.; Chen, N.; Zeng, Q.; Fan, M.; Rao, J. Effect of PVC film pretreatment on performance and lamination of wood-plastic composite plywood. *RSC Adv.* **2019**, *9*, 21530–21538. [CrossRef]
23. Ye, C.X.; Yang, W.B.; Xu, J.Y.; Chen, Z.J.; Liao, R.; Zhong, Z. Hot-pressing Technology of Formaldehyde-free Plywood Made by Veneer with Rolling Holes and High-density Polyethylene Film. *J. Northwest A F Univ. (Nat. Sci. Ed.)* **2015**, *43*, 76–82. [CrossRef]
24. Zhang, L.; Chen, Z.; Dong, H.; Fu, S.; Ma, L.; Yang, X. Wood plastic composites based wood wall's structure and thermal insulation performance. *J. Bioresour. Bioprod.* **2021**, *6*, 65–74. [CrossRef]
25. Tian, F.; Xu, X. Dynamical mechanical behaviors of rubber-filled wood fiber composites with urea formaldehyde resin. *J. Bioresour. Bioprod.* **2022**, *7*, 320–327. [CrossRef]
26. Bekhta, P.; Mueller, M.; Hunko, I. Properties of thermoplastic-bonded plywood: Effects of the wood species and types of the thermoplastic films. *Polymers* **2020**, *12*, 2582. [CrossRef] [PubMed]
27. Chang, L. Formation Mechanism and Interface Status Evaluation of High Density Polyethylene Poplar Composite Plywood. Ph.D. Thesis, Chinese Academy of Forestry, Beijing, China, 2014.
28. Fang, L. Interfacial Modification Methods and Mechanism of High Density Polyethylene Film/Poplar Veneer Plywood. Ph.D. Thesis, Chinese Academy of Forestry, Beijing, China, 2014.
29. Song, W.; Wei, W.; Li, X.; Zhang, S. Utilization of polypropylene film as an adhesive to prepare formaldehyde-free, weather-resistant plywood-like composites: Process optimization, performance evaluation, and interface modification. *Bioresources* **2017**, *12*, 228–254. [CrossRef]
30. Chen, Z.H.; Wang, C.C.; Cao, Y.; Zhang, S.B.; Song, W. Effect of adhesive content and modification method on physical and mechanical properties of eucalyptus veneer-poly-beta-hydroxybutyrate film composites. *For. Prod. J.* **2018**, *68*, 419–429. [CrossRef]

31. *GB/T 17657-2013;* Test Methods of Evaluating the Properties of Wood-Based Panels and Surface Decorated Wood-Based Panels—Part 4: Test Method. China National Standardization Administration: Beijing, China, 2013.
32. *GB/T 9846-2015;* Plywood for General Use—Part 5: Require. China National Standardization Administration: Beijing, China, 2015.
33. Gu, P.; Zhang, J. Vinyl acetate content influence on thermal, non-isothermal crystallization, and optical characteristics of ethylene-vinyl acetate copolymers. *Iran. Polym. J.* **2022**, *31*, 905–917. [CrossRef]
34. Yang, G.; Li, Q.C.; Li, Y.M.; Lu, W.; Xu, L.P.; Wang, Y.X. Capability and application of poly(ethylene-co-vinyl acetate)(EVA). *Chin. J. Colloid Polym.* **2009**, *27*, 45–46.
35. Zhou, D. Production progress and apply of ethylene-vinylacetate. *Guangdong Chem. Ind.* **2015**, *42*, 99–100, 104.
36. Tian, J.J.; Jiang, H.; Su, T.T.; Zhang, X.T.; Sun, Z.L. Study on the Thermal Decomposition of EVA Based on the TGA-FTIR. *J. Instrum. Anal.* **2003**, *22*, 100–102.
37. Yao, F.; Wu, Q.; Lei, Y.; Guo, W.; Xu, Y. Thermal decomposition kinetics of natural fibers: Activation energy with dynamic thermogravimetric analysis. *Polym. Degrad. Stab.* **2008**, *93*, 90–98. [CrossRef]
38. Bekhta, P.; Sedliacik, J. Environmentally-friendly high-density polyethylene-bonded plywood panels. *Polymers* **2019**, *11*, 1166. [CrossRef]
39. Bekhta, P.; Chernetskyi, O.; Kusniak, I.; Bekhta, N.; Bryn, O. Selected properties of plywood bonded with low-density polyethylene film from different wood species. *Polymers* **2022**, *14*, 51. [CrossRef]
40. Bekhta, P.; Pizzi, A.; Kusniak, I.; Bekhta, N.; Chernetskyi, O.; Nuryawan, A. A comparative study of several properties of plywood bonded with virgin and recycled LDPE films. *Materials* **2022**, *15*, 4942. [CrossRef]
41. Fang, L.; Yin, Y.H.; Han, Y.L.; Chang, L.; Wu, Z.H. Effects of number o film layers on properties of thermoplastic bonded plywood. *J. For. Eng.* **2016**, *1*, 45–50. [CrossRef]
42. Peng, X.; Zhang, Z. Hot-pressing composite curling deformation characteristics of plastic film-reinforced pliable decorative sliced veneer. *Compos. Sci. Technol.* **2018**, *157*, 40–47. [CrossRef]
43. Chang, L.; Tang, Q.; Gao, L.; Fang, L.; Wang, Z.; Guo, W. Fabrication and characterization of HDPE resins as adhesives in plywood. *Eur. J. Wood Wood Prod.* **2018**, *76*, 325–335. [CrossRef]

MDPI
St. Alban-Anlage 66
4052 Basel
Switzerland
Tel. +41 61 683 77 34
Fax +41 61 302 89 18
www.mdpi.com

Polymers Editorial Office
E-mail: polymers@mdpi.com
www.mdpi.com/journal/polymers

www.ingramcontent.com/pod-product-compliance
Lightning Source LLC
Chambersburg PA
CBHW040811250926
51850CB00007BA/1141
* 9 7 8 3 0 3 6 5 7 5 0 9 4 *